T0396643

Integrating Agriculture, Conservation and Ecotourism: Examples from the Field

Issues in Agroecology – Present Status and Future Prospectus

Volume 1

Series Editors
W. Bruce Campbell and Silvia López Ortíz

For other titles published in this series, go to
http://www.springer.com/series/8794

W. Bruce Campbell • Silvia López Ortíz

Editors

Integrating Agriculture, Conservation and Ecotourism: Examples from the Field

Editors
W. Bruce Campbell
Cardno ENTRIX
701 University Avenue
Suite 200
Sacramento, California
95825 USA
bruce_campbell3@hotmail.com

Silvia López Ortíz
Colegio de Postgraduados
Campus Veracruz
Km. 86.5 Carretera Federal Xalapa-Veracruz
via Paso de Ovejas
entre Paso San Juan y Puente Jula
Tepetates, Veracruz, Mexico
C.P. 91690
silvialopez@colpos.mx

and

Colegio de Postgraduados
Campus Veracruz
Apartado Postal 421
Veracruz, Mexico, C.P. 91700

ISSN 2211-2405   e-ISSN 2211-2413
ISBN 978-94-007-1308-6   e-ISBN 978-94-007-1309-3
DOI 10.1007/978-94-007-1309-3
Springer Dordrecht Heidelberg London New York

Library of Congress Control Number: 2011930868

© Springer Science+Business Media B.V. 2011
No part of this work may be reproduced, stored in a retrieval system, or transmitted in any form or by any means, electronic, mechanical, photocopying, microfilming, recording or otherwise, without written permission from the Publisher, with the exception of any material supplied specifically for the purpose of being entered and executed on a computer system, for exclusive use by the purchaser of the work.

Printed on acid-free paper

Springer is part of Springer Science+Business Media (www.springer.com)

Foreword

Finding Ways Forward, Together...

Agroecology encompasses not only aspects of ecology and agriculture, but the ecology of sustainable food production systems, including the technology and related societal and cultural values (e.g. Gliessman 1998; Altieri and Hecht 1990; Altieri 1989, 1987, 1983) to better promote healthy and functional environments for a sustainable quality of life (see also Castillo et al. 2005). To provide effective communication regarding the status and advances in this burgeoning field, connections must be established with many disciplines including (but not limited to) ecology, agriculture, sociology, anthropology, environmental sciences, ethics, rural development, policy and management, and economics, to provide integrated points of view that will help lead to a more sustainable construction of values than conventional commercial economics alone. Such designs are inherently complex and dynamic, and go beyond the individual farm to include landscapes, communities, and biogeographic regions by emphasizing their unique agricultural and ecological values, and their biological, societal, and cultural components and processes. This multifaceted perspective provides immense insight on dealing with systems level issues and contributing to the development of sustainable societies.

The concept of agroecology is not new; early agrarian and small-scale agroforestry-oriented populations knew what plant and animal species would coexist well and have sufficient yields in given local environments. However, as societies grew, they became spatially larger and more fixed in space and time; no longer migrating or moving, and requiring increasingly greater resource inputs to survive. As such, many (if not most) traditional and local-scale farming practices succumbed to the pressures of societal advances by occupying more space, intensively using more resources and existing spaces, and becoming more monoculturally and commercially oriented. Such practices were promoted because they not only continued to supply the dietary needs of a growing population, but they also were seen as unlocking the hidden potential in nature for greater food production and economic growth; often focusing on the evolving power of technological advances to help do so. Indeed, improvements in food storage and food delivery systems permitted agricultural products to be available for longer periods of time and for international distribution and trade. No longer present were the natural, diverse, and coevolved local ecological systems. Landscapes became more homogeneous and intensively managed with tremendous energy inputs to satisfy growing market demands, even for species not suited to the area. Left in place to operate continuously, the earlier low-level and chronic impacts to ecosystems have now become acute, and the more familiar means of managing or ameliorating such problems no longer function effectively. These legacies of accumulating impacts, in concert with those from modern society, have led us to the realization that our current systems are not sustainable in their present form; many strong changes and fresh perspectives are needed.

During the early 1900s, a union between agronomy and ecology was, in fact, promoted (Gliessman 1998; Klages 1928), and scientists explored the local (e.g. soil) and regional (e.g. climate) environmental conditions promoting crop adaptations. After World War II, however, the resultant economic and population surges caused a rift to develop between ecologists (who experimented in natural systems) and agronomists (who worked in cultivated agricultural systems). It wasn't until the 1970s that either group began to officially recognize the value of and work in the other, such that the academic and scientific literature began to incorporate the term *agroecology* and the concept of *agroecosystem* with increasing frequency. Key among these was the work of Mexican ethnobotanist Efraim Hernández Xolocotzi who researched indigenous cultivation systems (Hernández Xolocotzi 1977), and recognized that as socioeconomic forces became more influential on food production systems, ecological connections within these systems would decline. Although we cannot eliminate commercial production because of societal needs, Gliessman (1998) suggested that as the similarity in structure and function between agroecosystems and natural systems grew in their respective biogeographic regions, so did the capacity for sustainability. Is it possible then, to create a more sustainable, secure and equitable future for food production by recovering knowledge from centuries of traditional agricultural practices and modifying it based on what we know of natural systems so that it can be blended with the development of new technologies and societal needs? Within the pages of this and succeeding volumes of *Issues in Agroecology* exist numerous and diverse examples of how various solutions have been found, how diverse conditions affect possible outcomes, and how some conditions can be changed to improve progress toward sustainable objectives. Within each contribution is the constant presence of scale; whether temporal, geographic, social or economic in scope, as this factor is one of the most difficult to engender, implement or manage, and may sometimes be unpredictable. Political and socioeconomic issues also are common and sometimes overriding sources of influence.

Issues In Agroecology – Present Status and Future Prospectus is the result of 7 years of planning, gestation, and countless, tireless hours of discussion with colleagues, students and rural peoples regarding local, regional and larger scale issues. This review series was produced from the realization that since the 1980s there has been tremendous growth in the agricultural and agroecological literature at all scales of observation; growth that requires critical assessments and syntheses from the point of view of sustainability. Quite simply, are our collective efforts, irrespective of scale of analysis, region of application or topic of work, orienting us along paths toward improved sustainability, or are we still suffering from obscured vision by being 'lost in the forest through the trees'? Such a multifaceted question not only prompts a multidisciplinary approach in each invited review, but a multiscaled perspective in the assessment as well. Hence, the series approaches this question and the corresponding complex panorama of interrelated and integrated topics by presenting authoritative, comprehensive, and analytical reviews from leading scientists in all areas of agroecology worldwide. Authors for each review represent a collaborative mix sufficient to provide strong summaries and scholarly advances, and to identify inaccuracies, gaps, and needs to improve the foundations for discussion leading to novel routes of research activity, application of management methodologies, and education and outreach programs. Each review represents a concise and up-to-date synthesis of the rapidly growing quantity of scientific information in its respective topic within the highly interdisciplinary field of agroecology.

The first four Volumes of the review series (published twice per year) are oriented around particular themes that have arisen from growing interest in the literature, sufficiently enough to promote their own cohesive syntheses:

Volume 1 – Integrating Agriculture, Conservation, and Ecotourism: Examples from the Field
Volume 2 – Integrating Agriculture, Conservation, and Ecotourism: Societal Influences
Volume 3 – Sustainable Food Production Includes Human and Environmental Health
Volume 4 – Propagation for Consumption and Ornamental Trade

Later volumes will contain more individualized review topics, including guest edited volumes.

Given the interrelated and integrated nature of themes, issues and topics underneath the rather broad umbrella of agroecology, certain broad thematic considerations will always be fundamentally interwoven in reviews. In particular, the value of sustainability and the environment, the concern over the future of the world's food supply, and that institutional and political factors are very often more influential than technical ones with regard to dealing with sustainability issues. Coupled with the informed assessments of the routes to realize future potential, the review series is expected to be an essential part of the scientific method and a necessity for researchers, teachers, students, and field professionals when dealing with increasing global environmental and socioeconomic change. This format will make *Issues In Agroecology – Present Status and Future Prospectus* a highly citable review series that is guaranteed to enlighten researchers, technology users, educators, students, and the general public on the status and advances in agroecological topics around the world.

November 30, 2010   Dr. W. Bruce Campbell
                                                              Dr. Silvia López Ortíz
                                                              CoEditors-In-Chief

References

Altieri MA (1983) Agroecology. University of California Press, Berkeley

Altieri MA (1987) Agroecology: the scientific basis of alternative agriculture. Westview Press, Boulder

Altieri MA (1989) Agroecology: a new research and development paradigm for world agriculture. Agric Ecosyst Environ 27:37–46

Altieri MA, Hecht SB (eds.) (1990) Agroecology and small farm development. CRC Press, Boca Raton

Castillo A, Torres A, Velásquez A, Bocco G (2005) The use of ecological science by rural producers: a case study in Mexico. Ecol Appl 15(2):745–756

Gliessman SR (1998) Agroecology: ecological processes in sustainable agriculture. Ann Arbor Press, Chelsea

Hernández Xolocotzi E (ed.) (1977) Agroecosistemas de Mexico: contribuciones a la ensenanza, investigacion, y divulgacion agricola. Colegio de Postgraduados, Chapingo

Klages KHW (1928) Crop ecology and ecological crop geography in the agronomic curriculum. J Am Soc Agron 20:336–353

Acknowledgements

The realization of this review series is a large step forward for agroecology as a science, a movement, and a practice on an international scale, as well as for its sustainable evolution. Yet, such endeavors require a great deal of continuous and tireless collaborative effort from a diverse array of people. Hence, we are indebted to Dr. Maryse Walsh, Jacco Flipsen and Melanie van Overbeek of Springer Science and Business Media B.V., Dordrecht, and to the many reviewers of the initial series plans whose past and present commitment to the concept and publication of this series is invaluable. We thank the members of the international editorial committee, Dr. Alexander Wezel, Dr. Louise Jackson, Dr. Miguel A. Altieri, Dr. Ted Lefroy and Dr. Juan J. Villalba, who have given of themselves tremendously to promote the birth and continued production of this series.

November 30, 2010

Dr. W. Bruce Campbell
Dr. Silvia López Ortíz
CoEditors-in-Chief

Contents

Contributors

Series Editors

Dr. W. Bruce Campbell is a Visiting Professor at Colegio de Postgraduados, Campus Veracruz, Mexico, and a Senior Project Scientist with Cardno ENTRIX, Sacramento, California, USA. He is an aquatic and terrestrial ecologist focusing on the development of strategies to detect and interpret the changes in biological communities, assemblages, and populations resulting from environmental impact, habitat alteration, harvesting pressure, resource use, introductions of exotic species, and conservation and restoration practices. Such work is essential for the development and maintenance of functional foundations in ecological and agroecological research and management, as well as in the development of sustainable resource initiatives. To understand these changes and the foundations behind them, he also focuses on identifying the components and processes that define various systems and how these definitions change with observational scale. These objectives foster greater understanding of how to improve or help maintain natural productivity and ecological function while also benefiting human society and promoting sustainable practices. Email: bruce_campbell3@hotmail.com

Dr. Silvia López Ortíz is an Associate Professor at Colegio de Postgraduados, Campus Veracruz, Mexico. Her research focuses on plant-herbivore interactions, how body condition of livestock influences the ability of the animals to detoxify plant secondary compounds as well as affecting their ability to avoid those plants with higher concentrations of toxins, and how management practices can best be applied to reduce problems associated with consumption of toxic plants. She is involved in researching how different grazing management schemes alter pasture plant community structure, and how such changes translate to changes in ruminant dietary quality and pasture health and stability. She is currently researching native forage trees that can be used by farmers as a strategy to produce more and higher quality forage during the dry season. She has identified as many as 30 species of native trees that could be used as forage; in particular guacimo (*Guazuma ulmifolia* Lam.), which has shown strong success in outreach studies with local farmers who are planting the tree to create silvopastoral systems. Email: silvialopez@colpos.mx

Series Associate Editors

Dr. Alexander Wezel is an Associate Professor of Agroecology, and French Coordinator of the European Master of Science Program in Agroecology in the Department of Agroecosystems, Environment and Production at ISARA-Lyon, France. He is an agroecologist and landscape ecologist working with various topics related to land use and resource conservation in the tropics and subtropics. Over the last few years his research has focused on analysing and defining agroecology as a scientific discipline, as well as on research projects dealing with agroecosystem analysis and management and on relating effects from agricultural practices with water quality and biodiversity. He is actively engaged in work related to agroecology, landscape ecology, agroecosystems management, management of biodiversity, homegardens, and traditional agriculture in the tropics and subtropics. Email: wezel@isara.fr

Dr. Louise Jackson is Professor and Cooperative Extension Specialist in the Department of Land, Air and Water Resources at the University of California, Davis, California, USA. She holds the Orr Chair in Environmental Plant Science and is co-chair of the DIVERSITAS network on agrobiodiversity. Her research is focused on utilizing biodiversity to increase ecosystem services in intensive agricultural systems, ranging from the molecular to the ecosystem and landscape scales. She also is active in developing participatory processes for public involvement in biodiversity issues across agricultural landscapes, especially as relevant to climate change. She is a frequent contributor to regional, national, and international organizations on topics related to plant and soil ecology, agricultural sustainability, utilization of biodiversity, and land use change. Email: lejackson@ucdavis.edu

Dr. Miguel A. Altieri is a Professor of Agroecology at the University of California, Berkeley, California, USA, in the Department of Environmental Science, Policy and Management. He has served as a Scientific Advisor to the Latin American Consortium on Agroecology and Development (CLADES) in Chile, an NGO network promoting agroecology as a strategy for sustainable small farm development in the region. He also served for 4 years as the General Coordinator for the United Nations Development Program of Sustainable Agriculture Networking and Extension which aims at capacity building in agroecology among NGOs and the scaling-up of successful local sustainable agricultural initiatives in Africa, Latin America and Asia. He was the chairman of the NGO committee of the Consultative Group on International Agriculture Research whose mission was to ensure that the research agenda of the 15 International Agricultural Research Centers benefited the poor farmers. Currently, he is advisor to the FAO-GIAHS program (Globally Ingenious Agricultural Heritage Systems), which is devoted to identifying and dynamically conserving traditional farming systems in the developing world. He also is the general coordinator of the Latin American Scientific Society of Agroecology (www.agroeco.org/socla). His research focuses on the application of agroecological principles for the design of biodiverse, resilient and durable agricultural systems. Email: agroeco3@berkeley.edu

Dr. Ted Lefroy is Director, Centre for Environment, at the University of Tasmania, Hobart, Tasmania. He trained in agricultural science and spent 7 years working in rural development and extension in Queensland and Papua New Guinea before returning to southwestern Australia to work with watershed groups on a United Nations Man and the Biosphere Project on integrated resource management and conservation around the World Heritage Fitzgerald River National Park. He has since held research positions with state Department's of Agriculture, The University of Western Australia and CSIRO involving leadership of interdisciplinary research teams working with land managers to minimize the impacts of agriculture on natural resources and biodiversity. In 2005 he was appointed Professor of Environment and Director of the Centre for Environment at the University of Tasmania. Email: Ted.Lefroy@utas.edu.au

Dr. Juan J. Villalba is a Research Assistant Professor in the Department of Wildland Resources at Utah State University, Logan, Utah, USA. His research focuses on understanding the mechanisms that influence food selection and intake in herbivores, with the aim of creating efficient alternatives to manage animals and their environment. He also serves as Research Coordinator of the program BEHAVE (www.behave.net), a worldwide network of scientists, producers, land managers, and extension personnel committed to integrating behavioral principles and processes with local knowledge to enhance ecological, economic, and social values of rural and urban communities and landscapes. Email: jjvillalba@usu.edu

Authors for Volume 1

Dr. Jean-Claude Jauneau is a socio-economist at ISARA-Lyon, France. His work focuses on evaluating rural development programs (LEADER programs, assessment of services in mountain areas), working with agri-environmental policies (at local, regional or national levels), with programs concerning the remuneration of non-market functions in agriculture in Regional Nature Parks, and with the evaluation of small dairy farm sustainability in France. He is actively engaged in work dealing with agricultural economics, agricultural and rural policies, agri-environmental measures, and assessment of rural development policies and programs. Email: jauneau@isara.fr

Dr. David C. Weindorf is an Associate Professor of Soil Classification and Land Use, and SPESS Graduate Advisor at the Louisiana State University Agricultural Center, Baton Rouge, Louisiana, USA. He is a Fulbright Scholar with active international research collaborations with the Institute of Soil Sciences – Chinese Academy of Sciences (Nanjing, China) and Universitatea de Științe Agricole și Medicină Veterinară (Cluj-Napoca, Romania). His research is focused on new technologies in field soil surveys, land use/management, high resolution soil surveys, and international systems of soil taxonomy. He is a frequent contributor to regional, national, and international organizations on topics related to soil morphology/classification/

genesis, natural resource management, and field soil survey technologies. He serves as Editor of *Soil Survey Horizons* and is an active contributor to the National Cooperative Soil Survey (USDA-NRCS). Email: dweindorf@agcenter.lsu.edu

Dr. James P. Muir is a Professor in the Department of Soil and Crop Sciences at Texas AgriLife Research (Agricultural Experiment Station), Texas A & M University System, Stephenville, Texas, USA. He is actively involved in studying vegetative strategies for soil phosphorus reduction and the subsequent reduced phosphorus runoff from land receiving dairy manure, researching dairy composts in silage rotations to improve phosphorus cycling in dairies, and performing field trials to test annual and perennial grasses and legumes as manure-phosphorus sinks. Email: j-muir@tamu.edu

Dr. Cesáreo Landeros-Sánchez received his B.Sc. as an Agonomist Engineer specializing in irrigation in 1981 from the Universidad Autónoma de Chapingo, México, his M.Sc. in Water Management in 1987 from the Agricultural University of Wageningen, The Netherlands, and his Ph.D. in Water Management-Water Table Fluctuation and Solute Redistribution from Cranfield University, Silsoe College, UK. He is a Research Professor at Colegio de Postgraduados, Campus Veracruz (Tropical Agroecosystems), Mexico, where his work focuses on sustainable water management, water contamination, and soil and water conservation in tropical crops. As well, he works on the agricultural management of sugar cane with regard to irrigation and drainage, nitrogen fertilization, harvest residue utilization (trash), and filter-cake compost application. Email: clandero@colpos.mx

Dr. Ron J. Johnson is a Professor of Wildlife Ecology and BioSustainability in the Department of Forestry and Natural Resources at Clemson University, Clemson, South Carolina, USA. Dr. Johnson has worked with birds in agricultural systems for over 35 years, including bird damage management in agriculture and bird conservation and function in agroecosystems, especially the role of birds in the suppression of insect pests. He has evaluated the wildlife habitat interface between cropland and a variety of non-crop habitats including windbreaks, wooded and herbaceous riparian corridors and fencerows, grassland habitats, and wildlife in organic and non-organic agricultural systems. Recently, he led the initiation of the development of the Healthy Farm Index, assessment of interest and potential for wildlife-friendly farm certification, and an evaluation of relationships among farm birds, climate, and insect food resources. Email: ronj@clemson.edu

Dr. Julie A. Jedlicka is, at the time of this writing, a near-completion doctoral student at the University of California-Santa Cruz, Santa Cruz, California, USA. She has over 10 years of experience researching avian behavior and conservation within forest and agricultural systems. Her ornithological research includes studying avian microhabitat use in Costa Rican cloud forest, documenting foraging niche-shifts of warblers in Mexican shade coffee farms, and conducting meta-analyses on the effects of natural enemy biodiversity in suppressing arthropod herbivores. She is interested in the roles birds play at reducing insect pest populations, including research merging avian conservation with insect pest management in California vineyards. Email: jenvs@ucsc.edu

Dr. John E. Quinn is a post-doctoral researcher in the School of Natural Resources at the University of Nebraska, Lincoln, Nebraska, USA. His research focuses on birds in organic agroecosystems, particularly their breeding success and functional roles. His broad research interests emphasize concerns within biodiversity and conservation; in particular avian ecology, agroecology, restoration of working landscapes, and the role of birds as a medium for environmental awareness and education. Email: jquinn2@unl.edu

Dr. James R. Brandle is a Professor of Agroforestry, Applied Ecology, and Shelterbelt Ecology in the School of Natural Resources at the University of Nebraska, Lincoln, Nebraska, USA. Over the past 30 years his research has focused on the influence of shelterbelts on crop production with particular interest in the mechanics of wind movement in shelter, the microclimate created in shelter, the resulting yield benefits and the economics associated with shelterbelt performance. As part of these efforts he has evaluated various sustainable agriculture practices including conversion issues related to organic production. Current projects are related to vegetation diversity and the resulting biodiversity associated with predator-prey relationships and control of crop pests, feeding behavior of bird species, and management options for optimizing lady beetle foraging in alfalfa. Email: jbrandle@unl.edu

Dr. Shalene Jha is a UC President's Postdoctoral Fellow at the University of California, Berkeley, California, USA, and her primary research interest is to understand how local and regional agroecological landscapes influence plant and pollinator communities, their gene flow processes, and their ability to provide ecosystem services. Broadly, this research program covers topics in landscape genetics, pollination ecology, native bee community ecology, and foraging ecology. By integrating ecological research with molecular evolution analyses and GIS tools, her research aims to reveal how best to manage landscapes in order to support robust native pollinator and seed-disperser communities, conserve key dispersal processes for native plants, and provide valuable ecosystem services to agricultural communities. Email: s.jha@berkeley.edu

Dr. Christopher M. Bacon is a political ecologist, and at the time this volume was written, was a S.V. Ciriacy-Wantrup Fellow in the Department of Geography at the University of California, Berkeley, California, USA. Currently, he is an Assistant Professor in the Environmental Policy Institute at Santa Clara University, California, USA. He has conducted participatory action research in Mesoamerica for more than a decade, taught university courses, consulted, and contributed to developing a campus-wide hands-on sustainability curriculum. His work focuses on the role of smallholder cooperatives and alternative agri-food networks in enabling more sustainable livelihoods and reducing hunger. Email: CBacon@scu.edu

Dr. Stacy M. Philpott is an Assistant Professor in the Department of Environmental Sciences, at the University of Toledo, Ohio, USA. She received her Ph.D. in Ecology from the University of Michigan and completed a postdoctoral fellowship at the Smithsonian Migratory Bird Center. Her research interests include community ecology, ecosystem services, and linkages between agroecology and conservation.

Her research focuses on coffee agroecosystems and she has experience working in such areas and surrounding landscapes in Mexico, Costa Rica, and Indonesia. She has a background in theoretical ecology and human rights and dedicates her efforts towards putting ecological knowledge into practice to improve farmer livelihoods while conserving biodiversity. She currently works on understanding how coffee agroecosystems and urban gardens contribute to biodiversity conservation. Email: stacy.philpott@utoledo.edu

Dr. Robert A. Rice is a geographer at the Smithsonian Migratory Bird Center (SMBC), Washington, DC, USA, conducting research and addressing policy issues related to migratory bird habitat and conservation. His research interests focus upon the intersection between agricultural activities and the natural environment, as well as the socioeconomic benefits derived from diverse agroforestry systems. He currently is looking at soil carbon in agroforestry systems with different shade management regimes. Email: ricer@si.edu

Dr. V. Ernesto Méndez is an Assistant Professor of Agroecology and Environmental Studies in the Environmental Program and the Department of Plant and Soil Science at the University of Vermont, Burlington, Vermont, USA. His work analyzes interactions between agroecology, rural livelihoods, and environmental conservation in agricultural landscapes and agro-food systems. He also is interested in applying and critically evaluating participatory action research (PAR) approaches. A native of El Salvador, he holds a B.S. in Crop Science from California Polytechnic State University, an M.S. in Tropical Agroforestry from the Tropical Research and Education Center in Costa Rica, and a Ph.D. in Environmental Studies from the University of California at Santa Cruz. Email: emendez@uvm.edu

Dr. Peter Läderach is the Central America Leader of the Decision and Policy Analysis Program (DAPA) of the International Centre for Tropical Agriculture (CIAT). He has 10 years of experience in the development and implementation of spatial analyses and supply chain tools for coffee and cocoa, including analyses of the impact of climate change on the agricultural sector in Central America and the Caribbean. Email: p.laderach@cgiar.org

Dr. Travis Idol is an Associate Professor of Tropical Forestry and Agroforestry at the University of Hawaii-Manoa, Hawaii, USA. He was trained in forest soils and biogeochemistry at Purdue University. He worked as a postdoctoral researcher in mycorrhizal ecology at the University of California, Davis. His research interests include studying the productivity, silviculture, and biogeochemistry of restored native forests in Hawaii, evaluation of coffee agroforestry systems, and the design of agroforestry systems with native plants as a means for restoring degraded and alien-dominated forests. Email: idol@hawaii.edu

Dr. Jeremy Haggar is Leader of the Tree Crops in the Agroforestry Program at the Tropical Agricultural Centre for Research and Higher Education (CATIE). He was trained in tropical agroecology at the University of Cambridge. He worked as a postdoctoral researcher for the University of Florida studying the sustainability of tropical agroecosystems, and as Forestry Coordinator for the Organization for Tropical Studies, both based at La Selva Biological Station, Costa Rica. Subsequently he worked for the World Agroforestry Centre, ICRAF in Mexico studying agroforestry as an alternative to slash and burn agriculture on the Yucatan Peninsula. Currently, he is based in Managua, Nicaragua, where he coordinates regional projects to reinforce the production and business capacity of producer organizations in Guatemala, Honduras and Nicaragua to produce and sell sustainable and quality coffees. He is working on the evaluation of the environmental services that coffee production can provide and how those services may be compensated by socially and environmentally responsible markets. He is developing and validating processes to facilitate adaptation to climate change for coffee producing families and actors in the value chain. Email: jeremyhaggar@gmail.com

Dr. Linda J. Cox is a Community Economic Development Specialist in the Department of Natural Resource and Environmental Management at the University of Hawaii-Manoa, Hawaii, USA. Her areas of expertise include entrepreneurship, sustainable natural resource management, sustainable development, business management, and human resource development on domestic and international scales. Email: lcox@hawaii.edu

Dr. Sabine Weizenegger is a regional manager for the LEADER-Program in the Allgäu in Germany. She has worked on actor-oriented management of protected areas in sub-Saharan Africa at the University of Munich, on projects dealing with tourism and sustainable development in Benin, and with European cultural landscapes and the economic effects to protected areas. She works on introducing methods of regional development and LEADER to non-European countries on a free-lance basis. Her work involves geography, tourism and sustainable development, protected areas, rural development, regional development, cultural landscapes, and developing countries. Email: weizenegger@regionalentwicklung-oberallgaeu.de

Agroecology – Interpretations, Approaches and Their Links to Nature Conservation, Rural Development and Ecotourism

Alexander Wezel and Jean-Claude Jauneau

Abstract Different interpretations and definitions of agroecology are currently used world-wide. They vary from agroecology as a practice, agroecology as a movement, and varying approaches to agroecology as a scientific discipline, which are the plot/field, the agroecosystem, and the food system approach. The evolution of the interpretations and definitions are often closely linked to differences in the historical development of agroecology in different countries and regions of the world. More and more topics have become related to agroecology in recent years. In the second part of this review, we analyse and discuss the integration and linking of rural development, nature conservation and ecotourism, both within and to the framework of agroecology.

1 Introduction

In recent years, agroecology has increasingly become a topic of global interest and concern. This rise in popularity is due to the need to respond to the diverse challenges facing agriculture such as sustainable production, food security, climate change, conservation of biodiversity in agroecosystems, and rural development. These challenges involve global and systems aspects and cannot be attacked using

A. Wezel (✉)
Department of Agroecosystems, Environment and Production, ISARA Lyon,
23, rue Jean Baldassini, 69364 Lyon cedex 07, France
e-mail: wezel@isara.fr

J.-C. Jauneau
Department of Social Science and Management, ISARA Lyon,
23, rue Jean Baldassini, 69364 Lyon cedex 07, France
e-mail: jauneau@isara.fr

W.B. Campbell and S. López Ortíz (eds.), *Integrating Agriculture, Conservation and Ecotourism: Examples from the Field*, Issues in Agroecology – Present Status and Future Prospectus 1, DOI 10.1007/978-94-007-1309-3_1,
© Springer Science+Business Media B.V. 2011

only disciplinary approaches. Hence, agroecology as a scientific discipline might be particularly suited, because interdisciplinary and systems approaches are major foundations of many present agroecology interpretations. However, to understand the issue better, we must start from the beginning; the origins of agroecology. Since the first use of the term 'agroecology' in the early twentieth century, its meanings, definitions, interpretations and approaches have changed enormously up to the present. Thus, this review will summarise the evolution of the topic starting with an historical overview of agroecology as a scientific discipline and the actual definitions and approaches used. The review will then present different utilisations of the term *agroecology* as a science, a movement, and a practice. As well, our synthesis will analyse and discuss the integration and linking of rural development, nature conservation and ecotourism, both within and to the framework of agroecology.

2 Historical Overview of Agroecology

2.1 Initial Phase: 1930s–1960s

The history of agroecology began with Bensin (1928), who first used the term *agroecology*. According to López i Gelats (2004), it was Bensin who traced the term *agroecology* to 1928 from the Czechoslovak Botanical Society. Bensin (1930) suggested the term *agroecology* to describe the use of ecological methods on commercial crop plants. *Agroecology* would hence be preliminarily defined as the application of ecology in agriculture. Some years later, Bensin (1938) dealt with agroecology as a basic science of agriculture.

In the 1950s, several articles were written by the German ecologist/zoologist Tischler (e.g. Tischler 1950). In these papers, he presented the results of his agroecological research, in particular on pest management, and discussed unsolved problems concerning soil biology, insect biocoenosis interactions and plant protection in agricultural landscapes, including non-cultivated ecosystems such as hedgerows. His book, published in 1965, was probably the first to be entitled 'Agroecology' (Tischler 1965). He analysed different agroecological components (plants, animals, soils and climate) and their interactions within an agroecosystem as well as the impact of human agricultural management on these components. This approach combined ecology (interactions among biological components at the field level, or agroecosystem) and agronomy (integration of agricultural management). Further publications in the 1950s and 1960s dealt with pest management and zoology (e.g. Heydemann 1953) or field crops (Vavilov 1957).

Between the 1930s and 1960s other works on agroecology were published. The first book was published by the German zoologist Friederichs (1930) on agricultural zoology and related ecological/environmental factors for plant protection. This book also presented different pest management strategies, including biological control and the role of natural ecosystems for pest management, and evaluated the

Fig. 1 Major interpretations of agroecology from a historical perspective

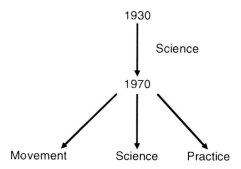

economic impact of pest damage. His approach was very similar to that of Tischler. A second important book was published by the American agronomist Klages (1942) in which he dealt with the distribution of crop plants on a physiological basis. He also analysed the ecological, technological, socio-economic and historical factors influencing their production. In addition, Klages (1928) is one of the first papers dealing with agroecology, but without using the term explicitly.

At the end of the 1960s, the French agronomist Hénin (1967) defined agronomy as being 'an ecology applied to plant production and agricultural land management' – which is very close to Bensin's definition – without actually using the word *agroecology*. Something similar can be stated for the Italian author Azzi (1956), who defined *agricultural ecology* as the study of the physical characteristics of environment, climate and soil, in relation to the development of agricultural plants. The foundations of his work were already laid 30 years earlier (Azzi 1928). More details about the first phase of the history of agroecology as well as the roots of agroecology before the year 1928 can be found in Wezel and Soldat (2009). This first phase in the history of agroecology concerned only agroecology as a science (Fig. 1). Later interpretations of agroecology were enlarged to gradually include interpretations of agroecology as a movement and as a practice. More details about this evolution will be presented after the historical overview.

2.2   Expansion of Agroecology as a Science: 1970s–1980s

As for the starting phase, very few publications can be found during the 1970s which used the term *agroecology* (Fig. 2). Since the 1980s, this quantity changed significantly with an increasing publication rate up to the present. During the 1980s, many publications dealt with agroecological zones or zoning (e.g. Henricksen 1986), or agricultural production related to different crops or to livestock (e.g. Moss 1980). At the end of the 1980s, sustainability and sustainable development became topics within agroecology (e.g. Altieri 1989; Dover and Talbot 1987) as well as alternative agriculture (Altieri 1987).

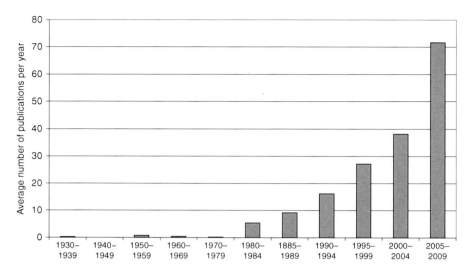

Fig. 2 Average number of publications using the word *agroecology* or *agroecological* in the title or in the author keywords for different periods from 1930 to 2009 (Note: 10-year periods from 1930 to 1979, afterwards 5-year periods)

Since the beginning of the 1980s, agroecology has emerged as a distinct methodology and conceptual framework for the study of agroecosystems (e.g. Puia and Soran 1984). *Agroecology* at that time was defined as the global study of agroecosystems protecting natural resources, with a view to design and manage sustainable agroecosystems (Altieri 1989). The key concept of *agroecosystem* emerged in the 1970s. The term was formerly suggested by the ecologist Odum (1969, quoted in Altieri 1995), who considered agroecosystems as 'domesticated ecosystems', intermediate between natural and fabricated ecosystems. Another new orientation in agroecology at the time was research into traditional farming systems and agroecosystems in tropical and subtropical developing countries (e.g. Arrignon 1987; Conway 1987; Altieri et al. 1983). Cox and Atkins (1979) was another important agroecology publication in the 1970s and 1980s. They provided a very broad overview and in-depth analyses of different factors and dynamics in agroecosystems, but also raised political, economic and energy-related questions regarding agricultural systems in developing and developed countries.

2.3 Institutionalisation and Consolidation of Agroecology: 1990s

During the 1990s, agroecological research expanded and consolidated, and several important textbooks were published (e.g. Gliessman 1997, 1990; Altieri 1995; Carroll et al. 1990). During this period, the number of publications dealing with agroecological zones, characterization, zoning or land-use classification, as well as with sustainability and sustainable agriculture increased enormously (e.g. Thomas

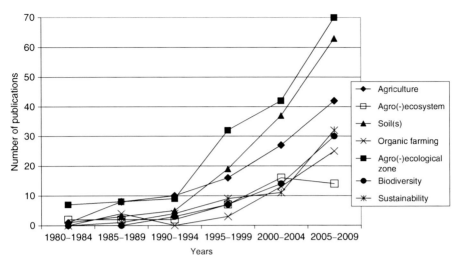

Fig. 3 Evolution of selected title words and author keyword clusters in publications from 1980 to 2009 for 5 year periods

and Kevan 1993). Consequently, the theme of biodiversity also emerged in the 1990s within agroecology-related publications (e.g. Alard 1994; Altieri 1993) (Fig. 3). At the end of the 1990s the word *soil* started to be used increasingly in agroecology publications under various topics such as soil fertility, conservation, productivity or zonation. As in the 1980s, the term *agroecosystem* continued to be present in the title or the keywords of different publications (e.g. Altieri 1999; Johns 1998), but to a lesser degree in comparison to the words *sustainability* or *biodiversity*.

2.4   New Dimensions in Agroecology: 2000 to Present

At the beginning of the twenty-first century, new definitions for *agroecology* appeared. For some authors agroecology moved beyond agroecosystems toward food systems. The first definition was provided by Francis et al. (2003) with *agroecology* defined as "the integrative study of the ecology of entire food systems, encompassing ecological, economic and social dimensions", or more simply the "ecology of food systems". Gliessman (2007) provided a similar definition for *agroecology* giving it as "the science of applying ecological concepts and principles to the design and management of sustainable food systems", but with certain emphasis on practical application. Ten new dimensions of agroecology, compared to traditional agronomic approaches, are presented by Clements and Shrestha (2004): new philosophy of agriculture, systems thinking, local adaption, non-crop biota, crop autecology, encompassing the agricultural landscape, closing the materials cycle, technology and ecology, human ecology, and the natural dimension.

Since 2000, publications dealing with sustainability and sustainable agriculture increased significantly, but now focused more often with sustainable development, biodiversity, and the inclusion of new topics such as organic farming/agriculture and agrobiodiversity (Wezel and Soldat 2009).

3   Current Variations in Definitions and Scales in Agroecology

In looking at the different definitions and descriptions in publications, it is evident that agroecology has changed from the plot or field scale (1930s–1960s) to the farm or agroecosystem scale (1970s–2000s) (Fig. 4), although the smaller scale approaches also are still used today. At present, the definitions of *agroecology* given by Francis et al. (2003) and Gliessman (2007) go beyond this context by leaving the concrete spatial scale and entering the dimension of the food system. This dimension includes local, regional, national and global geographical scales, as well as the food production systems, society, and the economics and politics that cannot be attributed directly to a specific scale, but which are connected and interwoven in different ways.

The change of definitions and scale can be related mainly to the evolution of the two basic disciplines from which agroecology is derived, agronomy and ecology. However, other disciplines such as zoology, botany/plant physiology, and their applications in agricultural and environmental issues, also play important roles (Fig. 5). Over time and with larger scales, more disciplines have become involved, and include geography, sociology, socioeconomics and anthropology (e.g. dealing with cultures, traditions, or indigenous knowledge). Main topics and objectives vary according to the different scales and where they are applied in agroecological research. In particular, with the food systems approach, many more new topics have become part of agroecology; providing more disciplines. While some are new, others such as sociology that were already in use for the agroecosystem approach (in certain cases), became more clearly visible from the 2000s onwards. The new topics of rural poverty, rural development, and biodiversity conservation have risen far more rapidly than research at the agroecosystem level (notice the apparent plateau of the keyword *agroecosystem* in Fig. 3), requiring more of a food systems approach.

The second major, but more restricted approach in agroecology is the agroecosystem approach. Here, ongoing research dominates the agroecosystem scale, including exchange with, and impact on the environment (e.g. Martin and Sauerborn 2006). Normally, interactions with society, politics and economy are not taken into consideration. A definition that summarizes this quite well is provided by the Department of Crop Science (Agroecology Section) at the University of Göttingen (2008): "Agroecological analyses focus on plant and animal communities, food web interactions, and conservation biology in temperate as well as tropical agricultural landscapes and agroecosystems", although agricultural production aspects are not clearly mentioned. Within agroecosystem approaches, the definitions and concepts might vary depending on the definition of an agroecosystem. Sometimes the farm is seen as equivalent to an agroecosystem, for others an agroecosystem is at the larger end of the scale; a local or regional landscape where

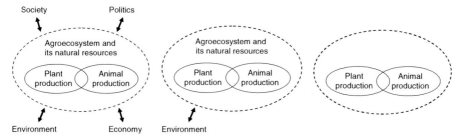

Fig. 4 The different definitions and views of agroecology in current research (*left*: food systems approach, *middle*: agroecosystem approach, *right*: plot or field approach) (From: Wezel and Soldat 2009, © Earthscan)

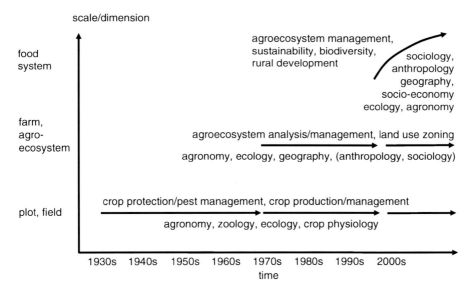

Fig. 5 Temporal changes in scale and dimension in the definitions of *agroecology* as well as related main topics and basic disciplines for applied research (*above the arrows* are main topics, *below the arrows* are basic disciplines) (Adapted from: Wezel and Soldat (2009))

agriculture is practiced (Conway 1987). Figure 6 illustrates an agroecosystem which is characterised by a mixture of viticulture, cereal and livestock production in the southern part of the Vercors Regional Natural Park in southeastern France. This agroecosystem will be presented and discussed in the second half of the chapter.

The third agroecological approach is restricted to the plot or field scale. Here, research almost exclusively analyses crop–pest and crop–weed interactions with a particular emphasis on natural processes (Fig. 7). In some cases, the impact of pesticides on crops and natural flora and fauna also is analysed. Research on animal production within this more restricted approach focuses often on the single animal, or the resources from single or several pastures, but does not really consider the interactions and implications for the agroecosystem or the environment.

Fig. 6 Viticulture and cereal production in the southern part of the Vercors Regional Natural Park in southeastern France

Fig. 7 Agroecological research at the plot scale: ladybird beetles are natural predators of aphids on organic wheat in southeastern France

4   Interpretation of Agroecology as a Science, a Movement and a Practice

As already mentioned in the historical overview, interpretations of agroecology grew and diversified from the 1970s onwards. Besides the scientific discipline, agroecology as a movement gradually emerged in the 1970s, and as a set of practices beginning in the 1980s.

In general, the basis for the agroecological movements was set in the environmental movements of the 1960s which often emerged as a consequence of the unexpected impacts of industrialised agriculture after the Green Revolution. Researchers with a focus on short-term yields and economic returns considered environmental and social factors to be externalities. Public policies rarely considered the environmental impact of agriculture, nor the social consequences of uni-dimensional rural development focused on production and economics. This environmentalism was primarily concerned with the impacts of toxic substances, in particular pesticides, on the environment. Other non-agricultural topics of these environmental movements included industrial pollution, nature conservation, and distribution of benefits. Nevertheless, from the 1960s to the 1980s the term *agroecology* generally was not used to explicitly describe a movement. This started in the 1990s, especially in Latin America and in the USA, when the word started to be used to express a new way of considering agriculture and its relationships with society.

Almost within the same time period, a third word usage emerged, that of recognising a set of agricultural practices which aims at developing a more "environment-friendly" or "sustainable" form of agriculture. One of the origins of agroecology as a practice began during the middle to late 1970s in Latin America (Hernández Xolocotzi 1977). It was seen as the basis for an agricultural development framework, supported by ecologists, agronomists and ethnobotanists working especially in Mexico and Central America. Agroecology helped local farmers to improve their indigenous farming practices as an alternative to high input, chemical-intensive agriculture promoted by international corporations (see Gliessman 2007; Altieri 1995, 1989). Practices such as conservation of natural resources, adapted soil fertility management and conservation of agrobiodiversity are the practical bases for the different agroecological movements in Latin America. Another example of agroecology as a practice is described by Arrignon (1987), who illustrated technical, more adaptive methods in agriculture such as water and livestock management or anti-erosion measures as a basis for rural and sustainable development in arid and sub-humid areas.

Today, the three main interpretations of agroecology, as a movement, as a science and as a practice, can be further specified in relation to (i) definitions for the scientific discipline, (ii) major objectives of the movements, and (iii) different scales (Fig. 8). For example, the agroecosystem scale is pertinent for the agroecosystems ecology approach of the scientific discipline as well as for the environmentalism and rural development movements.

In addition, the term 'agroecology' as a movement, as a science and as a practice is used in many countries in a combined way, and in some situations they are even strongly intertwined. In Germany for example, agroecology has a long tradition as a scientific discipline, and the term is not associated with a movement or with practices (Wezel et al. 2009). In the USA and in Brazil, agroecology is used to describe all three activities, with predominance toward science in the USA and a stronger movement and/or practice emphasis in Brazil. In France, agroecology was mainly known until recently as a practice, but is now increasingly also seen as a scientific discipline. In countries where agroecological movements are well established, the

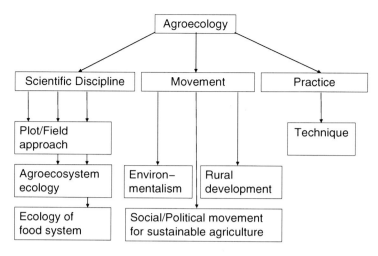

Fig. 8 Diversity of current meanings of *agroecology* and their relation to scale

idea of practice is strongly connected, or even incorporated, into these movements. Here, they merge for the development of the objective and to assist in the transition into sustainable agroecosystems (e.g. Gliessman 2007; Wojtkowski 2002) and with other models such as traditional, alternative or organic farming. There is large overlap in use of these several terms.

In this sense, agroecology encourages farmers and extension personnel to participate in the design of new systems, and also contribute to social or political movements. This is particularly the case for Brazil, and to a certain extent for the USA and France. In these situations, there is often a link between a political vision (the movement), a technological application (the practice) to achieve the goals, and a way to produce the knowledge (the science). A key point here for the scientists is to assess how these tight connections may influence the science of agroecology, where there will be application to meet a political vision using a set of technological practices. This association raises serious questions for some who have seen science more as an objective activity that is somewhat disconnected from practice. For example, when the science of agroecology is defined as the scientific basis of a sustainable development strategy which emphasises food sovereignty, conservation of natural resources and agrobiodiversity and empowers rural social movements, the science itself may appear as an advocacy activity that will be impacted by diverse goals and applications of results. Instead of considering agroecology as a general matrix including the wider range of disciplines (Caporal et al. 2006), collaborations between agricultural, natural and social scientists should help to clarify such embedded interpretations of agroecology. One must ask, of course, whether this connection between the science and the practice is any different from our accepted linkages between research and recommendation, such as studies of fertilizing rates, types of effective pesticides, or scheduling of irrigation.

5   Agroecology and Nature Conservation

In this section we will analyse and show where the links between nature conservation and agroecology exist. This will include the topics of species, ecosystem and landscape conservation, agrobiodiversity, functional biodiversity, biodiversity management, protected areas management, and international conventions.

Different aspects of nature conservation (e.g. diversity, biodiversity, conservation) have become gradually integrated into agroecology publications since the 1990s, in most cases linked to the agroecosystems approach within agroecology; and since the 2000s it also has been linked to the food systems approach (Fig. 5). The major starting point of this shift was probably the United Nations Conference on Environment and Development, held in Rio de Janeiro, Brazil in 1992, which raised awareness of the topics of conservation and management of biodiversity on the world's agenda. Today, it is widely accepted that to develop sustainable agriculture in different agroecosystems, nature conservation aspects cannot be ignored (e.g. Flade et al. 2006). In general, it is considered that conservation of biodiversity in agroecosystems is vital for the long-term functioning and stability of an agroecosystem (e.g. Altieri 1999; Collins and Qualset 1999). Although somewhat vague, it is more specific with the approaches of agrobiodiversity and functional biodiversity. Even though the ecological literature provides many insights (e.g. Hooper et al. 2005; Fonseca and Ganade 2001), much still needs to be learned about biodiversity as a natural capital for providing ecosystem goods and services for agriculture (Jackson et al. 2007).

According to Wood and Lenné (1999), agrobiodiversity is the total variation within and among species of living organisms related to agriculture. Agrobiodiversity includes all crops and livestock and their wild relatives and all interacting species such as pests, diseases, weeds, pollinators and biological control organisms, and the many organisms controlling nutrient cycling. Although Wood and Lenné (1999) do not include explicitly the habitats of these species (or more generally ecosystems) in their definition, agrobiodiversity cannot be seen without them as they are the necessary spatial areas where the organisms of agrobiodiversity are living (at least during parts of the life cycles). The importance of the functional role of agrobiodiversity has been stressed by Swift and Anderson (1994). The biotic components of agroecosystems can be divided up into three types: productive, resource (beneficial) and destructive. The productive biota includes crops and livestock. The resource biota contribute positively to the productivity of the system, e.g. via pollinators, plants of fallows, and much soil biota controlling nutrient cycling. Finally, the destructive biota includes weeds, pests and pathogens.

Functional biodiversity is defined as that part of the biodiversity composed of clusters of elements (at the gene, species or habitat level) providing the same (agro) ecosystem service, that is driven by within-cluster diversity (Moonen and Barberi 2008). The restoration of functional biodiversity of the agricultural landscape must be a key strategy in sustainable agriculture (Altieri 1994), although in practice it is probably much more difficult to achieve this as high levels of biodiversity (and with functional biodiversity) in managed landscapes are more likely to be maintained for reasons of intrinsic values or utilitarian (direct use) than for functional or ecosystem service values (Swift et al. 2004).

The functional biodiversity approach applies to the plot and agroecosystem scales as well. For example, which insect species, species groups or taxa should be conserved or promoted within fields to improve natural pest control, or which landscape elements or corridors (hedgerows, tree lines, thickets, herbaceous vegetation strips, natural grasslands, etc.) should be maintained as habitats for pollinators, predators of pests, or as physical elements for wind and water erosion control as well as for water regulation and purification. Nevertheless, the agrobiodiversity and functional biodiversity approaches do in practice often neglect these parts of biodiversity (species and ecosystems) which seem to not have obvious functions for agriculture or which cannot clearly be identified with the provision of ecosystem services (Jackson et al. 2007). For example, the values of rareness, uniqueness and aesthetic beauty, as well as non-agricultural uses of agroecosystems for recreation and hunting are not considered. The rareness or uniqueness of species or ecosystems and the question of how to protect them have been the foundation of "traditional" nature conservation over the last century. The concept of more "progressive" nature conservation was developed three decades ago by the IUCN (1980) in defining targets for nature conservation that went beyond the preservation of species and pristine ecosystems. Among them are the maintenance of essential ecological processes and live-support systems, protection of genetic diversity and the management of human use of the biosphere in a sustainable manner. The latter consequently targets agricultural landscapes as agriculture is the most important form of land use worldwide (Millennium Ecosystems Assessment 2005). Thus, the major question is how nature conservation can be incorporated into agricultural landscapes. Figure 9 shows that two major concepts exist: integration or segregation (Werner et al. 2006; Hampicke 1988). For the latter, areas for nature conservation and areas for crop production are separated, sometimes shielded by buffer zones. For the integration concept, nature conservation and agriculture are combined in the same area, or they are webbed onto separate areas, but close to each other.

All of these concepts have their pros and cons depending on the objectives and the areas where they are intended to be implemented. The concept of combining is a major challenge for many actual agroecological research projects working at the plot scale. Topics vary for example from (i) adaptive fertilization and mowing practices which allow plant species diversity to increase in pastures and meadows, (ii) to maintaining rare and endangered plant species, or (iii) to preserving habitats for certain nesting bird species. Topics related to cropping include, for example, (i) no/reduced tillage to increase soil organism abundance and diversity, and (ii) different crop rotations integrating cover crops (often legumes) to preserve diversity of soil organisms or to provide habitats for beneficial insects. For the agroecosystems approach to agroecology, the concept of webbing is probably the most attractive for areas with a relatively heterogeneous small-scale landscape matrix consisting of fields, pastures and different landscape elements as well as forests. These landscapes can still be found in many countries (at least in Europe!), often in hilly or mountainous areas

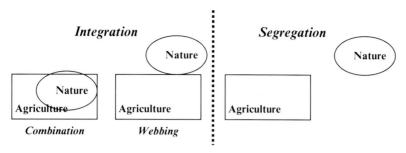

Fig. 9 Concepts in the spatial relationship between nature conservation and agriculture (Adapted from: Werner et al. (2006) and Hampicke (1988))

where industrialized agriculture has its limits. The non-productive areas in these landscapes are especially interesting as habitats for many different plant and animal species. Here, many research questions such as the required minimum area for different species, the necessary degree of habitat connectivity, the most adaptive type of management for the landscape elements as well as how to get better positive impacts from beneficial insects in hedgerows on the adjacent cropped field are not yet sufficiently answered. For monotonous, intensively used agricultural landscapes the concept of segregation might actually remain, in many cases, as the most feasible option because economic interests for many farmers or agricultural enterprises, as well as the general objective of feeding an increasing world population will probably prevent adopting more environmentally friendly farming practices in the near future. One important objective for agroecological research will be to develop agricultural practices in these intensively used landscapes which have less impact on biodiversity, or which might favor certain elements of biodiversity.

Protected areas have been established on approximately 10% of the world's land surface (Chape et al. 2003). Different agricultural activities are carried out in many of these protected areas, but especially in three out of the six major protected areas categories developed by the IUCN (UNEP-WCMC 2010; Chape et al. 2003): II National Parks, V Protected Land-/Seascape, and VI Managed Resource Protected Area. Different management strategies for the conservation of species and habitats in agricultural landscapes are implemented in many of these areas. Although the effectiveness of these strategies may vary considerably from country to country and from one area to another, they are generally expected to achieve better conservation results than outside the areas.

A particular world-wide strategy for practicing sustainable use is the biosphere reserve concept. It was initiated by the Man and Biosphere program (MAB) in 1970. Since then, 551 biosphere reserves in 107 countries (UNESCO 2010a) have been established, in which the major objective is to reconcile the conservation of biodiversity with its sustainable use. These biosphere reserves should function as model areas where sustainable use is commonly taught, practiced, and information about it communicated to other areas and regions. Thus, biosphere reserves have three interconnected functions: (i) conservation of landscapes, ecosystems, species and genetic variation, (ii) economic, human and culturally adapted development, and (iii) logistic support for research, monitoring, environmental education and training. This offers unique potentials to practice sustainable agriculture in the buffer and transition zones of the biosphere reserves, and link these projects with rural development (e.g. Shorfheide-Chorin Biosphere Reserve, Germany; Luberon Biosphere Reserve, France). The core zone of the biosphere reserves is restricted to strong nature conservation without any use (except for scientific research and monitoring).

Not to distant from the biosphere concept, but with a stronger focus on conservation, is that of the World Heritage Sites. Under the World Heritage convention (UNESCO 2010b), cultural and natural sites as well as mixed sites are protected, those that have outstanding value to humanity and which meet at least one out of ten selection criteria. The natural sites can partly include agricultural landscapes, for example the Alexander von Humboldt National Park in Cuba (Wezel and Bender 2002) as well as areas with very extensive agricultural use such as the Manú National Park, Peru (Ohl et al. 2008) or the Laponian Area, Sweden (UNESCO 2010b). Some mixed sites are cultural landscapes where traditional agriculture is still practiced, such as in the Pyrénées-Mont Perdu, France/Spain, or the Cliffs of Bandiagara (Land of the Dogons), Mali.

From the World Heritage Convention we come to two other conventions which are of special importance for nature conservation in agricultural landscapes and thus also for agroecological analyses: the CBD (Convention on Biological Diversity) and the CMS (Convention on Migratory Species). The CBD is presently the most well-known among international conventions. It was signed during the Earth Summit in Rio de Janeiro, Brazil in 1992, and entered into force in December 1993 (i.e., it was ratified by a sufficient number of countries) (UNEP, UN 2010). Important elements of the link between biodiversity, conservation and sustainable agriculture were already written under Article 6 (General Measures for Conservation and Sustainable Use) and Article 10 (Sustainable Use of Components of Biological Diversity) in the convention text. Thereafter, elemental links were worked out more explicitly in the thematic programme Agricultural Biodiversity. Since the ratification of the convention, many strategies and measures have been developed to conserve biodiversity and at the same time to sustainably use it. This especially concerns the use of different (old) crop and livestock varieties, their *in situ* conservation in agroecosystems, and the sharing of the benefits arising from commercial and other forms of utilisation of these genetic resources in a fair and equitable way. *In situ* conservation of this type of agrobiodiversity is an important part of nature conservation as not only species, but also their different habitats must be conserved. Moreover, the CBD aims to protect and encourage customary use of biological resources in ecosystems and agroecosystems in accordance with traditional cultural practices, among them different traditional agricultural practices.

The CMS also is an international convention (UNEP, CMS 2010), but less well-known than the CBD. As in the CBD, it offers different strategies for the conservation of biodiversity, but is restricted to species that migrate. Migrating species, as seen under the CMS, migrate from reproduction areas to summer or winter feeding areas and back, crossing over national boundaries. Many of these species (in particular migrating birds) feed or reproduce not only in natural areas, but often in different types of agroecosystems. Thus, management of these agroecosystems is of vital importance for their survival. This means that for protection of necessary habitats within the agroecosystem, low impact agricultural practices must be adopted.

6   Agroecology and Territorial Development

As we discussed previously, agroecology at the agroecosystem scale in most cases only focuses on agricultural activities from an agronomical and ecological point of view. Instead, the food system approach includes also the interactions, relations and influences from society, politics and economy with/on agroecosystems and their agricultural production. To analyse, explore and evaluate the relations and interactions between agriculture, other activity sectors and the society, the concept of territorial (or local) development will be useful. Thus, a close link between territorial development and the food system approach of agroecology can be assumed.

The concept of territorial development is generally used to define a set of social, cultural and economic processes that promote the economic dynamics and improvement of life quality of the population of a territory. This can be a metropolitan, rural (in this case the term territorial development can be used in an equivalent way to rural development) or peri-urban area. The keywords of this concept are valorisation of indigenous resources, inter-sectorial development, valorisation of local identity, self-control of development processes, solidarity, and democracy (Pecqueur 1989). Since the Rio Summit of 1992, environmental issues have been included, and in many cases it is now spoken of as territorial sustainable development (Ministère de l'Ecologie et du Développement Durable 2005).

This concept was a successful experiment in Europe within the framework of the LEADER programs which were designed to help rural stakeholders consider the long-term potential of their local region (European Commission 2010a). The territorial approach is described as a "bottom-up approach in the design and implementation of the programs, the integrated and pilot character of the activities, and the networking of all actors involved in the field of (rural) development" (European Commission 2010b). Often this applies to smaller regions (e.g. in France larger than one commune, but less than a district). The objective is to create a common place which is built by stakeholder strategies to valorise local resources in connection with the global society (Conseil Général du GREF 1999). Territorial development means in particular that development has to use a global approach in considering that the relations between the different sectors and the different stakeholders should be encouraged and reinforced to achieve synergies among them. From an agroecological perspective, the territorial development approach means that relations outside of agriculture, but which interfere with it, have to be taken into account. These relations can be technical, economical, social, human, or cultural. The relations between agriculture and the territory can impact each other in different ways:

- Agricultural impacts on the territory: farmers are generally the most important land users and their agricultural practices have effects on the territory, including impacts on landscape management, biodiversity and water quality through agricultural pollution.
- Territorial impacts on agriculture: there is often competition for land between different activities (tourism, forestry, settlements, protected areas); tourists or local populations have a demand for local agricultural products; local activities offer job supplies for farmers or their families (tourism, production sector, service sector) allowing multiple activities for farmers.
- Organisation at the territorial level impacts agriculture (farmer trade unions, cooperatives, farmer associations, agricultural services, extension services).

In looking at territorial development and the food system approach to agroecology, we must also take into consideration the issue of scale. The criteria for defining scales are not necessarily the same. For the food system approach it is difficult to attribute a clearly defined scale, as it can go beyond an agroecosystem scale and enter the dimension of the food system. Nevertheless, the agroecosystem is the pertinent scale for many analyses within agroecology of the food system when research questions are linked to a certain territory (this applies also in some cases for agroecology as a movement!). As mentioned before, depending on who defines an agroecosystem, it can be a farm, or a local or regional landscape where a certain type of agriculture is practiced. In contrast, the criteria used to determine a territory for territorial/local development can be geographical, economical or sociological, and in most cases in relation with administrative units (e.g. districts, sub-districts, association of communes, communes). In some cases there can be a good overlap between an agroecosystem and a territory, but often territories are larger than agroecosystems.

The role of agriculture within the concept of territorial development is closely linked with the promotion of multifunctional agriculture. Multifunctionality means that agriculture simultaneously carries out several functions. In general, these functions concern the three classic pillars/goals of sustainable development: economic sustainability, social sustainability and environmental sustainability. The functions of agriculture can be (i) economic (e.g. sale of products, income, services like clearing snow from roads, maintenance of ditches), (ii) societal (e.g. work place, employment, family living place, promotion of local identity, education), or (iii) environmental (e.g. preservation of a certain landscape type which is attractive for tourism, creating heterogeneous landscapes and often higher diversity of species and ecosystems). Different approaches for multifunctionality are used in the literature. The OECD (2001) for example, defines key elements of multifunctionality with (i) the existence of multiple commodity and non-commodity outputs that are produced by agriculture, and (ii) that some of the non-commodity outputs exhibit the characteristics of externalities, public goods or services, of which the public benefits without compensating the farmer because markets for these goods or services do not exist in most cases, or function poorly (OECD 2001; see also Lovell et al. 2010).

Mundler (2008, 2006) as well as Jauneau and Mundler (2007) indicate that accent must be put on global coherence (indeed, on the inseparable nature) of agriculture's different functions. The main challenge is to think of the role of agriculture and its relationships with other components in the local society when studying the multidimensional aspects of farm household activities and their contributions to social and economic development in general. From this perspective, multifunctionality concerns not only agricultural products, but also the range of practices and services of farmers and their families in the spaces they occupy, thus relatively closely linked to the food system approach in agroecology. In this sense, Lovell et al. (2010) propose that the fields of agroecology and multifunctionality, which have developed separately, might be integrated to form a more comprehensive offering for the sustainable design of agricultural landscapes.

To illustrate the theoretical concept of territorial development and multifunctionality of agriculture we will use the example of a cooperative research venture with the Regional Natural Park "Monts d'Ardèche" (Jauneau and Mundler 2007). In France, a Regional Natural Park is a protected area which constitutes a form of contractual protection between a local community and the Region (a federal state in France, in our case the Region Rhône-Alpes) where the Park is located. Regional Natural Parks in France are classified under IUCN category V (Protected Land/ Seascape). According to IFEN (1998), Regional Natural Parks are territories with rich and threatened natural assets, and are thus the subjects for development projects based on the protection and valorisation of these assets. A Regional Natural Park is created by decree for a maximum duration of 10 years. Its creation and the renewal of its classification are set in a contractual document defining the objectives of nature conservation as well as economic, social and cultural development. Regional Natural Parks are managed by a mixed syndicate which represents various local communities and concerned stakeholders. In general, agriculture plays an important role in a Regional Natural Park, which often consists of different agroecosystems (in most cases these are traditional agricultural landscapes) typical for the area and which the Park wants to conserve and manage. At the same time, the Park tries to develop other activities like tourism, services and nature conservation.

The Regional Natural Park of "Monts d'Ardèche" is located southeast of the Rhône-Alpes Region (Parc Naturel Régional Monts d'Ardèche 2010). It was created in 2001 on a chestnut producers' initiative (Dodelin and Pluvinet 2006). The Regional Natural Park covers 180,000 ha of a largely mountainous area (Fig. 10). The Park's objectives are to protect the region's patrimony as well as to sustainably manage the natural environment and the Park's landscapes (e.g. relicts of volcanoes, rivers, chestnut and blueberry production areas, vineyards, water mills, old agricultural terraces).

In the framework of the above mentioned research cooperation, four groups of functions, that can be managed by the farmers in addition to agricultural production, have been specified by a steering committee consisting of people elected from two communes, members of the Regional Chamber of Agriculture, persons from the Regional Natural Park, and researchers. Within these groups different themes are defined (Table 1).

Fig. 10 Typical landscapes in the Regional Natural Park "Monts d'Ardèche" in southeast France

Table 1 Functions and themes for agriculture and for territorial development in the Regional Natural Park "Monts d'Ardèche", France

Functions	Themes
Land management	Water management
	Biodiversity
	Landscape quality
	Waste management
	Fire prevention
Local cultural patrimony	Architecture of agriculture housing
	Terraces, low walls
Contribution to local employment	Quantitative and qualitative contribution to employment
Participation in local development	Training, social integration activities
	Tourism on the farm
	Contribution to local production and processing networks (slaughterhouses, dairies, small enterprises for livestock fodder)
	Open access to farm paths and parcels for other users (tourists, hikers)

After this first step, a methodology was proposed from the research side on how to evaluate and fulfill these different functions and how to remunerate the farmers for their efforts. Different propositions have been accepted by the Steering Committee, but have not yet been implemented because of local problems, such as a change in responsibility of different important stakeholders and lack of local willingness for immediate implementation.

7   Agroecology and Ecotourism

To illustrate the links between agroecology and ecotourism, we use the example of the Regional Natural Park "Vercors" (Vercors Regional Natural Park 2010). This Park also applies the concept of territorial development in focusing on touristic activities, among other sectors, and especially on ecotourism. Since their foundation in 1990, the International Ecotourism Society (2010) continues to use the following definition for *ecotourism*: "Responsible travel to natural areas that conserves the environment and improves the well-being of local people". According to Honey (2008), the seven defining points for *ecotourism* are:

1. Involves travel to natural destinations
2. Minimizes impact
3. Builds environmental awareness
4. Provides direct financial benefits for conservation
5. Provides financial benefits and empowerment for local people
6. Respects local culture
7. Supports human rights and democratic movements.

7.1   Ecotourism in the Regional Natural Park of Vercors (France)

The Vercors Regional Natural Park is a natural limestone citadel, extending over 186,000 ha between the Isère Valley to the north and the Diois Valley to the south (Vercors Regional Natural Park 2010). Over the centuries water has cut through its cliffs forming deep gorges, natural cirques, caves and chasms. Its wooded plateaus and valleys, shaped by generations of farmers, are home to a remarkable variety of wildlife. On the north side, the harsh rigour of the Alps prevails. On the gentler and lower south side, there is influence from a Mediterranean climate, and a heterogeneous landscape of vineyards, cereal and lavender fields, and pastures (Fig. 6). The Vercors Regional Natural Park was founded in 1970 to protect and promote these assets, maintain economic activity and promote harmonious human and environmental development.

In 2003, the Vercors Regional Natural Park received the certification "European Charter for Sustainable Tourism in Protected Areas" which is awarded by the Europarc Federation (2010). The reception of this certification allowed for recognition of the actions which had been realised or supported by the Park as well as to define a general frame of future objectives to reinforce different actions for sustainable tourism (in France, the term "ecotourism" is often replaced by the term "sustainable tourism"; whereas sustainable tourism is not necessarily carried out in nature!). Among the different actions are:

- Development of the label "Reception of the Vercors Park" for different types of tourist accommodations in the Park, but also to accompany hiking and other touristic activities as well as training and promotion. Presently, 70 hotels, restaurants and pensions (bed and breakfast) have received the label by assuring that they (i) respect and value nature by using environmental management measures (saving energy, using solar energy, etc.), (ii) help tourists to discover the territory's resources (landscape, traditions, leisure, culinary patrimony, etc.), and (iii) welcome guests personally and with a hearty touch.
- Support of development of touristic actions such as information trails or local museums for highlighting the nature, landscapes and cultural patrimony of the Park.

- Implementation of environmental management actions such as saving energy, and developing renewable energy in the touristic infrastructure.
- Implementation and organisation of open air activities (promoting sporting activities, protecting and monitoring of areas of concern, managing conflicts between use and conservation, establishing a long distance hiking trail network, developing youth holiday centres, etc.).
- Promoting local food products for tourists, including the possibility for local farmers to directly sell their products.

The link between ecotourism and agroecology in the Vercors Park concerns different points: management of agroecosystems, nature conservation in agroecosystems, multifunctionality of agriculture, local food products, and income for farmers. The intention of the Park is, for example, to manage the traditional agroecosystem of the valley pasture on the plateau of the Vercors and the alpine pastures and meadows in the surrounding alpine mountains in such a way that this attractive landscape is maintained for tourists (hiking and other sports activities), but also guarantees sufficient income for farmers and conserves special species or ecosystems which are associated with this agroecosystem (e.g. adapting mowing dates in valley pastures to conserve an endangered bird species such as the Corn Crake or Landrail, *Crex crex* [L.]). In addition, part of the milk produced in this agroecosystem is used to make a local variety of blue cheese, the "Bleu du Vercors-Sassenage", which is promoted with the label "Product of the Vercors Park" and a PDO label (Protected Denomination of Origin). A PDO is a European label which guarantees (i) the specificity/originality of the product (to distinguish it form standard products) and its link to the territory where it was produced, (ii) the quality of the product, and (iii) special regulations for production processes. In France, the PDO label is called AOP – Appellation d'Origine Protégée. The local blue cheese is sold either at the dairy cooperative or directly at farms. Similar initiatives are carried out for the agroecosystem at the southern feet of the Vercors Mountains. This agroecosystem, consisting of a mixture of vineyards, cereal fields, lavender fields and pastures (Fig. 6), also produces different local products with or without labels (e.g. different types of wine, lavender, cheese, meat, trout) which can also be sold directly to tourists.

8   Conclusions

Presently, different interpretations and definitions of *agroecology* are used. They vary from agroecology as a practice, agroecology as a movement, and varying approaches to agroecology as a scientific discipline. The evolution of the interpretations and definitions are often closely linked to differences in the historical development of agroecology in different countries and parts of the world. What all these approaches and interpretations have in common is the search for more sustainable agricultural systems and agroecosystems within the goal of sustainable development.

The latter demands that the more traditional agroecology approaches at the plot/field, and farm/agroecosystem scales be complemented by taking into consideration the social, political and cultural questions and dimensions for practical application. In this perspective, the concept of territorial development seems to be very promising as it considers social, cultural and economic processes, as well as environmental issues for improving the quality of life for the population of a territory. Closely linked to this is the concept of multifunctionality. From an agroecological perspective, multifunctionality of agriculture might be even more important as it focuses more on agriculture and as it includes more clearly natural resources and nature conservation aspects. In contrast, the topic of ecotourism might be better integrated into the territorial development concept. For Lovell et al. (2010), the farm level is the scale at which most overlap exists between agroecology and landscape multifunctionality approaches. This is where management activities occur and decisions regarding land use are made. According to Wilson (2008), it is at the farm level where the most direct expression of multifunctional action and thought are found, and where the most important level for mediation of multifunctional influences are exerted by other scales in the hierarchies of multifunctionality. Nevertheless, the actions of individual farmers at the farm scale can also have far-reaching impacts beyond the farm to local and even regional scales (Shellhorn et al. 2008, cited in Lovell et al. 2010).

Rural or territorial development is already present in different agroecology publications, particularly in interpretations of agroecology as a movement. In contrast, the link to nature conservation and ecotourism has so far not been clarified. In this chapter, we have approached the filling of this gap by illustrating this connection with different examples. Whereas different aspects of nature conservation seem to be more pertinent for the plot and agroecosystem approach of the scientific discipline of agroecology, ecotourism plays a more important role in the broader food system approach, and in most cases is closely linked to questions within rural or territorial development.

Acknowledgements We thank W. Bruce Campbell very much for his comments and corrections on the manuscript. The discussions on agroecology and inputs from Stephan Bellon, Christophe David, Thierry Doré, Charles Francis, Virginie Soldat and Dominique Vallod also are highly acknowledged as they laid the foundation for the first half of this review.

References

Alard D (1994) Grassland vegetation as an indicator of the main agro-ecological factors in a rural landscape: consequences for biodiversity and wildlife conservation in central Normandy (France). J Environ Manage 42(2):91–109

Altieri MA (1987) Agroecology: the scientific basis of alternative agriculture. Westview, Boulder, 227 pp

Altieri MA (1989) Agroecology: a new research and development paradigm for world agriculture. Agric Ecosyst Environ 27:37–46

Altieri MA (1993) Ethnoscience and biodiversity: key elements in the design of sustainable pest management systems for small farmers in developing countries. Agric Ecosyst Environ 46:257–272

Altieri MA (1994) Biodiversity and pest management in agroecosystems. Haworth, New York, 185 pp

Altieri MA (1995) Agroecology: the science of sustainable agriculture. Westview, Boulder, 433 pp

Altieri MA (1999) The ecological role of biodiversity in agroecosystems. Agric Ecosyst Environ 74(1–3):19–31

Altieri MA, Letourneau DK, Davis JR (1983) Developing sustainable agroecosystems. Bioscience 33:45–49

Arrignon J (1987) Agro-écologie des zones arides et sub-humides. Editions G.-P. Masonneuve & Larose et ACCT, Paris, 283 pp

Azzi G (1928) Agricultural ecology (in Italian). Edition Tipografia Editrice Torinese, Turin, 237 pp

Azzi G (1956) Agricultural ecology. Constable & Company, London, 424 pp

Bensin BM (1928) Agroecological characteristics, description and classification of the local corn varieties chorotypes. Book (Publisher unknown)

Bensin BM (1930) Possibilities for international cooperation in agroecological investigations. International review of agriculture. Mon Bull Agric Sci Pract (Rome) 21:277–284

Bensin BM (1938) Agroecology as a basic science of agriculture. Book (Publisher unknown)

Caporal FR, Costabeber JA, Paulus G (2006) Agroecologia: matriz disciplinar ou novo paradigma para o desenvolvimento rural sustentavel. In: 3rd Congreso Brasileiro de Agroecologia, Florianopolis, Anais: CBA, 25 pp

Carroll CR, Vandermeer JH, Rosset PM (1990) Agroecology. McGraw-Hill, New York, 641 pp

Chape S, Blyth S, Fish L, Fox P, Spalding M (2003) United Nations list of protected areas. In: IUCN – The World Conservation Union, IUCN, Gland/Cambridge and UNEP World Conservation Monitoring Centre, Cambridge, 45 pp. http://www.unep-wcmc.org/wdpa/unlist/2003_UN_LIST.pdf. Accessed Apr 2010

Clements DR, Shrestha A (2004) New dimensions in agroecology for developing a biological approach to crop production. In: Clements DR, Shrestha A (eds.) New dimensions in agroecology. Haworth, New York, pp 1–20

Collins WW, Qualset CO (eds.) (1999) Biodiversity in agroecosystems. CRC Press, Boca Raton/New York, 334 pp

Conseil Général du GREF (1999) Identification et structuration des territoires – recueil de recommandations méthodologiques. Ministère de l'Agriculture et de la Pêche, 35 pp

Conway GR (1987) The properties of agroecosystems. Agric Syst 24:95–117

Cox GW, Atkins MD (1979) Agricultural ecology: an analysis of world food production systems. W.H. Freeman and Sons, San Francisco, 721 pp

Department of Crop Science, Agroecology Section, University of Göttingen (2008) Introduction. http://wwwuser.gwdg.de/~uaoe/Agroecology.html. Accessed Jan 2008

Dodelin P-Y, Pluvinet P (2006) Valorisation des fonctions non marchandises de l'agriculture: Expérimentation sur deux communes du Parc Naturel Régional des Monts d'Ardèche. Masters thesis, ISARA Lyon, Lyon, 118 pp

Dover M, Talbot LM (1987) To feed the earth: agro-ecology for sustainable development. World Resources Institute, Washington, 88 pp

Europarc Federation (2010) European Charter for sustainable tourism in protected areas. http://www.european-charter.org. Accessed May 2010

European Commission (2010a) Introduction to Leader+. http://ec.europa.eu/agriculture/rur/leaderplus/index_en.htm. Accessed May 2010

European Commission (2010b) Leader+ Lignes directrices pour l'évaluation ex ante des programmes (Guidelines for the ex ante evaluation of programmes). http://ec.europa.eu/agriculture/rur/leaderplus/guidelines/eval_fr.pdf. Accessed Apr 2010

Flade M, Plachter H, Schmidt R, Wener A (2006) Introduction. In: Flade M, Plachter H, Schmidt R, Wener A (eds.) Nature conservation in agricultural ecosystems. Results of the Schorfheide-Chorin Research Project. Quelle & Meyer, Wiedbaden, pp 1–30

Fonseca CR, Ganade G (2001) Species functional redundancy, random extinctions and the stability of ecosystems. J Ecol 89:118–125

Francis C, Lieblein G, Gliessman S, Breland TA, Creamer N, Harwood R, Salomonsson L, Helenius J, Rickerl D, Salvador R, Wiedenhoeft M, Simmons S, Allen P, Altieri M, Flora C, Poincelot R (2003) Agroecology: the ecology of food systems. J Sustain Agr 22(3):99–118

Friederichs K (1930) Die Grundfragen und Gesetzmäßigkeiten der land- und forstwirtschaftlichen Zoologie. Vol 1: Ökologischer Teil, vol 2: Wirtschaftlicher Teil. Verlagsbuchhandlung Paul Parey, Berlin, 417 and 443 pp

Gliessman SR (ed) (1990) Agroecology: researching the ecological basis for sustainable agriculture, vol 78, Ecological studies. Springer, New York

Gliessman SR (1997) Agroecology: ecological processes in sustainable agriculture. Chelsea, Ann Arbor, 384 pp

Gliessman SR (2007) Agroecology: the ecology of sustainable food systems. CRC/Taylor & Francis, New York, 384 pp

Hampicke U (1988) Extensivierung der Landwirtschaft für den Naturschutz: Ziele, Rahmenbedingungen und Maßnahmen. Schriftenreihe Bayerisches Landesamt für Umweltschutz 84:9–35, Germany

Hénin S (1967) Les acquisitions techniques en production végétale et leurs applications. Economie Rurale, SFER, Paris, pp 31–44

Henricksen BJ (1986) Determination of agro-ecological zones in Africa: ILCA activities and expectations. ILCA Bull 23:15–22

Hernández Xolocotzi E (ed.) (1977) Agroecosistemas de Mexico: Contribuciones a la Ensenanza, Investigacion, y Divulgacion Agricola. Colegio de Postgraduados, Chapingo

Heydemann B (1953) Agrarökologische Problematik, dargetan an Untersuchungen über die Tierwelt der Bodenoberfläche der Kulturen. Ph.D. thesis, University of Kiel, Kiel

Honey M (2008) Ecotourism and sustainable development: who owns paradise? 2nd edn. Island Press, Washington DC, 416 pp

Hooper DU, Chapin FS III, Ewel JJ, Hector A, Inchausti P, Lavorel S, Lawton JH, Lodge DM, Loreau M, Naeem S, Schmid B, Setälä H, Symstad AJ, Vandermeer J, Wardle DA (2005) Effects of biodiversity on ecosystem functioning: a consensus of current knowledge. Ecol Monogr 75(1):3–35

IFEN (Institut Français de l'Environnement) (1998) L'Environnement en France. Editions La Découverte et Syros et Institut Français de l'Environnement, Paris, 480 pp

International Ecotourism Society (2010) What is ecotourism? http://www.ecotourism.org/. Accessed Apr 2010

IUCN (International Union for Conservation of Nature) (1980) World conservation strategy: living resource conservation for sustainable development. IUCN, Gland, 44 pp

Jackson LE, Pascual U, Hodgkin T (2007) Utilizing and conserving agrobiodiversity in agricultural landscapes. Agric Ecosyst Environ 121:196–210

Jauneau J-C, Mundler P (2007) Remuneration of non-market functions in agriculture – experimentation in two communities of the Regional Natural Park of "Monts d'Ardèche". In: Proceedings of the 3rd international scientific conference on rural development, Lithuanian University of Agriculture, Kaunas, 6 pp

Johns ND (1998) Conservation in Brazil's chocolate forest: the unlikely persistence of the traditional cocoa agroecosystem. Environ Manage 23(1):31–47

Klages KHW (1928) Crop ecology and ecological crop geography in the agronomic curriculum. J Am Soc Agron 10:336–353

Klages KHW (1942) Ecological crop geography. Macmillan, New York, 615 pp

López i Gelats F (2004) A discursive approach to agricultural and rural policy in Europe. Paper presented at the 3rd global conference – environmental justice & global citizenship, Copenhagen, 12–14 Feb 2004. http://www.inter-disciplinary.net/critical-issues/ethos/environmental-justice-and-global-citizenship/project-archives/2004/session-6a-organic-agriculture-ecology-and-sustainability/. Accessed July 2009

Lovell ST, DeSantis S, Nathan CA, Breton Olson M, Méndez VE, Kominami HC, Erickson DL, Morris KS, Morris WB (2010) Integrating agroecology and landscape multifunctionality in Vermont: an evolving framework to evaluate the design of agroecosystems. Agric Syst 103(5):327–341

Martin K, Sauerborn J (2006) Agrarökologie. Ulmer, Stuttgart, UTB 2793, 297 pp

Millennium Ecosystem Assessment (2005) Ecosystems and human well-being: biodiversity synthesis. World Resource Institute, Washington, DC

Ministère de l'Ecologie et du Développement Durable (2005) Territorial sustainable development projects and local agenda 21, France, 6 pp

Moonen AC, Bàrberi P (2008) Functional biodiversity: an agroecosystem approach. Agr Ecosyst Environ 127:7–21

Moss SR (1980) The agro-ecology and control of black-grass, *Alopecurus-myosuroides huds*, in modern cereal growing systems. ADAS Quart Rev 38:170–191

Mundler P (2006) La multifonctionnalité de l'agriculture: enjeux théoriques et d'arbitrage entre politiques sectorielles et politiques territoriales. In: Jean B, Lafontaine D (eds.) La multifonctionnalité de l'agriculture et des territoires ruraux: enjeux théoriques et d'action publique. CRDT – GRIDEQ, Rimouski

Mundler P (2008) L'agriculture en Rhône-Alpes, une multifonctionnalité évidente? In: Cornu P, Mayaud JL (eds.) Nouvelles questions agraires. Exploitants, fonctions et territoires. Collection pluridisciplinaire du Laboratoire d'études rurales/La Boutique de l'Histoire éditions, Lyon/Paris, pp 275–298

Odum EP (1969) The strategy of ecosystem development. Science 164:262–270

OECD (2001) Multifunctionality: towards an analytical framework. OECD Publications Service, Paris, 28 pp

Ohl J, Wezel A, Shepard GH Jr, Douglas WY (2008) Swidden agriculture in a protected area: the Matsigenka native communities of Manu National Park, Peru. Environ Dev Sustain 10:821–843

Parc Naturel Régional des Monts d'Ardèche (2010) Bienvenue dans le Parc naturel régional des Monts d'Ardèche. www.parc-monts-ardeche.fr. Accessed Apr 2010

Pecqueur B (1989) Le développement local: mode ou modèle? Syros, Paris, 149 pp

Puia I, Soran V (1984) Agroecologie: ecosistem si agroecosistem. Agronomia, Cluj-Napoca

Shellhorn NA, Macfadyen S, Bianchi FJJA, Williams DG, Zalucki MP (2008) Managing ecosystem services in broader landscapes: what are the appropriate spatial scales? Aust J Exp Agr 48:1549–1559

Swift MJ, Anderson JM (1994) Biodiversity and ecosystem function in agricultural systems. In: Schulze E-D, Monney HA (eds.) Biodiversity and ecosystem function. Springer, Berlin, pp 15–41

Swift MJ, Izac A-MN, van Noordwijk M (2004) Biodiversity and ecosystem services in agricultural landscapes – are we asking the right questions? Agric Ecosyst Environ 104:113–134

Thomas VG, Kevan PG (1993) Basic principles of agroecology and sustainable agriculture. J Agric Environ Ethics 6(1):1–19

Tischler W (1950) Ergebnisse und Probleme der Agrarökologie. Schriftenreihe der Landwirtschaftlichen Fakultät Kiel 3:71–82

Tischler W (1965) Agrarökologie. Gustav Fischer, Jena, 499 pp

UNEP, CMS (2010) Convention on migratory species. www.cms.int. Accessed Apr 2010

UNEP, UN (2010) Convention on biological diversity. www.cbd.int. Accessed Apr 2010

UNEP-WCMC (2010) Defining protected area management categories. www.unep-wcmc.org/protected_areas/categories/index.html. Accessed Apr 2010

UNESCO (2010a) Biosphere reserves. http://portal.unesco.org/science/en/ev.php-URL_ID=4801&URL_DO=DO_TOPIC&URL_SECTION=201.html. Accessed Apr 2010

UNESCO (2010b) World Heritage list. http://whc.unesco.org/en/list. Accessed Apr 2010

Vavilov NJ (1957) Agroecological survey of the main field crops. Academy of Sciences of the USSR, Jerusalem/Monson, 442 pp

Vercors Regional Natural Park (2010) Introduction page of the Vercors Regional Natural Park. www.parc-du-vercors.fr/eng/. Accessed Apr 2010

Werner A, Roth R, Zander P, Meyer-Aurich A, Jarfe A (2006) Scientific background for nature conservation agriculture. In: Flade M, Plachter H, Schmidt R, Werner A (eds.) Nature conservation in agricultural ecosystems. Results of the Schorfheide-Chorin Research Project. Quelle & Meyer, Wiedbaden, pp 529–572

Wezel A, Bender S (2002) Agricultural land use in the coastal area of the Alexander von Humboldt National Park, Cuba, and its implication for conservation and sustainability. GeoJournal 57(4):241–249

Wezel A, Soldat V (2009) A quantitative and qualitative historical analysis of the scientific discipline of agroecology. Int J Agric Sustain 7(1):3–18

Wezel A, Bellon S, Doré T, Francis C, Vallod D, David C (2009) Agroecology as a science, a movement or a practice. A review. Agron Sustainable Dev 29:503–515

Wilson GA (2008) From 'weak' to 'strong' multifunctionality: conceptualising farm-level multifunctional transitional pathways. J Rural Stud 24:367–383

Wojtkowski PA (2002) Agroecological perspectives in agronomy, forestry and agroforestry. Science Publishers, Enfield, 356 pp

Wood D, Lenné JM (1999) Why agrobiodiversity? In: Wood D, Lenné JM (eds.) Agrobiodiversity: characterization, utilisation and management. CABI, Wallingford, pp 1–13

Organic Compost and Manufactured Fertilizers: Economics and Ecology

David C. Weindorf, James P. Muir, and Cesáreo Landeros-Sánchez

Abstract Compost is a highly diverse group of organic soil amendments which provides substantial nutritive fertility to soils. The benefits of compost addition to soils are vast and have been well documented by a growing body of research. Composts are manufactured in a variety of methods and scales from simple localized plots to large scale commercial operations. This review examines the role of organic matter in soils, the process of composting, and the physical, chemical, and biological properties of compost. The global use of compost and its agro-ecological implications is explored. The review concludes with appropriate uses of compost, its comparison to traditional commercial fertilizer, as well as some limitations for its proper use.

1 Introduction

A variety of soil amendment products and potential nutrient sources provide flexibility for agricultural and horticultural systems. However, comparing the cost and value of these different soil amendments is not as simple as it might seem.

D.C. Weindorf (✉)
Louisiana State University AgCenter, 307 MB Sturgis Hall, Baton Rouge, LA 70803, USA
e-mail: dweindorf@agcenter.lsu.edu

J.P. Muir
Department of Soil and Crop Sciences, Texas Agrilife Research, 1229 North US Highway 281, Stephenville, TX 76401, USA
e-mail: j-muir@tamu.edu

C. Landeros-Sánchez
Colegio de Postgraduados, Campus Veracruz, Apartado Postal 421, Veracruz, Mexico, C.P. 91700

Colegio de Postgraduados, Campus Veracruz, Km 86.5 Carretera Federal Xalapa-Veracruz, vía Paso de Ovejas, entre Paso San Juan y Puente Jula, Tepetates, Veracruz, México. C.P. 91690
e-mail: clandero@colpos.mx

W.B. Campbell and S. López Ortíz (eds.), *Integrating Agriculture, Conservation and Ecotourism: Examples from the Field*, Issues in Agroecology – Present Status and Future Prospectus 1, DOI 10.1007/978-94-007-1309-3_2,
© Springer Science+Business Media B.V. 2011

Dairy manure compost, for example, supplies not only the major nutrients (N, P, and K), but also a broad range of secondary nutrients, micronutrients and organic matter. These plant nutrients have economic value, which can be used to estimate compost value for comparisons with traditional fertilizer materials. Organic matter applications, such as dairy manure, can also improve water and nutrient holding capacity of the soil, reduce erosion, and reduce fluctuations in soil pH.

Nutrients in compost products are more stable and are typically released gradually over three or more years; whereas inorganic fertilizers are generally formulated to release nutrients within a year of application. Thus, a realistic assessment of compost value requires at least a 3-year time frame. Also, since compost nutrient ratios and release rate may not be optimal for crop needs, some supplemental inorganic fertilizer (particularly N) may be necessary. The following information provides steps to determine the economic feasibility of using compost as an alternative or a supplement to inorganic fertilizers.

Currently, the need for reducing environmental impact requires diverse processes that permit an integrated reuse of solid residues, from which products or commodities of industrial importance can be obtained. The waste generated by communities is composed of diverse materials that vary according to climate, urbanization, and socio-economic stratum. Approximately 38% of all trash produced is biodegradable organic matter that does not have a market. This organic trash generates a serious environmental problem, even when it can be used for the production of compost or other uses (Garcia 1993).

The solution to the problem of urban and agro-industrial solid waste is to process them, but adequate techniques should comply with the following requirements:

- To provide a cost that is accessible by the community that will use it.
- Have a capacity to eliminate risks to human and environmental health, and not generate additional unforeseen waste as a part of the processing technology.
- Be able to consistently process the waste that is generated, which implies a general capacity to process high volumes and the flexibility to absorb fluctuations in the quantity of daily waste produced.

There are three techniques available for the treatment of urban solid waste:

- Sanitary landfills
- Composting
- Incineration

In most cases, the application of these techniques individually or combined, permits a satisfactory economic and sanitary solution.

Studies conducted around the world have used different types of agricultural and agro-industrial residues such as straw, stubble, cane chaff, and pineapple pulp, as well as the biodegradable fraction of urban solid waste for the production of antibiotics, enzymes, detoxified feeds for cattle, biofertilizers and substrate for cultivation. In all cases, employing solid fermentation techniques is an efficient technological alternative (FAO 1991).

Due to the impact from contamination by solid waste, investigations have been carried out to transform the organic matter into products of utility for agriculture. Processes such as composting arose from this idea which, when studied scientifically, can contribute to the solution of important problems affecting modern society (e.g. the sanitary disposal of organic waste) to provide humus for field application, to maintain fields in adequate condition for cultivation, and to induce the destruction of pathogenic microorganisms (Guerrero 1993).

This review defines the important characteristics of compost, describes its various manufacturing methods, and presents factors that directly influence the economic viability of using compost versus industrial fertilizer for agricultural production.

2 Soil Organic Matter

Soil organic matter (SOM) is generally defined as the organic fraction of the soil exclusive of undecayed plant and animal residues. Organic constituents within the soil have dramatic impacts on a range of soil properties, the use and management of the soil, and soil taxonomic classification.

Frequently, SOM is defined by its level of degradation from its original plant/animal sources. The level of SOM degradation varies widely and is characterized by a number of different organic substances within the soil to include humic and fulvic acids. Humic acid (humus) is aptly described as dark in color (black or dark brown), colloidal, and negatively charged given the high amount of oxygen (O^{2-}) within its macromolecular structure. Fulvic acid is a biologically stable, highly oxidized, water soluble complexing agent with a general chemical formula of $C_{20}H_{12}(COOH)_6(OH)_5(CO)_2$ (Schnitzer 1969). It is ubiquitous in nature and dramatically affects plant nutrient uptake. Humic and fulvic acids are functionally differentiated by their solubility; the former representing material that can be extracted from soil with dilute alkali and other reagents and precipitated by acidification to pH 1–2 (Soil Science Society of America 2010). Collectively, humic and fulvic acids are known as humic substances. The highly negative charge of humus gives it a large cation exchange capacity, on the order of 200 cmol kg^{-1} (Havlin et al. 2005; Brady and Weil 2002). Thus, a wide variety of cationic plant essential elements (Cu^{2+}, Fe^{2+}, Mg^{2+}, Mn^{2+}, K^+, Zn^{2+}, and Mo^{2+}) and metals (Al^{3+}, Cd^{2+}, Cr^{3+}, and Pb^{2+}) are sorbed to humus. The sorption of polyvalent cations is especially important as these can serve as bridges between negatively charged electrostatic clays. As these particles are joined together by cationic bridging, submicroaggregates of individual particles begin to form. Submicroaggregates grow and combine into microaggregates, where the foundations of soil structure start to emerge. Annabi et al. (2007) studied the influence of three urban composted materials (municipal solid sludge compost, sewage sludge/green waste, and biowaste compost) at two different stages of decomposition (immature and mature) on soil aggregate stability. They concluded that composts at both stages of decomposition enhance soil aggregate stability through fungal biomass stabilization and an improved resistance to slaking. Soil structure represents a feature of temporal pedogenic development where soil aggregates into one or more of six soil structural units (subangular blocks, angular blocks, plates, columns, prisms, or granules). The degree of structure development and expression are a function of developmental time under optimal (undisturbed) conditions. Soil structure dramatically impacts a number of soil properties including porosity, bulk density, water infiltration and percolation.

In many areas of the world, soil organic matter represents an essential nutrient source for agronomic production where commercial fertilizers are not available. It also plays a key role in governing the form of nutrients available for plant uptake. For instance, a common indicator of organic matter nitrogen content is the carbon to nitrogen (C:N) ratio. Organic materials with a C:N ratio of less than 25:1 are considered N rich. Nitrogen in this system is subject to mineralization, the conversion of plant-unavailable N to plant-available forms (NH_4^+ and NO_3^-). Brady and Weil (2002) describe the process as follows: Organic N (unavailable to plants) largely exists as amine groups ($R-NH_2$) in proteins or as part of humus. Microbial degradation of these compounds leads to the formation of simple amino acid compounds such as lysine (CH_2NH_2COOH) and alanine (CH_3CHNH_2COOH). Hydrolysis of the amine groups on these compounds leads to the formation of NH_4^+ which can finally be oxidized to NO_3^-. Organic materials with a C:N ratio of greater than 25:1 are considered N poor. Nitrogen in this system is subject to immobilization, the conversion of plant available N to plant-unavailable forms. In this process, microbes consuming dead organic material effectively incorporate the available N into their cellular structure. In doing so, free ionic species of N (NO_3^- and NH_4^+) are removed from the system and made unavailable for plant uptake.

The application of organic matter to soils has the potential to alter N dynamics of an ecosystem. If N is not bound by organic or mineral sources, excessive concentrations of free ionic species can pose water quality problems and health risks. Soil systems overwhelmed with nutrients (either via organic matter or inorganic fertilizer application) lose the potential to sorb those nutrients from soil solution. The electrostatic attraction of cations to the negatively charged surface of many clays and humus or the attraction of anions to structural cations along clay particle edges are finite. Ions in direct sorptive contact with the surfaces of electrostatically charged particles constitute the stern layer; the layer most strongly bound to the particles. Beyond the stern layer, diffuse double layer theory defines the inverse relationship between distance from the charged particle surface and attraction to that surface. If the charged particle surface is fully saturated with ions, the addition of more ions via organic matter application or fertilizer will allow such ions to remain in soil solution and will be prone to leaching through deep percolation into the water table or surface water runoff. If N enters surface waters via runoff, eutrophication can occur. Eutrophication is defined as the accumulation of nutrients that support a dense growth of algae and other organisms, the decay of which depletes shallow waters of oxygen. Human consumption of nitrate laden waters (either from surface or aquifer sources) can lead to a serious health condition known as Methemoglobinemia, where the hemoglobin of blood fails to properly bind oxygen, causing hypoxia (Kross et al. 1992).

2.1 Composting

Compost is generally defined as a mixture of various decaying organic substances used for fertilizing the soil. Colloquial claims concerning the virtues of compost are widespread and have led to some skepticism by the scientific community. The difficulty in quantifying the benefits of compost use stems from its dynamic nature; specifically, its variable source materials changes over time and methods of application. Nonetheless, the use of compost has become more widespread in recent years given new concerns over environmental sustainability and recycling.

Essentially, the process of composting involves accelerating the degradation of organic materials by optimizing conditions for microorganisms. Depending on its intended use, compost can manifest itself in a variety of products including general use compost, erosion control compost, and compost manufactured topsoil. Each of these products has unique properties and will be independently discussed. Furthermore, a broad array of methods for producing compost exists. Misra and Roy (2002) categorize a wide variety of composting methods into (a) traditional methods and (b) rapid composting methods. Traditional methods of facilitating anaerobic digestion include the Indian Bangalore Method and Passive Composting of Manure Piles. Traditional methods of facilitating aerobic decomposition through passive aeration include the Indian Indore Method (pit and heap methods) and Chinese Rural Composting (pit and high temperature methods). Large scale passive aeration is accomplished via turned windrows or passively aerated windrows. Rapid composting methods include the use of shredding and frequent turning, mineral N activators, effective microorganisms, cellulolytic cultures, forced aeration, in-vessel composting, and vermicomposting (Misra and Roy 2002). In the United States, commercial compost production is most commonly accomplished via turned windrows and will be the focus of the discussions that follow.

2.2 Origins of Compost: Feedstocks and Processing

The source materials of compost can come from a variety of origins and are often referred to as *feedstocks*. Common compost feedstocks include animal manures (cow, chicken, swine, horse, goat, and rabbit), bagasse, bonemeal, citrus waste, cottonseed meal, cotton gin trash, grass clippings, leaves, paper, rice hulls, sawdust, and sewage sludge (Fig. 1) (Martin and Gershuny 1992; Rynk 1992). In some instances, feedstocks may be purchased as byproducts of other industrial processes. In other instances, the feedstocks may be provided to a composter free of charge, saving the feedstock generator disposal fees. However, the transportation and storage of compost feedstocks can be cumbersome as the materials often contain appreciable water or emit foul odors from manure or slaughter waste. Feedstocks are often heterogeneous in their chemical composition and physical size, necessitating further processing to produce high quality, uniform compost.

Fig. 1 Feedstocks from common landscape operations in Dallas, Texas, USA (*clockwise from upper left*): wood, woody debris, fallen leaves, and grass clippings (Photos courtesy of Lawns of Dallas)

When compost is not an adequate mix of organic waste, the process of composting is slow and the final product is of low quality. To avoid this loss of quality, other materials can be added to improve the chemical composition and structure of the piles. According to Dalzell et al. (1991), these materials are:

Activators. Substances that enhance decomposition, and contain a large quantity of proteins and amino acids, as in manures and organic waste in general.

Inoculants. These are special bacterial cultures or media containing the agents responsible for the decomposition of organic matter. They include bacteria of the genus *Azotobacter*, mature compost, ground phosphorite, calcium phosphate and soil. Presently, many products exist on the market that can be used as biological inoculants, such as Ultrazyme® and Bio-Compost®. These products increase the rate of decomposition and reduce the time to obtain mature compost.

Enrichers. These are commercial fertilizers that can be incorporated into the composting process to increase the nutrient content.

Large feedstock materials such as tree and shrub waste (leaves or wood) are typically processed with a tub grinder (a large diesel powered grinding machine mounted on a tractor-trailer). Tub grinders are fed with an articulated loader and effectively reduce materials to a size of <5 cm. To ensure that the ground products are adequately processed, a set of large sieve shakers is used to separate the ground material into different size fractions for specific job requirements. Large objects retained by the sieves can be re-ground. For composts that seek to mix a variety of feed-stocks, grinding is the ideal time to combine them.

When grinding has produced a material of the desired size, the materials are placed into long, linear rows termed *windrows*. Several key factors govern the composting process including aeration, moisture, C:N ratio, pH, temperature, and particle size. Once ground material is placed into windrows, it may remain as static piles for passive aeration, be aerated artificially through turning of the windrows, or be actively aerated through a system of aeration pipes running through the pile. Functional degradation of organic feedstocks is accomplished by a number of aerobic fungi, bacteria, and actinomycete species which operate under two different thermal ranges: mesophilic (10–40°C) and thermophilic (>40°C) (Rynk 1992). Thermophilic composting is preferred as pathogens, weed seeds, and fly larva are destroyed at >63°C. Heat and CO_2 are generated as the microorganisms begin to degrade freshly added feedstock. As the degradation proceeds, heat can begin to limit microorganism activity. Similarly, as O_2 is consumed by the aerobic organisms, degradation slows as windrows turn anaerobic, generating H_2S, NH_3, and CH_4. For this reason, regular aeration or turning of the windrows is critical. Another critical factor in composting is moisture. For optimal degradation, windrows should contain 40–65% moisture (Rynk 1992). Below 40%, microbial activity is inhibited and above 65% moisture displaces oxygen causing anaerobic conditions within the windrow. Moisture rate reduction can be accomplished via the incorporation of cellulosic bulking agents such as bagasse, paper, peanut shells, and sawdust (Iqbal et al. 2010). The C to N (C:N) ratio is critical for facilitating organic matter degradation. Rynk (1992) found that C:N ratios of 25:1–30:1 were ideal for active composting, but ratios of 20:1–40:1 produced acceptable results. The C:N ratio of feedstocks varies widely with green, tender vegetation and sawdust having ratios of 12:1 and 400:1, respectively (Martin and Gershuny 1992; Rynk 1992). Optimal and acceptable conditions for composting are given in Table 1.

As compost reaches the end of active degradation, heat generation will decline, even after turning or aeration. The original volume of feedstocks can be reduced up to 50% by the composting process (Rynk 1992). Finished compost is said to be *cured* and should not contain foul odors. Cured compost need not be completely homogenous in its composition, but it should not be undergoing active degradation.

Windrows are one of the most utilized composting techniques and are utilized under aerobic conditions. This technique is also known as biopiles, biocells, or composting piles (Iturbe-Argüelles et al. 2002). The biopiles are a form of composting in which piles are formed. The system can be opened or closed, permits the addition of nutrients and water, is placed in a treatment area, and may include systems for the collection of leachates and some form of ventilation (Eweis et al. 1998).

Choosing the type of biopile system depends chiefly on the climatic conditions and the structure of the volatile organic compounds in the organic material. Generally, the biopiles are designed as closed systems, because they maintain temperature and avoid saturation with rainwater. As well, they reduce the evaporation of water and volatile organic compounds. Two of the most used biopile systems are extended biopiles (Fig. 2) and static biopiles (Fig. 4). The difference between these

Table 1 Optimal conditions
for composting in windrows
(Rynk 1992)

Factor	Optimal	Acceptable
C:N ratio	25:1–30:1	20:1–40:1
Moisture (%)	50–60%	40–65%
Oxygen content (%)	20%	>5%
Particle size (cm)	Variable[a]	0.3–1.3 cm
pH	6.5–8.0	5.5–9.0
Temperature (°C)	54–60°C	43–65°C

[a]Depends on intended use

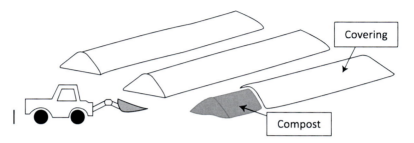

Fig. 2 Schematic representation of an elongated biopile system

Fig. 3 Windrow/biopile turning machines in Texas, USA (Photos courtesy of Saqib Mukhtar)

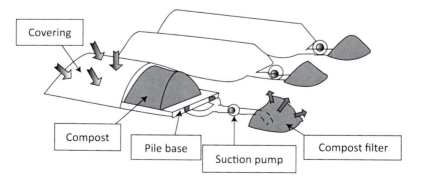

Fig. 4 Schematic representation of a static biopile system

technologies lies in the method of ventilation that provides oxygen to the composting process (Eweis et al. 1998). The system of elongated biopiles (windrows) is the most economic and simple composting process.

The material to compost is stacked on a platform in extended piles (Fig. 2) and ventilation is carried out by manually or mechanically mixing the compost (Fig. 3), a process that at the same time permits homogenization of temperature. The mixing of the compost provides for equitable material distribution (nutrients, water, air, contaminants and microorganisms) and facilitates biodegradation of the pathogens. The frequency of mixing the pile depends on the microbial activity which can generally be determined from the temperature profile of the compost; typically measured daily (EPA 1995) or monthly (Sellers et al. 1993).

In contrast, static biopiles do not need to be mixed mechanically since ventilation and equalization of heat in the compost is carried out via a system that injects (compressor) or extracts (suction) air using pipes placed in the base that are aligned in parallel along the pile (Fig. 4). In static biopiles, an air extraction system is normally employed that permits the capture of a certain fraction of the volatile organic compounds so that they can be removed from the organic material during the ventilation process. These vapors are sent to a biofiltration system or for catalytic oxidation processing (Eweis et al. 1998). The use of an injection or extraction system for air in this type of biopile permits manual or automatic control of the velocity of airflow to provide oxygen to the composting process. Thus, a temporal relationship between airflow and microbial activity can be established.

Important factors in the design and operation of a compost biopile (Dalzell et al. 1991) include economic (commodity) factors, materials cost, availability and durability, commodity reproduction, and appearance. Chemical materials or variables used in processing include pH (degree of acidity or alkalinity), capacity for cationic exchange, nutrient content, and soluble salt content. Physical and structural aspects important in evaluating final compost quality include particle size, density, porosity, ventilation, and water retention capacity. According to Garcia (1993), the feasibility of the composting process is determined by the degree of control over the percentage of humidity, since this process can be completed in a relatively short time (2–4 months).

2.3   Compost Properties

In an effort to standardize characterization and analysis of composts, the United States Composting Council (USCC) in partnership with the US Department of Agriculture (USDA) (2002) established *Test Methods for the Evaluation of Composts and Composting (TMECC)*. Guidelines from TMECC have become the industry standard for quantifying physical, chemical, and biological properties of compost in the USA. Application of the TMECC protocols is facilitated through the Seal of Testing Assurance (STA) program (administered by the USCC), whereby certified laboratories provide analysis of composted products.

2.3.1   Physical

Finished compost is typically dark in color and described as very dark gray (10YR 3/1), very dark brown (10YR 2.5/2) or black (10YR 2.5/1). However, feedstocks can influence compost color. Particle size is a function of processing and is variable according to the product's intended use. The Texas Department of Transportation (2004) specifies the following particle size limits:

- *Compost Manufactured Topsoil (CMT)*. Consists of 75% topsoil blended with 25% compost measured by volume. For use, CMT is either blended on-site (BOS), blended in-place (BIP), or pre-blended (PB), as specified on the plans.
- *Erosion Control Compost (ECC)*. Consists of 50% untreated wood chips blended with 50% compost measured by volume. Wood chips must be less than or equal to 12.7 cm in length with 95% passing a 5.1 cm screen and less than 30% passing a 2.5 cm screen.
- *General Use Compost (GUC)*. Consists of 100% compost, with 95% passing a 1.6 cm screen and 70% passing a 1 cm screen.

The bulk density of compost is known to vary widely based on feedstock particle density, moisture, and porosity. However, bulk density is important for calculating loading rates and transportation costs. Weindorf et al. (2006) found the bulk density of compost derived from grass clippings and leaves to be 0.70 g cm^{-3}. Van Ginkel et al. (1999) evaluated the bulk density of chicken manure/wheat straw compost and found ranges of 150–950 kg m^{-3}. They linked such wide variation to moisture content and compaction stemming from the height of the compost piles.

2.3.2   Chemical

The chemical properties of finished compost are essential to its use as a viable soil amendment. Typical chemical properties evaluated include compost pH, salinity, nutrient (elemental) content, and heavy metal content. As an amendment promoting soil fertility, the pH of compost should ideally serve to facilitate a pH of ~6.5 (slightly acidic). Slightly acidic conditions allow for the best overall availability of both soil macro- and micronutrients. Similarly, compost salinity must be carefully monitored so as not to exacerbate soil conditions where salinity can be harmful. Plant tolerance of salinity is highly species specific. For example, onions, oranges, beans, carrots, broccoli, corn, grapefruits, and tomatoes are moderately sensitive or sensitive to salinity (Maas and Grattan 1999). Sorghum, oats, soybeans, beets, asparagus, and artichokes are moderately tolerant or tolerant of salinity (Maas and Grattan 1999).

Elemental analysis of composts focuses on two key parameters: nutrient content and trace metals. The total quantity of plant essential nutrients within composts varies widely based on feedstock and composting methods. The C:N ratio is of particular importance to agronomic and horticultural applications and is sometimes adjusted to the ideal 25:1–30:1 (Table 1) by the addition of fertilizer N. Other commonly evaluated elements include Ca, Mg, K, P, Fe, Cu, Mn, Zn, Mo and Cl. Trace metal content is of particular concern where the application of composts could pose threats to surface water quality or environmental degradation. The US EPA 40 CFR § 503.13 sets forth the ceiling concentrations, cumulative pollutant loading rates, monthly average concentrations, and annual pollutant loading rates permissible for land application of organic materials (US EPA 2010). Trace elements covered under all or part of these regulations include As, Cd, Cu, Pb, Hg, Mo, Ni, Se, and Zn.

2.3.3 Biological

While a range of different biological organisms are known to reside within composted products, they are generally classed as microscopic and physical decomposers. The former concerns bacteria, actinomycetes, protozoa, and fungi within compost. Up to 25% of the mass of finished, stable compost is comprised of living and non-living cellular material from microbes (US Composting Council-USDA 2002). The population dynamics of this group vary considerably based on feedstock, aeration, moisture, and heat within the compost. Bacteria are single celled organisms which reproduce via binary fission. They typically produce enzymes which functionally degrade the material on which they reside, serving as a food source for their life and propagation. In doing so, bacteria and fungi generate up to 90% of the CO_2 produced by living organisms on the earth (Nardi 2003). However, bacteria are generally less mobile than other microorganisms and thus, unable to escape unfavorable environments. This causes bacteria populations to proliferate and then die in cyclical patterns. Actinomycetes are vital to humus formation and are known to produce rudimentary antibiotics (Nardi 2003). As they decompose organic substances, they liberate C, N, and NH_3. Protozoa are single celled organisms that consume large amounts of bacteria as food (Nardi 2003). However, they have limited persistence to high temperatures of the thermophyllic phase of composting (Martin and Gurshuny 1992). Fungi represent one of the final stages of microscopic degradation. They essentially act as primitive plants, but lack chlorophyll and depend on organic substrates for survival. Various forms of fungi thrive in compost from 21°C to 49°C (Martin and Gurshuny 1992).

Macroscopic physical decomposers include mites, millipedes, centipedes, sow bugs, snails, slugs, spiders, springtails, beetles, flies, ants, nematodes, and earthworms. The presence of these decomposers in compost forms a complex web of interdependence with microscopic organisms, which form their primary food source.

Typical assessment of biological properties of compost includes pathogen testing (fecal coliforms) and a measure of biological activity via some form of respirometry. Adani et al. (2003) evaluated the dynamic respiration index (DRI), static respiration index (SRI), and specific oxygen uptake rate (SOUR) and found that the three methods were well correlated and aptly characterized biological stability of organic materials. A commonly employed field technique utilizes the Solvita Maturity Test; a colorimetric test for qualitatively assessing CO_2 and NH_3 generation from a given quantity of compost. Changa et al. (2003) concluded that such tests provided useful information

Table 2 Class A and B
biosolids limits (US EPA
2010)

Pathogen	Density limits
Class A biosolids	
Salmonella	<3 MPN $(4 \text{ g})^{-1}$ TS
OR	
Fecal coliforms	<1,000 MPN g^{-1} TS, and
Enteric viruses	<1 PFU $(4 \text{ g})^{-1}$ TS, and
Viable helminth ova	<1 $(4 \text{ g})^{-1}$ TS
Class B biosolids	
Fecal coliforms	<2,000,000 (MPN or CFU)g^{-1} TS

for identifying potential toxic plant responses to excessive NH_3 in a simple, broadly applicable field test. However, they concede that the Solvita test is no replacement for actual lab respirometry. Pathogen testing typically focuses on salmonella and/or fecal coliforms. The US EPA 40 CFR § 503.13 distinguishes two classes of compost products: Class A biosolids and Class B biosolids (Table 2) (US EPA 2010). It is important to note that proper thermophyllic composting typically results in finished compost which meets Class A biosolids limits. However, if composting temperatures are minimal, fecal coliforms may remain viable in manure-based feedstocks, posing potential health risks to humans.

According to Guerrero (1993) and Coronado (1997), the incorporation of compost into soils as a source of organic matter produces several positive effects in its biological, physical, and chemical properties including:

- Contributing essential nutrients (e.g. N, P, K, S, B, Cu, Fe, Mg) for plant growth during the process of decomposition (Koepf 1965).
- Contributing to the biological activity of soils by incorporating organic acids and alcohols during their decomposition such that they serve as sources of C for the microorganisms and N fixers which produce substances for growth such as tryptophan and indole-acetic acid.
- Providing food for the microorganisms that are active in the process of decomposition, and that produce antibiotics that protect plants against disease, thus contributing to plant health (Koepf 1965).
- Incorporating intermediate metabolites produced during decomposition that can be absorbed by the plants to increase their growth. When organic matter is in the form of humus it provides more benefits (Guerrero 1993).
- Incorporating segregated substances that favor soil structure to improve water and air transport, diminish compaction, and favor the development of the plant roots and plowing of the soil (Crovetto 1992).
- Buffering against abrupt modifications of pH (Buchanan 1993).
- Providing metabolites such as phenols that contribute to plant respiration, improved P absorption, and plant health (Guerrero 1993).
- Increasing soil organic material to improve retention of soil humidity (Crovetto 1992).
- Reducing inorganic fertilizer requirements.
- Improving water infiltration and drought tolerance.

- Reducing soil compaction and crusting.
- Improving root growth and yields.
- Increasing populations of microbes and earthworms in the soil.
- Improving plant resistance to disease.
- Slowly releasing nutrients to plants.
- Improving nutrient holding capacity.
- Increasing ease of cultivation.
- Increasing pollution prevention and remediation.

Specifically, the application of organic matter positively influences the soil microbial community of bacteria and fungi, enlarging their abundance and diversity. The application of organic fertilizers increases the production of cultivation and increases resistance against pests and diseases. Due to the large reservoir of N in the soil, the application of nitrogenous fertilizer only favors plant vegetative growth and not that of the soil microbial fauna (bacteria, fungi, nematodes). As well, the exclusive and continuous use of chemical fertilizers leads to the reduction and disappearance of organic matter, favoring the loss of soil structure and the increase of soil compaction (Córdoba 2009; Neely et al. 1991).

Also important is that the quality of the compost can be considered as a 'fertilizer' or 'soil conditioner', depending on its effect on plant nutrition. 'Fertilizers' are a source of quickly available nutrients that have a direct effect, reflecting a short time in plant growth. 'Soil conditioners' affect plant growth indirectly by improving the physical properties of the soil by improving water retention, aeration, structure and drainage, properties that are intimately related to the prevention of soil erosion, the recovery of degraded soils (López-Martínez et al. 2001; Castellanos et al. 1996), and the favoring of diversity and microbiological activity (Neely et al. 1991). That being said, composted materials in some countries are not specifically labeled for sale as 'fertilizer' due to requirements in uniformity of material (guaranteed analysis) and testing.

3   Global Compost Dynamics

The type of materials composted worldwide is expansive. However, composting is most often carried out to provide disposal of unwanted organic refuse, reuse/capture of a nutrient stream where resources are limited by availability or financial constraints, or to protect environmental quality as a nutrient management practice. Certain conditions serve to assure the effectiveness of composting operations. First, the supply of feedstocks must be continuous and located physically near to the composting operations. As such, local organic waste streams often govern the types of compost produced in a given area. Transportation of many feedstocks and composted products is difficult since the appreciable water content of the products makes them heavy. Large scale operations require heavy equipment for loading, mixing, and moving the compost. Large trucks must be utilized to carry the finished product to end-users, requiring fuel and labor. Second, the compost should be uniform, consistent in its properties, and free from foreign inorganic matter such as plastics or glass. Organic products are inherently variable, but thorough mixing and processing will provide a consistent, appealing product. Last, effective composting requires 'buy-in' by end-users; they must appreciate the benefits of compost and believe in its proper use.

Worldwide, one of the most commonly composted feedstocks is manure from livestock. The proliferation of confined animal feeding operations (CAFOs) for industries such as dairies and feedlots has exacerbated the need for environmentally responsible manure management. Huang et al. (2008) evaluated the nutrient content of 120 manures from composting and farm operations across 22 Chinese provinces and documented the levels of K, Ca, Mg, Fe, and Zn using near infrared spectroscopy. They found that the near infrared spectroscopy technique is a potential method for predicting nutrient metal content of animal manure compost products. Mupondi et al. (2006) studied the inclusion of goat manure in pine bark compost in South Africa. They concluded that the addition of goat manure enhanced cabbage seedling growth compared to pine bark compost with no manure. In central Texas, USA, Butler et al. (2008) compared corn yield from fields supplied with dairy manure compost versus inorganic fertilizer. They found comparable yield performance between the two nutrient supply strategies, but noted that some accumulation of salinity and adjustment in soil pH were evident with repeated compost application. They concluded that the combined use of some manure compost and some inorganic fertilizer would be a feasible strategy for optimal corn production.

Another major feedstock for composting operations is municipal sewage sludge or urban wastes. In some areas, solid urban wastes are applied directly to soils as fertilizers for crops. For example, Ouagadougou in Burkina Faso, Africa, is a city of >1.2 million residents generating 300,000 tons of solid urban waste annually (Kabore et al. 2010). Traditionally, waste products have been applied to soils directly, providing high crop productivity of cereal and legume crops. However, in an effort to reduce pathogen prevalence, pit composting has been employed to process solid urban waste prior to agricultural use. Kabore et al. (2010) recommend mixing household waste, slaughter house waste, and tree leaves to accelerate organic matter stabilization and produce compost with higher available N content. In India, urban populations are expected to reach 341 million by 2010, generating 65 million tons of municipal solid waste (Kumar and Gaikwad 2004). Bhattacharyya et al. (2003) compared municipal solid waste compost to cow dung manure with and without the addition of urea and fertilizer for rice production in West Bengal. Rice production was greater with cow dung manure + urea, and municipal solid waste compost + urea compared to fertilizer. Furthermore, they noted that rice uptake of heavy metals (Zn, Cu, Pb, and Cd) was still within safe limits. Soumare et al. (2003) compared the use of mineral fertilizer and municipal solid waste compost as soil amendments supporting the growth of ryegrass in Mali. They found that mineral fertilizers and 50 T ha^{-1} municipal solid waste compost increased dry matter yields by 69.7%, 65%, 10% and 17.5% for the Gao and Bgda soils, respectively. While inorganic fertilizer provided the most production, increases in soil organic carbon, available P, Fe, Mn, Zn, Cu, K, and pH were linked to compost, confirming its appropriateness as a soil amendment. Farrell and Jones (2009) argued that even after composting of municipal solid waste, caution must be applied to its prudent agricultural use. Risks from sharp objects like glass shards, organic pollutants, and heavy metals remain, though they conclude that the latter poses limited risks for plant uptake and environmental degradation. Nonetheless, they advocate careful investigation of contaminant levels and detailed risk assessment prior to the application of municipal solid waste compost.

Other examples of composted products around the world include mushroom waste in Ireland (Courtney and Mullen 2008), cabbage waste and sawdust in South Africa (Manungufala et al. 2008), and sweet sorghum bagasse combined with pig slurry and sewage sludge in Spain (Negro et al. 1999). While composted products are widely heralded as beneficial soil amendments for agricultural production, deleterious results also are possible. Levy and Taylor (2003) evaluated the effects of four composted products (horse manure/bedding, mink farm waste, municipal solid waste/sewage sludge, and pulp mill waste) on the growth and establishment of tomatoes, cress, and radish. They found that horse manure/bedding and mink farm waste dramatically stimulated vegetative growth, but municipal solid waste/sewage sludge and pulp mill waste were strongly inhibitory, producing vegetative deformity and stunted growth.

3.1 Cost and Scale of Application

Generally, conventional costs for technologies like incineration or the construction and management of controlled confinements oscillate between $250 and $1,000 USD/m^3 (Van Deuren et al. 1997). For the particular case of biopiles, the estimated costs are between $25 and $150 USD/m^3 (Semple et al. 2001; Potter 2000). These costs vary according to the quantity and type of soil to treat, the volume of agent availability, the type of contaminants, the type of process to employ, the need of prior and subsequent processing, the need of equipment for the control of volatile organic compounds, and climatic conditions.

According to Echeverry (2002), in a comparative study between organic and inorganic fertilization in the cultivation of bananas in Colombia, the cost of organic fertilization was approximately $80 USD ha^{-1}, equivalent to 33% of the cost of chemical fertilization ($240 USD ha^{-1}), which is clearly favorable in terms of cost. There was no statistical difference in the weight between racemes produced with chemical or organic fertilizers. The advantages of employing organic fertilizers are their lower cost and contribution to the improvement and conservation of long-term soil fertility.

In Cuba, where the tendency is to develop solutions and techniques of fertilization to avoid the destruction of the environment and to eliminate high dependence on imported chemical fertilizers, investigations have focused on filter-cake compost obtained from waste generated by sugarcane production. This material contributes a high quantity of nutrients for the production of compost on a large scale. According to Rodriguez (2002), 35 tons ha^{-1} of filter-cake applied to soils provides:

- 312 kg urea with a 2002 value of $84–$106 USD per ton
- 282 kg triple superphosphate with a 2002 value of $129–$138 USD per ton
- 70 kg potassium chloride with a 2002 value of $112–$116 USD per ton

Organic compost application represented a savings of $2,980 USD for 13.4 ha and had a residual effect for 5 years, guaranteeing increments in performance of 6–15%. In the preparation of biological compost, only enhancers were used to expedite the process.

3.2   Scale of Application

In order to choose the size of the site required for composting, the following factors should be taken into consideration: the anticipated volume of raw materials, the technology to be used (the higher the level, the less space required), the equipment to be used (which depends on the method and raw materials), and the projections for growth. Also important is accessibility (roads suitable for traffic and convenient to feedstocks, or raw materials), population density (no houses within half a mile), and type of neighbors (some industries require a clean atmosphere and no flies). Some characteristics of a desirable site include slightly sloped land (for drainage), a firm soil type that packs well, not located in a flood plain, convenient utilities, and a rectangular or square site, which is more efficient than a circular or irregularly shaped site. Key to the success of any composting operation is a marketing or distribution program for compost products. The compost must be of consistently high quality so as to develop long-term markets.

The application of compost for agricultural production, as a means of recycling green waste that is produced by communities and agricultural and livestock activities, may be a sustainable and inexpensive solution. However, the large quantities produced necessitate the development of education and organization of compost producing infrastructures and equipment programs, particularly in developing countries.

When dealing with compost application at a relatively small spatial scale, the need for infrastructure, financial resources and a labor force may not represent a major concern because the compost volumes that are to be transported and the costs involved are smaller compared to those for large scale compost application. It is well known that a significant fraction of the solid waste generated in the world is organic material that can be recycled through small scale composting (Fig. 5). There are many advantages to this strategy of waste management. For instance, households, businesses and institutions may save money by composting items such as food scraps and yard trimmings while sending less waste to landfills and incinerators. In addition, small scale composting is often the most environmentally sound way of recycling organic materials.

Fig. 5 Small scale composting in Malawi and Costa Rica (Photos courtesy of David C. Weindorf)

However, for big projects that run at a large scale, the economics and the infrastructure requirements will be two of the key factors taken into consideration, since the cost in creating the required infrastructure for production and distribution may be considerable. In both cases, the environmental benefits justify the investments involved.

From the social and economic perspectives, small scale projects can be more suitable for developing countries or individual households. Large scale projects normally involve great financial investment and the establishment of a network of compost production sites and centralized sites, which must be properly equipped to compost the increasing volumes of waste produced and to meet a growing demand from the agricultural sector. This large scale approach can be of greater applicability in developed countries, although its cost of implementation can be much greater than at smaller scales.

The application of compost at both small and large scales allows farmers to minimize the use and cost of commercial fertilizers, replacing them locally with an economical and sustainable alternative. Thus, farmer agricultural productive activities become more environmentally friendly and competitive. The compost can be used to mulch landscaping, enhance crop growth, enrich topsoil, and provide other benefits. Reduction in the need for inorganic fertilizers and pesticides when using compost is highly beneficial to the aquatic ecosystems, flora, fauna, and human health.

4   Organic and Inorganic Fertilization: Culture, Economics, and Sustainability

Presently, modern agriculture bases its productivity, to a large extent, on the use of inorganic fertilizers, including urea, nitrates and its by-products. Such use has yielded deteriorations in field productive capacity, problems with soil hydricity and erosion, soil compaction, salinization, loss of soil structure, and water contaminated with chemical compounds such as nitrates and insecticides (Pérez 2008).

According to the Mexican Association of Ecological Farmers (1992), composting is a fertilizer technique founded in the larger topic of organic agriculture. The use of compost, as opposed to inorganic fertilizers, is characterized by its low solubility because it delivers nutrients more slowly to the plants, has a greater duration, and reduces nutrient loss through leaching. As well, the varied nutrient composition of compost responds to the needs of the plants (Narea and Valdivieso 2002). According to Gross (1986), most improvements in farming occur with soil fertility and productivity. Investigations in Germany and the Netherlands (Table 3) have shown that nitrate filtration levels are significantly lower with organic agriculture than in traditional farming systems. The purpose for using compost in agriculture is to reduce contamination and prevent environmental degradation by using more sustainable methods of cultivation (FAO 2003).

In the framework of sustainable development, the process of composting presents important perspectives for resolving many problems produced by contamination in Mexico. Composting technologies particularly and bioremediation in general, are viable processes for application since most of the country has adequate climatic conditions for farming, with annual average temperatures that oscillate between 18°C and 26°C, temperatures favorable for implementing the aforementioned types of composting (Cooperband 2002).

Nevertheless, before using the process of composting for remediation of any given site, it is necessary to include complete local information (origin of the contamination, characterization of the soil and of the contamination to be treated) and to establish tests of contaminant biodegradation by indigenous and exogenous microorganisms to select the type of technology based on the costs and the availability of materials and equipment to carry out the treatment (Zechendorf 1999).

According to Soil and More, Mexico (2010), a private company dedicated to the production of compost in Mexico, the use of compost improves the economic

Table 3 Reduction of nitrate filtration indices with organic agriculture compared to traditional inorganic agriculture (From Stolze et al. (2000); cited by FAO (2003))

Percentage filtration	Authors
>50	Smilde (1989)
>50	Vereijken (1990)
57	Paffrath (1993)
50	Reitmayr (1995)
40	Berg et al. (1997)
64	Haas (1997)

situation of the agricultural producers in the area, as well as the social development and environmental conditions of the region. Application of high quality compost increases crop production, and reduces the cost of chemical fertilizer and pesticide applications, considerably improving the economic situation of the agricultural producers. Compost application is a sustainable means of developing the fertility of soils degraded by agricultural activities in the region. The practice more efficiently uses irrigation water because it increases the retention of humidity in the soil. Because of the natural microbiological community, the compost also acts as a natural filter for removing many agricultural pathogens.

In sustainable agriculture, the application of organic materials to the soil is indisputably necessary, since they are a vital source for reconstructing its organic matter and for supplying nutrients (Álvarez et al. 2006). The employment of compost in agriculture unites aspects of cultivation, ecology, economy and society in an integrated manner, to substitute for or complement the use of traditional fertilizers at the farm level (Echeverry 2002).

5 Limitations of Composts as Fertilizers

Composts rarely provide nutrients to plants in exactly the right balance. This is especially the case for macro-nutrients such as excess P or deficient N. For example, bovine manure usually has a higher P:N ratio than what non-leguminous crops require. Once composted, that P:N ratio is even greater after N loss to volatilization (McDowell and Sharpley 2004).

As a result of nutrient imbalances as well as improper management, composts have been identified as potential environmental threats. In some cases, especially with animal manure, compost-N can overwhelm soil capacity to hold it until plants can effectively utilize it (Daliparthy et al. 1994). This is particularly the case in soils with shallow water tables where leaching quickly carries soluble nitrates to those tables or in cold climates when crops are absent or dormant. Composting those manures lowers N concentrations and mitigates this problem but results in deficient soil-N for most crops. Where water tables are further from composts on the soil surface, excessive P contribution to surface water runoff is more likely to be problematic (McGechan et al. 2005). When these composts are surface-applied to perennial forage fields where incorporation into the soil via tillage is not possible, negative impacts of P on downstream surface water quality have been identified (Sharpley and Syers 1979). In such cases, quantities applied to the crop may be limited by environmental considerations rather than crop requirements.

6   Manufactured Fertilizers Versus Composts as Fertilizers

Industrially manufactured fertilizers are used throughout the world and are generally credited, along with genetic manipulation and selection of key crops, for the huge increases in food production known as "The Green Revolution" (De Datta et al. 1968). Before the widespread production and use of manufactured fertilizers, crop productivity, especially from non-legume grain crops, was limited by inherent soil fertility. Once nutrients were "mined" from the soil by years of cropping, the production rates of those soils declined along with the capacity for a reasonable return on labor and seed invested. Basically, nutrients in soil organic material were converted into crop products and once these were exhausted, soils lost their fertility. Pre-industrial farmers mitigated this decline by various means, including:

- Moving on to other virgin soils
- Resting the land via fallows for several years
- Rotating with green manure crops (usually legumes) grown specifically for organic matter production
- Rotating with fertility-enhancing food crops such as legume pulses
- Collecting and incorporating animal wastes such as cattle manure or bat/bird guano
- Incorporating composts created from human and animal waste

As modern human population increased, demand for food production (i.e., mining soil fertility) climbed. The age-old methods for maintaining soil fertility simply could not keep up with market demand as society moved away from farms into urban areas. More people needed to be fed from less land. Mining and concentrating nutrients such as P or fixing atmospheric N into plant-available forms became possible using fossil fuels. The advantages of manufactured fertilizers compared to compost fertilizers were various, including:

- Ease of transport due to high nutrient concentration
- Low costs, reflecting low fossil-fuel costs
- Ease of incorporation into soils
- Near total nutrient availability
- Nearly unlimited raw material
- Precise nutrient balance reflecting varied crop and soil fertility needs
- Immediate availability to plant roots

It is easy to see, then, why compost fertilizers lost traction to what became known as industrial fertilizers. But soils, and eventually the environment, may have paid a price for this switch (Lappé et al. 1998). Many of the advantages manufactured fertilizers brought to agricultural production also carried dangers. These include:

- Changes in soil chemistry, especially pH
- High nutrient availability (solubility) making them easily leached into the environment
- Application of primary nutrients (mostly N, P and K) depletes or masks minor elements
- Soil cation exchange capacity (CEC) and/or OM is unable to hold nutrients as efficiently until plants need them

As the cost of fossil fuels rises and human population continues to grow, farmers are faced with a dilemma: starve populations by reducing crop yields or run the risk of damaging the environment by the continued heavy use of industrial fertilizers. The "dead zone" in the Gulf of Mexico is considered a prime example of the latter (USGS 2010), due at least in part to agricultural runoff into the Mississippi River of North America. The first option is politically and socially unacceptable while the latter inevitably will cost future generations.

A third option may be to join the two approaches by making old soil fertility methods more productive and new fertilizer uses more sustainable. Improving soil organic matter by using greater incorporation of composts, crop rotations and green manures, while boosting crop yields with judicious use of industrial fertilizer regimens may be the best compromise. In this manner, farmers realize dual benefits via:

- Stretching limited soil, compost, green manure, and organic matter resources
- Correcting nutrient imbalances/deficiencies of composts/green manures with industrial fertilizers
- Binding pesticides long enough to allow them to decompose before causing environmental concerns
- Improving industrial fertilizer nutrient delivery and balance
- Improving industrial fertilizer retention and slow release by association with soil OM

Numerous investigations have verified that the productive and ecological benefits of using compost as organic fertilizer are greater than those obtained from the use of chemical fertilizers alone (Bizzozero 2006; Barzaga et al. 2004; FAO 2003). As well, the use and application of chemical fertilizers is presently limited, not only by their effects on the environment, but because their price has grown rapidly, nearly 105% during 2007–2008 alone (Seceña 2010). Hence, the production of compost is a highly beneficial alternative, not only for producing good agroecological conditions, but also because the waste utilized can be acquired at a very low cost (Sandoval and Stuardo 2001).

In a comparative analysis carried out by the Cuban sugar company "Dos Rios", with only one application of filter cake compost due to the slowness of its decomposition (and therefore applied for the entire life cycle of the crop), a low cost of only $131.00 ha^{-1} would be incurred against a cost of $562.25 ha^{-1} by using inorganic fertilizers (chemical) for sugarcane cultivation. This translates to a savings of $431.25 ha^{-1}, and only for the fertilizer (Barzaga et al. 2004). According to these

Table 4 Changes in soil
sustainability indicators
produced by using organic
fertilizers in Chile (Bizzozero
2006)

Indicators	1994	1998
Soil erosion	60 tons ha^{-1}	12 tons ha^{-1}
Organic matter	2.1%	3%
Water retention/humidity	8%	11%
Biodiversity (Shannon index)	1	2.28
Aluminum saturation	4.8	2.4
Sum of magnesium, calcium, sodium	5.75	8.8

authors, for those companies capable of producing a quality product, the production of this biofertilizer constitutes an important source of income.

In China, in a comparative study between conventional and organic berry production systems, the supplies, products and net income of the organic system were higher than from the inorganic system. The greater supplies for the organic system consisted mostly of labor, especially for the task of fertilizer application, but costs of purchasing chemical fertilizers and insecticides were lower. Given that yield and net income was higher, the high cost of manual labor is offset by the high revenue from the product (FAO 2003).

In Chile, where the most important problem was soil erosion, compost was used in an agricultural fertilization project to recover some soil properties. Toward the end of the project in 1998, significant changes were observed. Organic management not only controlled erosion, but improved the structure and fertility of the soil (Table 4). The improvement in humidity retention, the reduction of erosion, and the introduction of rotational cropping resulted in a variety of food and forage with greater productivity (approximately 20% in the case of cereals, and between 20% and 60% for horticulture). As well, the area destined for horticulture grew by 260%, significantly increasing income by approximately $1,300 USD in 1994, and more than $6,000 USD annually in 1998 (Bizzozero 2006).

According to Bizzozero (2006), yield is the quantity (in kg ha^{-1}) of product obtained from a current production system with regard to the surface area utilized to provide financial gain. This parameter does not consider the form of the product obtained, the ecological impacts generated during its production, the supplies contributed or the cost of the same, nor the social impacts. Positive impacts also have been observed from organic fertilizers on crops, such as increasing the number of seedlings, shortening the cultivation cycle by 7–10 days, increasing flowering and fruition, and increasing performance between 5% and 20%, as well as obtaining fruits with greater commercial quality (appearance and size).

In Europe, even the water treatment plants favor the employment of organic fertilizers in areas of water resource protection. This is an economically efficient solution to reduce the costs of drinking water purification and to minimize groundwater contamination with nitrates and insecticides. When imposed as a regulation in organic agriculture, it has resulted in the low presence of N in organic operations. That implies lower costs, since the cost of production on the farm for 1 kg of N in organic operations can surpass 7–16 times the cost of the inorganic or mineral fertilizers. Therefore, contrary to what occurs on conventional farms where fertilizers and sewage sludge are a general waste problem, organic farmers develop efficient strategies for the management of N. For example, intercropping, cover crops, the optimum incorporation of legumes in the land or the limited use of liquid manure to avoid the volatilization (loss to the atmosphere) of N are common practices (FAO 2003).

7 Conclusions

Composts are dynamic substances generated worldwide as technological alternatives for the bioremediation and organic fertilization of soils. They permit improvement and conservation equilibrium of nutrient flows and minimize the use of external resources. Composting is based on the same system that is used naturally to maintain nutrient recycling (Granados and López 1996). It is a process of solid phase aerobic fermentation which takes advantage of automatic heat production by the different native microbial populations for the total or partial biodegradation of organic matter to obtain organic compost that is black, stable, homogeneous, and nutrient-rich (Semple et al. 2001). Composting is employed as an alternative to the use of industrial fertilizers for soils supporting a wide variety of crops across the world. The source materials (feedstocks) and composting methods employed vary with geographic location and available resources, but play an important role in sustainable agricultural production.

Acknowledgements To Colegio de Postgraduados, Campus Veracruz (LP12 Agroecosistemas Sustentables) for the facilities provided to CLS in the preparation of the contribution to this review. Thanks also are due to M.C. Itzel Galaviz Villa for contributing to the report.

References

Adani F, Gigliotti G, Valentini F, Laraia R (2003) Respiration index determination: a comparative study of different methods. Compost Sci Util 11:144–151

Álvarez SE, Vázquez AA, Castellanos JZ, Cueto WJ (2006) Efectividad biológica de abonos orgánicos en el crecimiento de trigo. Terra 24:261–268

Annabi M, Francou C, Poitrenaud M, Bissonnais YL (2007) Soil aggregate stability improvement with urban composts of different maturities. J Soil Sci Soc Am 71:413–423

Barzaga UJ, Pérez JA, Santana S (2004) Producción de compost y resultados preliminares de su aplicación en áreas de la empresa azucarera "Dos Rios". Jornada Científica por el 40 Aniversario del Instituto Nacional de Investigaciones de la Caña de Azúcar Sede Tecnológica 11 y 12 de Noviembre de 2004. http://www.santiago.cu/hosting/etica/Sede40/index.htm. Accessed July 2010

Berg M, Haas G, Köpke U (1997) Grundwasserschonende Landbewirtschaftung durch Organischen im Vergleich zu Integriertem und Konventionellem Landbau. VDLUFA-Schriftenreihe 46:615–618

Bhattacharyya P, Chakraborty A, Bhattacharya B, Chakrabarti K (2003) Evaluation of MSW compost as a component of integrated nutrient management in wetland rice. Compost Sci Util 11:343–350

Bizzozero F (2006) Tecnologías apropiadas. Biofertilizantes. Nutriendo cultivos sanos. CEUTA, Montevideo, 47 pp

Brady NC, Weil RR (2002) The nature and properties of soils, 13th edn. Prentice Hall, Upper Saddle River

Buchanan M (1993) Agricultural markets for yard waste compost. Biocycle 34(9):33–36

Butler TJ, Han KJ, Muir JP, Weindorf DC, Lastly L (2008) Dairy manure compost effects on corn silage production and soil properties. Agron J 100:1541–1545

Castellanos JZ, Marques OJ, Etchevers JD, Aguilar SA, Salinas JR (1996) Long-term effect of dairy manure on forage yields and soil properties in an arid irrigated region of northern Mexico. Terra 14:151–158

Changa CM, Wang P, Watson ME, Hoitink HAJ, Michel FC (2003) Assessment of the reliability of a commercial maturity test kit for composted manures. Compost Sci Util 11:125–143

Cooperband LR (2002) Paper mill residuals and compost effects on soil physical properties in an irrigated vegetable rotation. J Environ Qual 31:2086–2095

Córdoba VCA (2009) Efecto de dos tipos de compost y un biofertilizante sobre poblaciones microbianas edáficas en un cultivo de hortalizas en Risaralda, Colombia. Revista Brasiliana de Agroecología 4:1159–1162

Coronado M (1997) Agricultura orgánica versus agricultura convencional. Primer encuentro: Abonos orgánicos vs. Fertilizantes químicos. CIED. http://www.ciedperu.org/articulos/organico.htm. Accessed Oct 2003

Courtney RG, Mullen GJ (2008) Soil quality and barley growth as influenced by the land application of two compost types. Bioresour Technol 99:2913–2918

Crovetto C (1992) Rastrojos sobre el suelo. Una introducción a la cero la branza. Editorial Universitaria. Santiago, Chile, 312pp

Daliparthy J, Herbert SJ, Veneman PLM (1994) Dairy manure applications to alfalfa: crop response, soil nitrate and nitrate in soil water. Agron J 86:927–933

Dalzell HW, Biddlestone AJ, Gray KR, Turairajan K (1991) Manejo del suelo: producción y uso del composte en ambientes tropicales y subtropicales, vol 56. Boletín de Suelos de la FAO (FAO), Roma, 178 pp

De Datta SK, Tauro AC, Balaoing SN (1968) Effect of plant type and nitrogen level on growth characteristics and grain yield of indica rice in the tropics. Agron J 60:643–647

Echeverry NE (2002) Fertilización orgánica vs fertilización inorgánica de plátano Cachaco común en Colombia. Agronomía INFOMUSA 10(2):7–10

EPA (1995) An analysis of composting as an environmental remediation technology. EPA530-R-98-008. http://mie.esab.upc.es/ms/informacio/usos_compost/Composting%20as%20remediation%20technology%20EPA%201.pdf. Accessed Aug 2010

Eweis JB, Ergas SJ, Chang DP, Schroeder ED (1998) Bioremediation Principles. McGraw-Hill, Boston, London, 296 pp.

FAO (1991) Manejo del suelo. Producción y uso del composte en ambientes tropicales y subtropicales. Organización de las Naciones Unidas para la Agricultura y la Alimentación, Roma, 178 pp

FAO (2003) Agricultura orgánica, ambiente y seguridad alimentaria. Serie: Environment and natural resources management. ISBN: 9253048190 Y4137/S. 4: 262. http://www.fao.org/documents/pub_dett.asp?lang=es&pub_id=152711. Accessed Aug 2010

Farrell M, Jones DL (2009) Critical evaluation of municipal solid waste composting and potential compost markets. Bioresour Technol 100:4301–4310

García MI (1993) Elaboración de compostas a partir de los desechos sólidos biodegradables. Reporte No.153907. UAM-Iztapalapa, 11 pp

Granados D, López G (1996) Agroecología. Universidad Autónoma Chapingo, México, 54 pp

Gross A (1986) Abonos. Guía práctica de la fertilización. Edición Mundi-Prensa, Madrid, 560 pp

Guerrero J (1993) Abonos orgánicos: Tecnología para el manejo ecológico de suelos. RAAA, Lima, 189 pp

Haas G (1997) Leistungen des Ökologischen Landbaus zur Verringerung von Gewässerbelastungen. In: Dohmann M (Hrsg.) 30. Essener Tagung für Wasser- und Abfallwirtschaft, Aachen, 13/1–13/13

Havlin JL, Beaton JD, Tisdale SL, Nelson WL (2005) Soil fertility and fertilizers, 7th edn. Prentice Hall, Upper Saddle

Huang G, Han L, Yang Z, Wang X (2008) Evaluation of nutrient metal content in Chinese animal manure compost using near infrared spectroscopy. Bioresour Technol 99:8164–8169

Iqbal MK, Shafiq T, Ahmed K (2010) Characterization of bulking agents and its effect on physical properties of compost. Bioresour Technol 101:1913–1919

Iturbe-Argüelles R, Flores-Torres C, Chávez-López C, Roldán-Martín A (2002) Saneamiento de suelos contaminados con hidrocarburos mediante biopilas. Ingeniería, Investigación y Tecnología III 1:25–35

Kabore TWT, Houot S, Hien E, Zombre P, Hien V, Masse D (2010) Effect of the raw materials and mixing ratio of composted wastes on the dynamic of organic matter stabilization and nitrogen availability in composts of Sub-Saharan Africa. Bioresour Technol 101:1002–1013

Koepf H (1965) Compost. Reprinted from Bio-Dynamic issue N°77. Bio–Dynamic farming and gardening association, Inc. 18 pp

Kross BC, Ayebo AD, Fuortes LJ (1992) Methemoglobinemia: nitrate toxicity in rural America. Am Fam Physician 46:183–188

Kumar S, Gaikwad SA (2004) Municipal solid waste management in Indian urban centres: an approach for betterment. In: Gupta KR (ed) Urban development debates in the new millennium – studies in revisited theories and redefined praxes. Atlantic Publishers, New Delhi, pp 100–111

Lappé FM, Collins J, Rosset P (1998) World hunger: 12 myths. Grove, Greenwich

Levy JS, Taylor BR (2003) Effects of pulp mill solids and three composts on early growth of tomatoes. Bioresour Technol 89:297–305

López-Martínez JD, Díaz EE, Martínez RE, Valdez CD (2001) Abonos orgánicos y su efecto en propiedades físicas y químicas del suelo y rendimiento en maíz. Terra 19:293–299

Maas EV, Grattan SR (1999) Crop yields as affected by salinity. In: Skaggs RW, van Schilfgaarde J (eds) Agricultural drainage, Agronomy Monograph 38. ASA, CSSA, SSSA, Madison

Manungufala TE, Chimuka L, Maswanganyi BX (2008) Evaluating the quality of communities making compost manure in South Africa: a case study of content and sources of metals in compost manure from Thulamela Municipality, Limpopo province. Bioresour Technol 99:1491–1496

Martin DL, Gershuny G (eds) (1992) The Rodale book of composting. Rodale, Emmaus

McDowell RW, Sharpley AN (2004) Variation of phosphorus leached from Pennsylvanian soils amended with manures, composts or inorganic fertilizer. Agric Ecosyst Environ 102:17–27

McGechan MB, Lewis DR, Hooda PS (2005) Modeling through-soil transport of phosphorus to surface waters from livestock agriculture at the field and catchment scale. Sci Total Environ 344:185–199

Mexican Association of Ecological Farmers (1992) Decisions research on sustainable agriculture. Agronomy News, January 1992, pp. 26–29

Misra RV, Roy RN (2002) On-farm composting methods. Food and Agriculture Organization of the United Nations, Rome

Mupondi LT, Mnkeni PNS, Brutsch MO (2006) Evaluation of pine bark or pine bark with goat manure or sewage sludge composts as growing media for vegetable seedlings. Compost Sci Util 14:238–243

Nardi JB (2003) The world beneath our feet – a guide to life in the soil. Oxford University Press, New York

Narea G, Valdivieso C (2002) Agricultura orgánica: Situación actual, desafíos y técnicas de producción. Servicio Agrícola y Ganadero. Departamento de Protección de Recursos Naturales Renovables. Santiago, Chile. 150pp

Neely CL, Beare MH, Hargrove HW, Coleman DC (1991) Relationships between fungal and bacterial substrate-induced respiration, biomass and plant residue decomposition. Soil Biol Biochem 23:947–954

Negro MJ, Solano ML, Ciria P, Carrasco J (1999) Composting sweet sorghum bagasse with other wastes. Bioresour Technol 67:89–92

Paffrath A (1993) N-Dynamik auf ausgewählten Flächen des Boschheide Hofes und des konventionellen Vergleichsbetriebes. In: MURL (ed), Abschlußbericht Forschungs- und Entwicklungsvorhaben "Alternativer Landbau Boschheide Hof" 1979–1992. Forschung und Beratung 49: 56–66

Pérez ER (2008) Contaminación agrícola y políticas públicas en México. XI Jornada de economía crítica. 27 al 29 Marzo. Bilbao, V. España, pp 1–23

Potter CL (2000) Biopile treatment of soils contaminated with hazardous waste. Seminar Series on Bioremediation of Hazardous Waste Sites 10:1–5

Reitmayr T (1995) Entwicklungen eines rechnergestützten Kennzahlensystems zur ökonomischen und ökologischen Beurteilung von agrarischen Bewirtschaftungsformen–dargestellt an einem Beispiel. Agrarwirtschaft Sonderheft, 147 pp

Rodríguez CS (2002) El potencial de desarrollo de la agricultura cañera cubana sobre bases sustentables. http://www.nodo50.org/cubasigloXXI/economia/castellon3_311202.pdf. Accessed July 2010

Rynk R (ed) (1992) On-farm composting handbook. Natural Resource, Agriculture, and Engineering Service – Cooperative Extension, Ithaca

Sandoval A, Stuardo A (2001) El compost; una buena alternativa. Centro de Semillas de Árboles Forestales. Facultad de Ciencias Forestales, Universidad de Chile. http://www.uchile.cl/facultades/cs_forestales/publicaciones/cesaf/n13/2.html. Accessed May 2003

Schnitzer M (1969) Reactions between fulvic acid, a soil humic compound and inorganic soil constituents. J Soil Sci Soc Am 33:75–81

Seceña MJC (2010) Evaluación del manejo del nitrógeno en el agroecosistema caña de azúcar. Tesis de Doctorado. Colegio de Postgraduados, Campus Veracruz, 114 pp

Sellers K, Pederson TA, Fan C (1993) Review of soil mound technologies for the bioremediation of hydrocarbon contaminated soil. In: Calabrese E, Bell C (eds) Hydrocarbon contaminated soil, vol 3. Lewis Publishers, Boca Raton

Semple KT, Reid BJ, Fermor TR (2001) Impact of composting strategies on the treatment of soils contaminated with organic pollutants. Environ Pollut 112:269–283

Sharpley AN, Syers JK (1979) Phosphorus inputs into a stream draining an agricultural watershed. II. Amounts contributed and relative significance of runoff types. Water Air Soil Pollut 11:417–428

Smilde KW (1989) Nutrient supply and soil fertility. In: Zadoks JC (ed) Development of farming systems. Pudoc, Wageningen, The Netherlands, p 25–31

Soil and More, Mexico (2010) http://www.soilandmore.nl/node/28. Accessed Aug 2010

Soil Science Society of America (2010) Glossary of soil science terms. https://www.soils.org/publications/soils-glossary. Accessed Mar 2010

Soumare M, Tack FMG, Verloo MG (2003) Effects of municipal solid waste compost and mineral fertilization on plant growth in two tropical agricultural soils of Mali. Bioresour Technol 86:15–20

Stolze M, Piorr A, Häring A, Dabbert S (2000) The environmental impacts of organic farming in Europe, vol 6, Organic farming in Europe: economics and policy. University of Hohenheim, Department of Farm Economics, Stuttgart

Texas Department of Transportation (2004) Standard specifications for construction and maintenance of highways, streets, and bridges. Texas Department of Transportation, Austin

US Composting Council-USDA (2002) Test methods for the evaluation of composts and composting. Composting Council Research and Education Foundation, Holbrook, CD-ROM computer file

US EPA (2010) Electronic code of federal regulations – Title 40: Protection of environment – Part 503 – Standards for the use or disposal of sewage sludge. 503.13 Pollutant limits. http://ecfr.gpoaccess.gov. Accessed Mar 2010

USGS (2010) The Gulf of Mexico hypoxic zone. United States Geologic Survey, Washington, DC. http://toxics.usgs.gov/hypoxia/hypoxic_zone.html

Van Deuren J, Wang Z, Ledbetter J (1997) Remediation technologies screening matrix and reference guide, 3rd edn. Technology Innovation Office, US EPA, Washington, DC. http://www.epa.gov/tio/remed.htm. Accessed Aug 2010

Van Ginkel JT, Raats PAC, Van Haneghem IA (1999) Bulk density and porosity distributions in a compost pile. Neth J Agric Sci 47:105–121

Vereijken P (1990) Integrierte Nährstoffversorgung im Ackerbau. Schweizerische Landwirtschaftliche Forschung 29:359–365

Weindorf DC, Zartman RE, Allen BL (2006) Effect of compost on soil properties in Dallas, Texas. Compost Sci Util 14:59–67

Zechendorf B (1999) Sustainable development: how can biotechnology contribute? Trends Biotechnol 17:219–225

Global Perspectives on Birds in Agricultural Landscapes

Ron J. Johnson, Julie A. Jedlicka, John E. Quinn, and James R. Brandle

I've wandered the world in search of life: bird by bird I've come to know the earth...

– Pablo Neruda

Abstract Earth is home for about 10,000 bird species. They inhabit all continents and interface with agroecosystems worldwide. Bird migrations across continents and nations make birds a truly global phenomenon of broad but complex conservation appeal. Global agricultural expansion during the past 200 years and intensification in the last 50 have been key drivers in global habitat loss and in declines of about 60% of the birds listed on the IUCN red list. Agricultural intensification is a continued concern as is expansion in tropical areas such as Latin America. Maintaining field-edge and set-aside habitats and using lower-intensity practices in production areas

R.J. Johnson (✉)
Department of Forestry and Natural Resources, Clemson University,
260 Lehotsky Hall, Clemson, SC 29634, USA
e-mail: ronj@clemson.edu

J.A. Jedlicka
Environmental Studies, University of California-Santa Cruz,
1156 High Street, Santa Cruz, CA 95064, USA
e-mail: jenvs@ucsc.edu

J.E. Quinn
School of Natural Resources, University of Nebraska-Lincoln,
408 Hardin Hall, 3310 Holdrege Street, Lincoln, NE 68583–0962, USA
e-mail: jquinn2@unl.edu

J.R. Brandle
School of Natural Resources, University of Nebraska-Lincoln,
407 Hardin Hall, 3310 Holdrege Street, Lincoln, NE 68583–0974, USA
e-mail: jbrandle@unl.edu

W.B. Campbell and S. López Ortíz (eds.), *Integrating Agriculture, Conservation and Ecotourism: Examples from the Field*, Issues in Agroecology – Present Status and Future Prospectus 1, DOI 10.1007/978-94-007-1309-3_3,
© Springer Science+Business Media B.V. 2011

are important options for sustaining bird populations globally. Many key threats to birds in agroecosystems are global but specific impacts and management options may differ among geographical areas. Global climate change creates uncertainties for agriculture and birds, including impacts on bird migration and nesting, and concerns about synchrony between birds, habitats, and food resources. Climate change adds to other existing challenges of habitat loss and fragmentation, urbanization, migration barriers, and uncertain food resources. The push for biofuels has resulted in production intensification and habitat losses, especially removal of set-aside lands. Wildlife-friendly farming approaches can facilitate bird movement in fragmented agroecosystems and can provide important habitat for agricultural species and migratory birds. Wildlife-friendly and land sparing approaches are currently being debated toward the goal of sustaining biodiversity and food production. Global influences from social and political systems affect agroecosystems, people, and birds.

Ecotourism may hold potential to benefit local economies, people, and biodiversity if proper and persistent attention is given to ensure these outcomes. Producing food and fiber while, at the same time, sustaining biodiversity and ecosystem services is a challenge for interdisciplinary research in collaboration with working farms and farmers. Research and decision-support tools are needed to facilitate development of policies and infrastructures to support sustainable agriculture and to facilitate conservation of biodiversity in agroecosystems. A conservation vision for the future is needed that embraces the realities of both natural resource limits and human desires for improved quality of life. The positive relationships between people, birds, and sustainable farms may be a key starting point to develop such a vision.

1   Introduction

Of the some 10,000 living bird species on earth (Gill and Donsker 2010), one in eight is threatened with global extinction (BirdLife International 2010; Baillie et al. 2004). Of special concern are farmland birds in Europe; grassland birds in North America; Nearctic-Neotropical migrants between South and North America; Palearctic-African migrants between Europe and Africa; waterbirds in Asia; raptors in Africa; and woodland, grassland, and wetland species in Australia (BirdLife International 2010; Attwood et al. 2009; Olsen 2008). Even many common species are in steep decline (Gaston 2010; Olsen 2008). For example, populations of the Northern Bobwhite (*Colinus virginianus*), a bird known in North America as the "farmer's friend", have plummeted 82% in the past 40 years, one of 20 species in the continental United States (US) that have declined ≥50% in that timeframe (Audubon 2010).

Agricultural expansion and land use change are leading drivers in the global decline of biodiversity (Norris 2008; Scharlemann et al. 2004; Krebs et al. 1999; Tilman 1999; Matson et al. 1997) and in the decline of about 60% of the birds listed on the IUCN (International Union for Conservation of Nature) Red List (Norris 2008). Today, agricultural croplands and pastures cover about 38% of the

Earth's total ice-free land (Ellis et al. 2010; FAO 2007b), and agriculture is the primary user of Earth's freshwater resources (FAO 2007b; Gleick 2000; Shiklomanov 2000; Postel et al. 1996). Birds face additional stressors from climate change (Mawdsley et al. 2009), land and energy development (Czúcz et al. 2010; McDonald et al. 2009), and biofuel expansion (Butler et al. 2010; Fargione et al. 2009). Coupled with these stresses are an increasing human population and inequitable use and distribution of food and resources (Godfray et al. 2010; May 2010). Currently estimated at 6.8 billion, the global population is predicated to level off near nine billion in about 2050 (Godfray et al. 2010). At the same time, people are searching for improved standards of living that include changes in diet (e.g. increases in meat consumption), energy consumption, and land use (Godfray et al. 2010; FAO 2009; Delgado 2003) ultimately seeking more from the limited land available. Thus, ensuring future food supplies while at the same time conserving biodiversity is a global issue and a pressing challenge for society (Wilson et al. 2010; Norris 2008; Krebs et al. 1999; Matson et al. 1997). Moreover, because loss of biodiversity and associated ecosystem simplification and homogenization result in loss of ecosystem resilience and services important to people (Laliberté et al. 2010; Tscharntke et al. 2005; Foley et al. 2005), this topic extends beyond protecting avian communities.

As you read this review about interactions of birds and agriculture, we encourage you to think not only about the current reality and juxtaposition of avian conservation in agricultural lands, but also the potential to create farming systems that provide food for humans and quality habitat for many species. In this review, we focus primarily on the more recent decades and consider how agricultural lands have and will both negatively and positively affect bird populations. We begin with an overview of agricultural change, especially expansion and intensification, and then address birds in agriculture in three geographic areas with different agricultural development patterns and history. Our review of birds and agriculture concentrates on Europe, the United States and Canada, and Latin America, covering primary topics that are experienced elsewhere. Finally, we scale up to provide an overview of current global topics being considered in the scientific literature, and conclude with suggested research and conservation needs.

2 Agricultural Change and Birds

2.1 Expansion and Intensification

Agricultural expansion into new areas and subsequent intensification of the production process brought associated impacts on bird habitats. Between 1700 and 1980, agriculture expanded globally with an estimated 466% increase in cultivated land, accounting for 12 million km^2 brought into cultivation (Meyer and Turner 1992). This expansion was higher in some areas than others, with North America

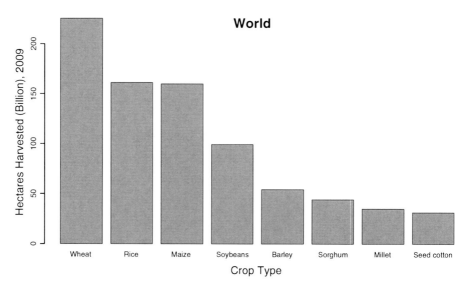

Fig. 1 Land area in the leading crops for World agriculture, arranged by crop from most land area (*left*) to lesser amounts (*right*). Similar graphs for agriculture in Europe, the United States and Canada, and Latin America are presented later in this review (From FAO (2008))

experiencing a 6,666% increase (Meyer and Turner 1992). Overall, cropland increased from about 2% of total ice-free land in 1700 to about 12% in 2000, while pastures/rangelands increased from 3% to 26% (Ellis et al. 2010) (Fig. 1). Expansion has mostly run its course in Europe and continues to a lesser extent in North American grasslands, but continues aggressively in the tropics (DeFries et al. 2010; Hansen et al. 2008; Jenkins 2003).

Intensification of agriculture followed expansion, facilitated by the increased availability and use of synthetic nitrogen fertilizers and pesticides after WWII (Aspelin 2003; Galloway et al. 2003). Intensification can be described as taking a maximum proportion of primary production for human consumption, which results in less being available for the rest of nature (Krebs et al. 1999; Vitousek et al. 1986). Another characterization is that agricultural intensification is the use of practices that increase yields in the short term but that also typically result in land simplification and homogenization (Donald et al. 2001; Krebs et al. 1999; Tilman 1999; Matson et al. 1997). Intensification normally means high-yielding crop varieties supported by applications of synthetic pesticides and fertilizers, mechanization, and often irrigation (Krebs et al. 1999; Matson et al. 1997). Intensification may also involve reduced crop diversity and simplified rotations, often monocultures, and land simplification through drainage and loss of non-crop areas (Filippi-Codaccioni et al. 2010; Donald et al. 2001; Krebs et al. 1999; Matson et al. 1997). The recent advent of genetically-modified (GM) crops is also considered a form of agricultural intensification (Groot and Dicke 2002; Hails 2002; Krebs et al. 1999). Overall, agricultural expansion and the more recent intensification are recognized as major drivers in loss of birds and other biodiversity globally through loss and degradation of habitats (Norris 2008; Krebs et al. 1999).

As agricultural expansion and intensification increased and the associated impacts on birds and other biodiversity became more apparent, research began to focus on understanding how bird populations might be conserved within farmland (Askins et al. 2007; Vickery et al. 1999; Martin and Finch 1995; Robbins 1979). Currently, a key global question facing society is how to produce needed food, fiber, and energy while sustaining bird populations and other biodiversity (Foley et al. 2005; Krebs et al. 1999; Matson et al. 1997). The need in part is to understand how birds interface with agroecosystems, how they are affected by associated management practices, and what conservation approaches might reduce negative impacts.

2.2   Bird Movements and Habitats Overlap Agriculture

Migrant and resident birds interact with agroecosystems in different spatial and temporal ways so land-use changes affect migrant and resident populations differently (Lima and Zollner 1996). Birds use agricultural habitats for food and cover resources during migration and other non-breeding periods, and for nesting and foraging during the breeding season (Fig. 2). While the former require steady food resources and protection from predation and weather events over a few consecutive months of the year, the latter require these amenities year-round in addition to meeting reproductive needs for raising young. Maintaining species richness of both residents and migrants is facilitated by planning land-use activities at large geographical (landscape) scales (Tscharntke et al. 2005).

Migration is complex, energy intensive, and varies by species, routes, and distances. Migrants encounter multiple challenges including natural obstacles such as mountains or bodies of water; and anthropogenic obstacles such as cell phone towers, tall buildings, wind farms, and fragmented landscapes that affect availability of stopover habitats (Faaborg et al. 2010; Kirby et al. 2008; Newton 2008). Moreover, recent global climate change appears to underlie the altered migration schedules documented for some birds in both Europe and North America, and concerns about potential increases in the frequency and intensity of storms that affect migrating birds (Faaborg et al. 2010). Some long-distance migrant birds fly between continents. Nearctic-Neotropical migrants fly between South and North America and Palearctic-Afrotropical migrants between Europe and Africa (Fig. 2b). Other birds migrate shorter distances by shifting closer to the equator during the non-breeding period. Examples include migrants in the United States and Canada that shift from northern latitudes southward to more southern

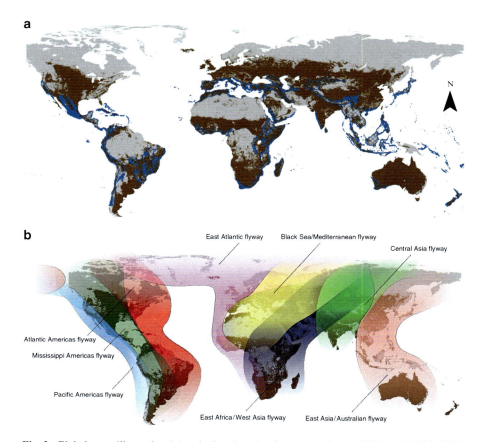

Fig. 2 Global maps illustrating (**a**) agricultural production areas in brown (Ellis et al. 2010, URL: http://www.ecotope.org) and biodiversity hotspots in blue (© Conservation International 2004, used by permission, URL: www.conservation.org) and (**b**) primary bird migration routes.

states, or austral (Southern Hemisphere) migrants in southern South America, Australia, and Africa that shift northward during the non-breeding season (Newton 2008; Jahn et al. 2004). Migration and the varied land use/land cover patterns encountered through seasons create challenges for bird conservation and sustainable agroecosystem management.

Land use changes associated with agriculture overlap globally with migratory routes (Kirby et al. 2008) and with biodiversity hotspots (Conservation International 2004). Scharlemann et al. (2004) found that the proportion of land in agricultural use is greater in Endemic Bird Areas (42%) than in other parts of the world (37%), a trend that continues from historical times and is expected to continue. The clear overlap between agriculture and bird conservation priorities highlights the need for agricultural practices that sustain biodiversity.

2.3   Landscape Legacy

The historical legacy of a landscape, which includes human traditions, patterns of disturbance, succession, and plant and animal use or introductions, strongly affects current biodiversity and prospects for restoration or change (With et al. 2008; Askins et al. 2007; Foster et al. 2003). In that regard, current bird populations in agroeco-systems reflect the land and agricultural history of the area. Agricultural history dates back over 10,000 years (Rowley-Conwy 2009; Pringle 1998) and humans have had today's domesticated crops, animals, and basic tools such as the plough, hoe, sickle, harrow, and axe for over 2,500 years (Burger 1994). In Europe, agriculture has been present for about 9,000 years (Rowley-Conwy 2009) and pigs and dairying for about 6,000 (Spangenberg et al. 2008; Larson et al. 2007). Much of Great Britain has been farmed for over 5,000 years with subsequent clearing events (~4,700 and 2,230 years ago) and most of the landscape has been organized for intense agricultural use for over 1,000 years (Wilson et al. 2009; Fyfe et al. 2003; Fowler 2002). In the Americas, maize and other crops were under cultivation in Mexico about 8,700 years ago (Piperno et al 2009; Ranere et al. 2009) and, in eastern North America, domesticated plants appeared about 5,000 years ago (Price 2009), crop complexes 3,800 years ago, and maize about 2,200 years ago (Smith and Yarnell 2009). Before European coloni-zation of the Americas, however, agriculture was mixed with hunting-gathering activities and overall land-use pressure was less intense than in Europe (Smith and Yarnell 2009; Flannery 2001; Burger 1994).

With the long-term intense use of land in Great Britain and parts of Europe, birds that remain in farming systems are now collectively known as farmland birds and conservation efforts are primarily through agri-environmental policies focused within and around farm fields (Wilson et al. 2009; Zeder 2008). In the Americas, birds in agroecosystems are still associated primarily with their native habitat types and consequently are managed by their primary habitat preference. In Europe, the United States, and Canada most clearing of new land has stabilized. There, conser-vation efforts associated with farming are primarily focused on soil and water and thus attempt to discourage planting row crops on highly-erodible land and encroachment on remaining natural habitats such as riparian areas or wetlands. Although benefits to wildlife are generally secondary outcomes, the habitat value of mid-term (10–15 years) set-aside areas to declining grassland birds is a recog-nized benefit of government programs (Herkert 2009; Johnson and Schwartz 1993a, b). Finally, in tropical areas where agricultural intensification is more recent, additional clearing of forest land for crops or pasture is a major concern, especially in parts of South America (especially Brazil), sub-Saharan Africa, and Indonesia (Hansen et al. 2008; Jenkins 2003).

The history and legacy of various global areas affect decisions about farming and bird habitats. These are reflected in the accounts of the geographical areas that follow below, beginning with Europe, followed by the United States and Canada, and then Latin America, three areas that represent a range of agricultural management patterns and impacts on bird populations.

3   Europe

Northern Lapwing (*Vanellus vanellus*)

3.1   Historical Overview

A patchy wooded environment covered much of the European landscape until the sixteenth century when a growing human population and implementation of frequent agricultural disturbances began to transform the region's land use and land cover patterns (Firbank 2005; Williams 2003: 102, 168). Ultimately, the outcome became a heterogeneous mix of crops, grassland, and linear cover personified by the three-field crop rotation of a fall crop, spring crop, and fallow (Firbank 2005; Williams 2003: 107). As a result, the continent's avian community shifted and many species thrived in the new agroecosystem that emerged (Kleijn et al. 2006). The remaining species are recognized today as farmland birds, considered the norm and of great conservation importance. Consequently, a substantial body of research, management, and policy is targeted towards maintaining an early successional landscape and the associated suite of bird diversity.

This emphasis is warranted. Currently, agricultural lands provide habitat to more at-risk species than any other habitat type in Europe (Wilson et al. 2005). Of the 173 priority species in agricultural habitats, 81 use arable fields and improved grasslands (Tucker and Evans 1997). Moreover, the decline (Fig. 3) and contraction of European farmland bird populations is well documented (Donald et al. 2006, 2001; Fuller et al. 2005a) and is estimated currently at 50% of 1980 levels and at the lowest point observed over the last 30 years (Butler et al. 2010; Wilson et al. 2010).

Since the 1970s, declines in farmland birds have been greater than those of woodland birds (Fig. 3), suggesting a greater impact from agricultural intensification (farmland birds) than from expansion and habitat loss (woodland birds) during this time period. Excellent reviews of birds in agricultural systems in Europe, with a focus on the UK are provided by O'Conner and Shrubb (1986) and Wilson et al. (2009).

Fig. 3 Index of change for farmland birds in the United Kingdom (From DEFRA (2010))

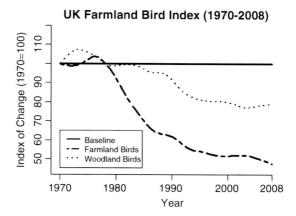

The causes of population decline stem from adopting industrial agricultural management practices following World War II (Wilson et al. 2005), including the intensification of both arable fields and grasslands. Particular changes in farm practices frequently identified as drivers of avian declines include increased field and landscape homogeneity, shorter crop rotations, loss of semi-natural or non-crop habitat, chemical use, a switch from spring to autumn sowing, land drainage, a switch from hay to silage along with earlier harvesting, and the decline in availability of habitat quality at the edge of ranges (Wretenberg et al. 2010; Wilson et al. 2009; Butler et al. 2007; Boatman et al. 2004; Benton et al. 2003; Atkinson et al. 2002; Donald et al. 2001). Demands for biofuel crops (Eggers et al. 2009), along with modernization, specialization, and land abandonment (Wilson et al. 2010) are considered to be ongoing drivers of change across the continent.

A finer examination of the continent highlights a more nuanced relationship between agricultural intensification and bird response (Butler et al. 2010; Donald et al. 2001). Today, 50% of Europe is managed as farmland (Butler et al. 2010), though this percentage varies from 82% in Ireland to 3% in Norway. Intensification of farmland is greatest in Western Europe (Stoate et al. 2009). Until recently, many farms in Eastern Europe remained small with mixed farming systems that included grass-based livestock and arable land. In this area, slower rates of species decline were observed (Reif et al. 2008); however, induction into the European Union has propelled the intensification of Eastern European farmland (Reif et al. 2008; Herzon and O'Hara 2007). Consequently, patterns and usage of farmland by birds may become more similar continent-wide in the future.

Further variations on the negative relationship between intensification and farmbird populations are reported in the literature. For example, farmland birds in Sweden and England exhibit similar population declines, despite Sweden's not following the same trend of increased intensification (Wretenberg et al. 2006). In contrast, patterns of agricultural intensification in Denmark follow continental trends, yet farmland bird populations remain stable (Fox 2004). Differing policy and cultural interests associated with the varied patterns of farming intensification among nations provides an opportunity to examine the impacts of intensification, perhaps gaining a better understanding of the importance of working land for avian conservation.

While the negative trend for farmland birds is well established, the general pattern of decline observed from broad studies does not always reflect information gained or change observed through local research and conservation efforts at the field and farm scale. Below we focus on research and conservation efforts in arable (tilled) cropland, improved grasslands, and semi-natural habitats such as field-edge hedgerows and grass buffers. Other farmland habitats in Europe, including vineyards and orchards (Genghini et al. 2006; Mols and Visser 2002), are important but beyond the scope of this review. Arable cropland, grassland, and associated margins provide important food and cover resources throughout the year. Below we review the use of each during the breeding, non-breeding, and migratory periods.

3.2   Tilled (Arable) Cropland

Arable land provides essential foraging opportunities to many European farmland birds (Bas et al. 2009; Atkinson et al. 2002; Robinson et al. 2001) (Fig. 4). Non-crop vegetation in arable fields provides an important source of seeds, but perhaps as importantly, it recruits insects (Marshall et al. 2003). Yet, recent changes in farming practices have reduced the value of arable cropland as a food source. A shift to fall planting (Evans and Green 2007) and increased nitrogen inputs (Billeter et al. 2008) resulted in increased density of crop vegetation, limiting many species' ability to forage. The increased use of pesticides and shift to fall planting lowers both seed and insect food resources (Butler et al. 2007; Boatman et al. 2004). Similarly, the loss of winter stubble, resulting from a shift to fall planting, reduced the availability of seeds for granivorous farmland birds (Evans and Green 2007; Evans 2003; Hole et al. 2002). The introduction of genetically modified crops is engineered to limit weed and insect populations, further impacting avian food resources (Wilson et al. 2009; Firbank 2005).

Including arable fields in conservation efforts is important because the needs of many farmland species are best met by arable fields that in the past provided sufficient food and cover but are now being lost to intensification (Butler et al. 2007). Foraging and nesting opportunities can be improved by providing both spatial and structural vegetative heterogeneity within a field (Wilson et al. 2005; Morris et al. 2004) such as incorporation of greater disturbance to produce an abundance of seeds (Wilson et al. 2010). Foraging opportunities presented by arable land are also important during non – breeding and migration periods. Specifically, European Golden Plovers (*Pluvialis apricaria*) and Northern Lapwings (*Vanellus vanellus*) in Britain during the non-breeding season selected arable cropland proportional to the availability of the land use type, despite the previously held belief that these species preferred pasture (Gillings et al. 2007). Small changes can have large effects. For example, a novel solution that balances crop production and conservation in working

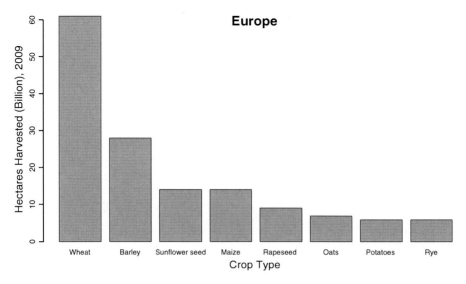

Fig. 4 Land area in the leading agricultural crops in Europe, arranged by crop from most land area (left) to lesser amounts (right) (From FAO (2008))

fields focused on improving Skylark (*Alauda arvensis*) nesting success. The results of this project suggest that Skylark chick abundance and adult density were greater in winter wheat fields that left two unplanted patches, approximately the width of a standard planter, per hectare (Morris et al. 2004).

3.3 Grasslands

Birds use managed grasslands and pasture as nesting habitat and, throughout the year, as foraging habitat. Like arable land, changes in cultural practices of grass-lands have reduced their value for biodiversity. Reduced plant diversity in forage mixes has lowered plant heterogeneity and increased density, both of which nega-tively affect bird forging ability (Whittingham and Evans 2004). Additionally, grassland intensification has limited insectivores in their preferred habitat (Atkinson et al. 2005, 2002). Increased frequency of mowing (or the change from hay to silage) has shortened the safe nesting window in managed grasslands. It also may limit access to needed food resources for young, an outcome that may be amplified by climate change (Kleijn et al. 2010). Delayed harvest (Magana et al. 2010) and reduced stocking rates (Bas et al. 2009) can be important adjust-ments to ensure the nesting success of many species. Increasing grassland hetero-geneity spatially and temporally with short and tall grasses will provide different foraging opportunities for different species, important to conservation efforts (Whittingham and Evans 2004).

Northern Wheatear (*Oenanthe oenanthe*)

Grassland grazing has variable effects on farmland birds depending on grazing intensity (stocking rate), bird species, and timing. For example, although moderate grazing improved habitat quality of managed grasslands for Northern Wheatears (*Oenanthe oenanthe*) in Sweden (Arlt et al. 2008), high intensity grazing limited the value for many species during the breeding season (Tichit et al. 2007). Similarly, intensification through higher nitrogen inputs may benefit farmland insectivores in grasslands during the non-breeding season because of increased soil invertebrate populations (Atkinson et al. 2005), but may reduce vegetation cover of winter set-asides that provide food and cover important for species survival (Whittingham et al. 2005).

3.4 Margins

As in other agroecosystems (Boutin et al. 1999; Best et al. 1990), field edges and margins in Europe play a key role in avian conservation by providing foraging and nesting habitat for many birds (Vickery et al. 2002). There is a strong observed correlation between continent-wide declines of farmland birds and loss of woody edges (Wilson et al. 2009). One quarter of the risk to farmland birds is attributed to the loss of margins and hedgerows (Butler et al. 2010). Moreover, in-field changes associated with greater intensification are so great that the magnitude of change in edge habitats may have been masked (Wilson et al. 2009), suggesting that the full impact of the loss of edge habitats remains uncertain.

Yet in contrast, Bas et al. (2009) report that only 17% of bird species that nest in shrubs or hedges had lower abundance in higher-yielding farmland, compared to 68% of ground nesters. Thus, improving the value of resources in cropped fields is considered a primary need (Butler et al. 2007). This conclusion that margins may not be the limiting factor for at least some farmland birds is demonstrated by the

Yellowhammer (*Emberiza citrinella*), a species that nests in margins and forages in cropland. Yellowhammers have maintained high breeding productivity yet their population continues to decline because of poor quality non-breeding habitat (Cornulier et al. 2009).

3.5 Bird Use of Multiple Land Types

Spatial and temporal heterogeneity is frequently identified as an essential landscape feature for forging and nesting European farmland birds (Gilroy et al. 2010; Tucker and Evans 1997). For example, species richness in Poland was greatest in areas with moderate amounts of arable land, high grassland abundance, and high availability of linear woody edges (Sanderson et al. 2008). In England and Wales, Atkinson et al. (2002) observed a use ratio of 2:2:1 of species using arable, mixed, and grassland landscapes respectively, though the ratio did shift more to arable lands within mixed landscapes in the non-breeding winter season. The current spatial isolation of farming plots from native grassland areas contributes to farmbird declines (Evans 2003; Atkinson et al. 2002). Unfortunately, this is a pattern observed globally as vegetative cover on farms becomes simplified, a result of crops in one region and livestock in another. Simplification of farm landscapes requires birds such as the Lapwing to forage further distances from nest sites (Breitbach et al. 2010) or to fly farther to move among habitat types for different needs (Berg 2008, Evans 2003; Söderström and Pärt 2000).

Spatial and temporal heterogeneity also includes landscape elements not used for crop production. Abundance of birds was positively correlated to areas of semi-natural habitat (Brambilla et al. 2008), highlighting the key point that, for some species, even the best agricultural crop or pasture land will not replace semi-natural or natural habitat in conservation efforts (Billeter et al. 2008). Modifications to include semi-natural habitat can be as simple as adding margins and in-field strips or setting aside a portion of land. However, these efforts have not always been effective because of low quality, quantity, and arrangement (Birrer et al. 2007). Low-intensity cropland can also be seen as a complement to natural areas, improving regional conservation efforts by providing additional habitat (Pino et al. 2000). However, conservation models from nature reserves may not be suitable for working farms because of discord between habitat types and management goals (Bignal and McCracken 1996).

3.6 Key Threats

The key threats in Europe revolve around the adoption of intensive agricultural management practices (Wilson et al. 2005). As described above, changes in farm practices frequently identified as drivers of avian declines include increased field and landscape homogeneity, shorter crop rotations, loss of semi-natural or non-crop habitat, chemical use, a switch from spring to autumn sowing, land drainage, a switch from hay to silage along with earlier harvesting, and the decline in availability of habitat quality at the edge of ranges (Wretenberg et al. 2010; Wilson et al. 2009; Butler et al. 2007; Boatman et al. 2004; Benton et al. 2003; Atkinson et al. 2002; Donald et al. 2001). In the future, demands for biofuel crops (Eggers et al. 2009), along with modernization, specialization, and land abandonment (Wilson et al. 2010) are considered to be ongoing drivers of change across the continent.

3.7   Conclusion

Avian conservation efforts in European policies and scientific publications focus on maintaining a semi-natural landscape shaped by agriculture. Mounting evidence links the decline in European farmbird communities to the loss of food resources caused by intensification of agricultural practices in cropland and improved grassland. For example, an estimated 76% of the risk to farmland birds is attributable to the loss of food resources driven largely by changes in cropped areas (Butler et al. 2010).

Management recommendations and subsequent conservation practices need to consider the interaction between local and landscape effects (Wilson et al. 2010; Wretenberg et al 2010; Söderström and Pärt 2000), as finer scale responses are often species and region specific (Siriwardena et al. 2000). Consequently, a particular challenge will be translating the success of local measures to larger landscapes and ensuring that policy makers and land managers are informed of the costs and benefits of different approaches (Stevens and Bradbury 2006).

The long history between agriculture and biodiversity in Europe has resulted in an avian fauna adapted to agriculture. Recent intensification and change raises concern about whether some of these farmland bird populations will be sustained. Europe's history with birds and farms demonstrates that conservation in agricultural systems can likely sustain species adapted to agriculture but cannot sustain the diversity of other species not so adapted, a special concern in areas such as the tropics where rapid land use change is now occurring.

4   United States and Canada

Compared to Europe, the United States and Canada are more recent in agricultural expansion and as yet lack the more-defined 'farmland bird' group recognized in Europe. Birds on farmlands in the United States and Canada are still classified primarily with their original woodland or grassland habitats and much of the bird research has been on these habitats rather than on crop fields or farming practices. As research tools such as GIS and remote sensing became available, research began to incorporate landscape-scale perspectives that included both crop and non-crop habitats in the larger landscape (Turner 2005; Perkins et al. 2003; Best et al. 2001; Freemark et al. 1995).

A variety of bird species use the diverse agroecosystems in the United States and Canada for nesting, foraging, or migratory stopover, with use varying by life history needs (Boutin et al. 1999; Koford and Best 1996; Rodenhouse et al. 1993; Best et al. 1990). Habitat availability and suitability for birds in agricultural landscapes vary by geographical region, farming practice, topography, farm history, and other factors. In more intensive agricultural areas, non-crop habitats that remain tend to be linear strips of grassy or woody vegetation along field boundaries (Mineau and McLaughlin 1996; Warner 1994; Best et al. 1990; Sugden and Beyersbergen 1984). Examples include wooded riparian corridors, windbreaks or tree rows, wooded or herbaceous fence rows, and grassed waterways. In addition some block-shaped habitats of various sizes occur. Examples include grasslands, primarily set-aside grasslands, small woodlots, and wetlands. Below is a brief historical overview of birds and agriculture in the region followed by descriptions of typical farmland habitats and how birds use them. Finally, there is a brief description of some issues and key threats, and a concluding summary.

4.1 Historical Overview

Expansion and settlement by Europeans across the United States and Canada was well advanced by the early 1900s. Extensive forested areas were cleared in the East for croplands by 1850, followed by high clearing rates of grasslands in the Midwest (1860–1880) and the Great Plains (1880–1900) (Ramankutty and Foley 1999). By 1920, the United States state of Illinois had gone from nearly two-thirds prairie to <1% native grasslands, and harvested hay from over half native vegetation in 1900 to entirely cool-season introduced species by 1920 (Warner 1994). Yet prior to the 1960s, farms generally had less-intensive management, smaller fields interspersed with non-crop habitats, and remained generally more wildlife-friendly than today (Warner et al. 2005; Koford and Best 1996).

Agricultural intensification patterns began to change markedly after the 1940s (Dimitri et al. 2005). Although the amount of land being farmed in the United States remained fairly stable in the 1900s, farm size increased by 67%, the number of farms decreased by 63%, and specialization reduced the number of commodities from about five to about one (Dimitri et al. 2005). The larger fields and crop monocultures were accompanied by effects on wildlife. For example, over half of the grassland bird species that bred in Illinois declined, and four species declined >85% between 1966 and 1991, declines that reflected similar regional and national trends (Herkert 1994; Warner 1994). Intensification also occurred in southern Canada during this time period. For example, assessment of land use in a 29,000 km² study area in Saskatchewan in the early 1980s found that 82.7% of the upland area was tilled annually, over half of the study plots were >90% tilled, and even two-thirds of the public road rights-of-way were used for private farming, leaving little or no potential for wildlife in the intensive agricultural landscape (Sugden and Beyersbergen 1984).

Agricultural land and bird populations in eastern North America have continued to shift. Extensive forest clearing prior to the mid-1800s was followed by farm abandonment (late 1800s–1950) and subsequently by forest regeneration (~1940–2000) (Litvaitis 1993). For example, forest cover in New Hampshire, estimated to be 95% before European settlement was reduced to 47% by 1880, and then increased back to 87% forest cover by 1980 (Litvaitis 1993). The forest clearing led to declines in some forest bird species but increases in grassland birds, a trend that reversed after the 1940s when habitats began to shift back to forests (Norment 2002; Vickery et al. 1994; Litvaitis 1993). Concurrent with this forest regeneration in the East were wide-spread increases in agricultural intensification that affected grassland and early-successional species in the Midwest and Great Plains (Warner 1994). Because migratory and resident bird species affiliated with grassland, shrubland, and the forest-interior have all experienced substantial overall declines (Askins et al. 2007; Litvaitis 1993; Robbins et al. 1989), there is debate about which historic period and land cover or which species group should be a management priority (Askins et al. 2007; Norment 2002). How can conservation plans best sustain range-wide bird populations in view of past and current land use and land cover, with complications from other factors such as habitat fragmentation, fire suppression, development, and intensification within managed landscapes (Askins et al. 2007; Vickery et al. 1994; Litvaitis 1993)?

Bird research reflects the needs and concerns of the time, and from the late 1800s to the 1930s, economic ornithology was a primary focus of bird research with combined support from the American Ornithologists' Union and the United States Department of Agriculture (Evenden 1995; Palmer 1900). Questions related to economic relationships between birds and people, with much focus on determining which birds were beneficial in biological control or as game birds and which were harmful. Although it was clear that birds consumed pest insects in agricultural systems, little guidance was available on how to apply the information effectively in management decisions (Evenden 1995). Following WWII, the advent of widely-available pesticides shifted commercial, farmer, and research interests toward this new technology and away from birds as natural predators (Evenden 1995). By the early 1950s, bird research in agriculture had turned from interest in beneficial functions to focus on controlling species that damaged crops (Evenden 1995). The subsequent agricultural intensification with little consideration for bird habitat or environmental externalities signaled hard times to come for birds in agroecosystems.

Sections below describe common habitats associated with and embedded in farmland and research related to how birds use them. Tilled (row crop or arable) fields can provide food resources such as insects and seeds, and nesting sites for some species (Warner et al. 2005; Best et al. 1995). Woody remnants can host a diverse and abundant bird fauna and are important to Neotropical migrants (Koford and Best 1996; Rodenhouse et al. 1993, 1995). In contrast, grasslands are also needed to sustain unique grassland species, which overall are in decline (Ribic et al. 2009b; With et al. 2008; Boutin et al. 1999; Warner 1994). The mix of habitats

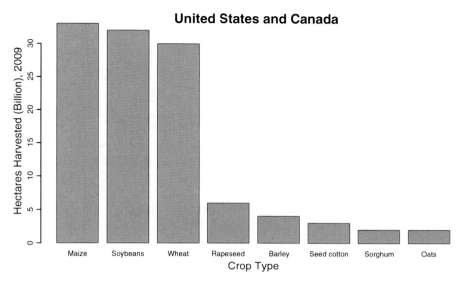

Fig. 5 Land area for the leading agricultural crops in the United States and Canada, arranged by crop with the most land area (*left*) to lesser amounts (*right*) (From: FAO (2008))

available on farms and in surrounding landscapes affects the value of the area for birds. Thus, conservation of bird diversity across agroecosystems should consider conservation needs and goals, the geographical location, habitats available, and management options to maintain unique or rare as well as common species.

4.2   Tilled (Row Crop) Fields

As agricultural intensification increased following the 1960s, fields became larger with fewer crop rotations and less crop diversity (Warner et al. 2005; Best 1983). Inputs of synthetic pesticides and fertilizers maintained monocultures and chemically isolated them from other biota, resulting in reduced habitat value for birds. Practices within fields that affect bird use include crop diversity, type (Fig. 5) and phenology, field size, and various management practices such as tillage methods and pesticide use (Best 2001; Koford and Best 1996; Rodenhouse et al. 1993; Best et al. 1990). Bird use of row crops is also influenced by the type and quality of habitats at field edges and in the surrounding landscape (Warner et al. 2005; Best et al. 2001).

In the Midwestern United States and southern Ontario, at least 94 bird species have been documented using crop fields, primarily for foraging, and at least nine species have been documented nesting in row crops (Hagy et al. 2007; Boutin et al. 1999; Best et al. 1998; Best et al. 1995). Most bird use of row crops is near field edges where non-crop edge habitats provide adjacent cover (Puckett et al. 2009; Best et al. 1990). Thus, smaller fields with relatively more edge habitat have more bird use than larger fields (Best et al. 1990). As crops grow and change, bird-use patterns also shift in relation to the crop structure, food resources, and habitat needs of various species (Best 2001). Moreover, not all bird use is near field edges. Some open-area species such as Killdeer (*Charadrius vociferus*), Horned Lark (*Eremophila alpestris*), and Mountain Plover (*Charadrius montanus*) nest in fields away from edges when crops are short (Best 2001; Knopf and Rupert 1999). Others will forage 200 m or more from field edges where seed and insect resources are available, and Red-winged Blackbirds (*Agelaius phoeniceus*) may perch in maize away from edges when the crop is high (Boutin et al. 1999; Johnson and Caslick 1982).

The frequency and type of soil disturbance in crop fields affects suitability as bird habitat. Under conventional tillage, fields are prepared for planting by inverting the surface, which removes food and cover resources used by birds during breeding and non-breeding seasons (Galle et al. 2009; Best 1985). Following planting, weed control is accomplished by tillage. In the 1980s, various conservation tillage or reduced tillage systems increased in use based in part on benefits of reduced soil erosion, fewer labor and energy inputs, and improved moisture conservation (Hobbs et al. 2008; Johnson 1986; Best 1985; Rodgers and Wooley 1983). These systems typically control weeds using herbicides rather than tillage, so more residue remains on the soil surface and tillage disturbance is reduced. The plant residues provide cover and food resources (insects, waste grains) that increase bird use during nesting, migration, and winter periods (Galle et al. 2009; Best 1985; Rodgers and Wooley 1983). No-till (without tillage) is one such alternative system where crops are planted directly into existing plant residues (Best 1985). Studies of no-till systems have documented more bird species and greater abundance or greater nesting densities compared to conventional tillage (Basore et al. 1986; Best 1985; Castrale 1985; Warburton and Klimstra 1984). Although nesting densities can be greater in no-till than in tilled systems, it may not translate to nesting success and more research is needed related to nesting success and long-term impacts of herbicides used to control weeds (Basore et al. 1986; Best 1986). Management that retains crop residues, however, also retains arthropods and waste grains, important food resources for resident and migratory birds, and reducing the number of tillage passes reduces nesting disturbance (Galle et al. 2009; Koford and Best 1996; Rodenhouse et al. 1993).

Crop field management practices that benefit birds are summarized by Rodenhouse et al. (1995, 1993), Koford and Best (1996), and Warner et al. (2005) and some points they list are included in this paragraph with additional more-recent or original sources. Pesticide impacts on birds, which can be both direct and indirect, can be reduced by using pesticides only when truly needed and by employing integrated pest management techniques (Ehler 2006). Maintaining a diversity of crop types and crop rotations, especially crops with different structures, helps meet habitat needs of different birds. Herbaceous strips such as grassed waterways through fields can protect soil and benefit birds (Bryan and Best 1994; Kemp and Barrett 1989) as can field-edge buffers and other field-edge habitats. Studies in the United States and Canada have found greater bird species richness and abundance in organic than in non-organic fields, largely related to greater food and cover resources (Beecher et al. 2002; Freemark and Kirk 2001). In fact, organic management resulted in greater bird species richness and abundance in the adjacent edge habitats as well as within the tilled field (Beecher et al. 2002). Organic agriculture is discussed further in the global topics section of this review. For the western Great Plains and eastern Colorado, Knopf and Rupert (1999) provide management options to reduce Mountain Plover nest and chick losses within crop fields. They suggest adjusting planting times within crop fields to a short time window and using weed control methods that minimize disturbance and, on adjacent grasslands, seeding only native warm-season grasses and managing for grazing intensity (Knopf and Rupert 1999).

4.3   Woody Habitats

Woody habitats in agricultural regions occur in a range of sizes and are typically recognized as woody riparian zones, windbreaks, fencerows, or small woodlots. While there are some differences in their function, they all potentially provide foraging, nesting, navigational aids, dispersal corridors, and migratory stopover sites for

Downy Woodpecker (*Picoides pubescens*)

woodland birds in highly fragmented agricultural areas (Bonter et al. 2009; Packett and Dunning 2009; Naiman et al. 2005; Skagen et al. 2005; Rodewald and Brittingham 2004; Haas 1995; Skagen et al. 1998). They also provide key winter habitat for resident birds (Knopf and Samson 1994; Knopf et al. 1988; Szaro 1980). Because woody habitats on farms, especially in the Midwest, are typically small in size, often in linear strips along field edges (Warner et al. 2005; Best et al. 2001), birds that benefit are predominantly species adapted to edges or smaller patches of habitat. Some of these woodland edge birds such as Downy Woodpecker (*Picoides pubescens*) Yellow-billed Cuckoo (*Coccyzus americanus*), Eastern Wood-Pewee

(*Contopus virens*), and Black-capped Chickadee (*Poecile atricapillus*) require the woody habitat for essentially all their needs including nesting and foraging (Perkins et al. 2003; Martin 1981). These species may also forage in crop field edges directly adjacent to the woody habitat but typically remain near the field edge where most bird foraging occurs (Puckett et al. 2009; Best 2001; Boutin et al. 1999; Best et al. 1990). Other woodland edge birds such as the American Robin (*Turdus migratorius*), Blue Jay (*Cyanocitta cristata*), American Goldfinch (*Spinus tristis*), and Eastern Kingbird (*Tyrannus tyrannus*) require woody habitat for nesting but forage in both the woody habitat and surrounding fields. Still others, such as the Mourning Dove (*Zenaida macroura*) nest primarily in woody habitat but forage primarily in surrounding fields (Cassel and Wiehe 1980).

Eastern Kingbird (*Tyrannus tyrannus*)

Studies in the United States and Canada have found that bird species richness and abundance are greater in woody habitats than in comparably-sized herbaceous or grassy habitats, but species composition varies and both habitat types are needed (Deschdenes et al. 2003; Best et al. 1990, 1995; Stauffer and Best 1980). Alternatively, in intensive agricultural row crop areas where no grasslands are present, windbreaks can provide crop protection (Brandle et al. 2009; Mize et al. 2008) and along with woody riparian areas, may be the only non-crop habitat available. Below we summarize some of the unique characteristics or impacts of woody habitats.

4.3.1   Riparian Corridors

Riparian refers to transition zones where water meets land, such as the river or streamside habitats that cross through agricultural landscapes (Naiman et al. 2005; Stauffer and Best 1980). Over 89% of riparian corridor area in North America has been lost in the past 200 years (Popotnik and Giuliano 2000; Naiman et al. 1993). Riparian habitats in the United States and at least in some watersheds of Canada cover <2% of the landscape but are among the most productive and valuable terrestrial habitats for birds and other wildlife at local to regional scales (Santelmann et al. 2006;

Jobin et al. 2004; Popotnik and Giuliano 2000; Knopf and Samson 1994). The value of riparian habitats for birds is especially evident in intensive agricultural or grazed areas (Jobin et al. 2004; Perkins et al. 2003; Popotnik and Giuliano 2000) and during migrations through arid regions where food and cover resources may be less predictable (Skagen et al. 2005, 1998).

Studies have documented 136 species of birds using the varied riparian habitats in the Great Plains of the United States (Tubbs 1980). Studies on farms in the Midwest, mostly with small streams, typically find about 30–70 bird species, depending on the study location, vegetation types, season, and other factors (Smiley et al. 2007; Perkins et al. 2003; Fitzmaurice 1995; Stauffer and Best 1980). Riparian habitats are important for post-fledging passerine birds (Akresh et al. 2009) and, in agricultural areas with few trees, may provide key winter habitat for resident birds (Knopf and Samson 1994; Knopf et al. 1988; Szaro 1980). Birds also benefit from riparian areas as dispersal corridors and, during migration, as navigational aids and stopover sites (Naiman et al. 2005; Skagen et al. 2005, 1998; Haas 1995).

Streamside habitats on farms are often extensively altered through encroachment and removal of edge vegetation, narrowing the corridor, corridor channelization, water withdrawals, and livestock trampling or overgrazing (Smiley et al. 2007; NRC 2002; Knopf et al. 1988; Szaro 1980). Such modifications damage riparian corridor functions and make them less suitable as wildlife habitat or in providing other eco-system services. The type of vegetation, width, composition, and related factors affect habitat characteristics in the stream (e.g. shading, temperatures, detritus), water quality (e.g. filtration effects), and streamside habitat available for birds and other biodiversity (Naiman et al. 1993; Gregory et al. 1991).

Generally, increased riparian corridor width and presence of adjacent buffer habitat correlate with increased benefits to birds in agricultural systems (Peak and Thompson 2006; Peak et al. 2004; Stauffer and Best 1980), and vegetation compo-sition (grass, shrub, or trees) strongly influences which bird species will be present (Smith et al. 2008). In agricultural areas in the Midwest, Stauffer and Best (1980) found that some species needed minimum widths of 100–200 m. Similarly, agricul-tural areas in Missouri had greater species richness in wide (400–530 m) forested riparian areas than in narrow (55–95 m), and in narrow riparian areas, birds bene-fited from adjacent grassland-shrub buffer strips (30–45 m wide) (Peak and Thompson 2006; Peak et al. 2004). The wider riparian areas attracted nesting forest interior species and the buffer strips increased richness and density of grassland-shrub species. Even the wider riparian zones with buffer strips, however, were insuf-ficient to sustain nesting populations of some species in the agricultural setting (72–82% row crops and grasses). Habitat in the surrounding landscape is another important variable (Freemark et al. 1995; Andrén 1994). In a landscape study of riparian areas with varying percentages of surrounding woody habitat, Perkins et al. (2003) found that nearby woody habitats may compensate some species for limited woody habitat in narrow riparian corridors.

Riparian areas are natural landscape features and part of the larger surrounding landscape, so it is important to consider watershed protection and the surrounding landscape in management planning (Richardson et al. 2005; Jobin et al. 2004;

Gregory et al. 1991). Management of riparian areas for birds should consider use through all seasons (Knopf and Samson 1994; Szaro 1980), vegetation structure (Seavy et al. 2009; Smith et al. 2008; Stauffer and Best 1986, 1980), needs of sensitive species, and minimizing grazing impacts on breeding birds (NRC 2002; Knopf and Samson 1994; Kauffman and Krueger 1984; Szaro 1980). Restoration of riparian zones for erosion control should include measures to ensure sufficient habitat area and vegetation to enhance birds (Smiley et al. 2007; Maul et al. 2005).

4.3.2   Windbreaks, Shelterbelts and Tree Rows

Windbreaks or shelterbelts are narrow linear strips of vegetation, usually trees and shrubs. In agricultural regions, they are planted to reduce wind speeds or alter wind flow patterns (Brandle et al. 2009; Kort 1988); to control erosion; and to protect crops, livestock, and homes. In the north-central region of the United States, <2% of the crop land is protected by windbreaks. While more extensive in the past, many have been removed to facilitate irrigation development (Brandle et al. 2009). A review of windbreaks in relation to bird communities found that, in the Great Plains of North America, at least 108 species of birds used windbreak habitats and, of these, 29 species benefited substantially, 37 moderately, and the remainder had only minor or accidental benefit (Johnson and Beck 1988).

Linear strips of trees and shrubs that grow naturally along fence lines or field boundaries may appear superficially like designed and planted windbreaks and may serve some of the same functions, but effects on wind and birds will vary with the vegetation type, spacing, and orientation. Woody plant composition may be a more diverse mix of local species and, because seeds are often brought there through bird diets, the species present typically include some with food value for birds.

Windbreaks provide nesting, foraging, non-breeding, and migratory stopover habitat for birds (Johnson and Beck 1988; Yahner 1981, 1982a, 1983; Martin 1980, 1981). Wider windbreaks, like wider riparian areas, typically have more species and individuals during both migration and breeding seasons, although density may be higher in more narrow windbreaks because agriculture-forest birds often concentrate in smaller or isolated habitat patches and forage in the surrounding landscape (Schroeder et al. 1992; Yahner 1983; Martin 1980, 1981). Complex vegetative structure, often associated with older windbreaks, fills wildlife needs more dependably, and the canopy and understory layers appear to be especially important (Schroeder et al. 1992; Yahner 1982b). For example, in a study of Minnesota shelterbelts, Yahner (1982b) found that 60.7% of the bird species primarily used the ground stratum, 28.6% the canopy, but only 10.7% the midstory. Similarly, Martin and Vohs (1978) reported that shelterbelts with the highest bird diversity, in their mid-Great Plains study, had a developed tree canopy and an open understory with a tall lush grass layer.

4.3.3 Fencerows

Vegetated fencerows were a common habitat feature on farms through the mid-1900s, used to separate farms and fields and to manage livestock. Although less common now, they provide benefits to birds that shift as plant succession shifts from herbaceous to brushy to woody tree rows. The fence functions to protect a linear area where vegetation can persist as habitat, and the fence itself functions as a perch site for many birds. Studies in Iowa found 62 species of birds using fencerows, 12 in those with herbaceous vegetation, 38 with scattered trees and shrubs, and 48 with continuous trees (Best 1983). Another study in Michigan evaluated bird nesting in grassy-herbaceous, shrub, or wooded fencerows and found 152 nests of 16 species with a high (58%) overall nesting success, likely because of the limited number and type of predators present (Shalaway 1985).

As agricultural intensification increased in the mid-1900s, however, fencerows were increasingly removed as mixed farms were consolidated into larger units with larger monoculture fields (Taylor et al. 1978; Vance 1976). An estimated 30–80% of such fencerows were removed between the 1930s and early 1990s (Koford and Best 1996; Rodenhouse et al. 1993). Although such fencerows are unlikely to be restored to earlier levels, many still remain and have value to birds. As with most management planning, the value of fencerows or fences varies with location and situation. For example, fences can become a detriment to some wildlife species in rangelands or areas where fences interfere with important wildlife movements (Hayward and Kerley 2009; Fleischner 2010).

4.3.4 Grazing and Management in Woody Habitats

Fencing to prevent unlimited livestock access to woody riparian or windbreak vegetation will generally benefit birds, particularly those that nest or forage near the ground (Popotnik and Giuliano 2000; Knopf and Samson 1994; Yahner 1983; Szaro 1980; Tubbs 1980; Dambach and Good 1940). Grazing livestock in wooded habitats alters vegetation structure and disturbs the nesting process, especially for shrub-nesting species; trampling compacts soil and, in riparian corridors, can increase erosion, reduce water quality, and destroy in-stream fish breeding sites (Krausman et al. 2009; Popotnik and Giuliano 2000; Kauffman and Krueger 1984). Studies in Pennsylvania comparing grazed to control (fenced) riparian areas found greater nest density (Hafner and Brittingham 1993) and increased bird species richness and abundance (Popotnik and Giuliano 2000) in fenced areas. Another Pennsylvania study evaluated riparian sites that had been fenced from livestock for 3–8 years and found that birds responded to the enhanced canopy, shrub, and herbaceous cover for both nesting (38% nest success) and migratory stopover (Argent and Zwier 2007). Studies in Colorado (Stanley and Knopf 2002; Sedgwick and Knopf 1991; 1987) concluded that moderate late-season (August–September, October–November) grazing in riparian areas with appropriate rest periods was compatible with restoration and use by breeding birds, but vegetation recovery was slower than without grazing. A meta-analysis in five western states of seven western riparian systems, vegetated mostly with deciduous trees and shrubs (Tewksbury et al. 2002), found that grazing effects varied among systems but that bird abundance and richness were lower at grazed survey sites. They recommended reducing cattle grazing in deciduous habitats and protecting the few remaining relatively pristine deciduous riparian areas. Where a riparian stream is used as a water source for adjacent livestock, providing livestock access to water in a way that minimizes trampling of the stream and stream bank will help protect the water and habitat quality (Hafner and Brittingham 1993). Other management practices used in riparian areas such as prescribed burning and disking may have value as tools to maintain open grassland in some areas and enhance habitat for some breeding birds, but much remains unknown about other species, appropriate burn frequencies, and long-term impacts (Benson et al. 2007).

Research has found that availability of food or other resources is an important management consideration for birds using shelterbelts (Capel 1988; Martin 1980; Podoll 1979), a point that likely also relates to other woody habitats. Proximity to other wooded habitats, oldfields (herbaceous vegetation interspersed with trees and shrubs), water sources (ditches, irrigation canals), and crop fields using organic management or no-tillage may have resource benefits for birds (Bernier-Leduc et al. 2009; Yahner 1983). In contrast, Schroeder et al. (1992) found no correlation between adjacent land use and bird species richness, possibly because of uniformity in resource availability in the adjacent agricultural fields. Food resources adjacent to shelter from prevailing winds can be important in winter. For example, fruit-bearing shrubs with wind protection can provide food, shelter, and potential sun exposure during stressful cold periods (Johnson et al. 1994; Capel 1988; Yahner 1983; Podoll 1979).

Specific habitat components may be needed by some species. For example, snags provide important nesting and foraging sites for 85 species of North American cavity-nesting birds, most of which are insectivorous species beneficial to forests and agriculture (Scott et al. 1977). Snags are standing dead or partly-dead (e.g. dead or broken top) trees (Thomas et al. 1979). Without suitable nesting sites, birds cannot persist, so the absence of snags in wooded habitats is a limiting factor for cavity-nesting species (Stauffer and Best 1980; Thomas et al. 1979; Scott et al. 1977). Where snags or natural cavities are limited, another option is to erect nest boxes properly sized for desired species that accept nest boxes (Steenhof and Peterson 2009; Willner et al. 1983; McComb and Noble 1981).

4.3.5   Landscape and Woody-Patch Perspectives

Avian response to the landscape differs between forest and agriculture-forest species (Andrén 1994; Martin 1981). Perkins et al. (2003) found that richness of the forest birds increased as woody habitat increased in the surrounding landscape, whereas abundance of the agriculture-forest species decreased. Forest birds apparently benefit from increased woody habitat that provides sufficient space for territories and resource needs, whereas agriculture-forest species appear to concentrate nests in small or limited woody habitat because they can forage in nearby fields (Perkins et al. 2003; Andrén 1994; Martin 1981).

Understanding how various bird species or guilds respond to the amount and arrangement of habitat will help clarify management options as more is learned about their various functions within an agricultural landscape. For example, management could perhaps encourage forest or agriculture-forest species by manipulating the percentage of woody habitat in the landscape. Moreover, when habitat is removed or degraded, the populations of some species will decline only in relation to the amount of habitat removed, but others may require a minimum amount of habitat and will disappear when the minimum threshold is crossed. For example, Perkins et al. (2003) found that the Great Crested Flycatcher (*Myiarchus crinitus*) and Eastern Wood-Pewee were present in woody patches (1.4–3.1 ha) only when there was ≥14.7% and ≥24% woody cover, respectively, in the surrounding 500 m area. In contrast, Yellow-billed Cuckoo (*Coccyzus americanus*), Black-capped Chickadee, and White-breasted Nuthatch (*Sitta carolinensis*) were present in the woody riparian patches regardless of the amount of surrounding woody cover. Thus, removal of vegetation on one farm may affect not only the local habitat, but also forest species on a nearby landscape scale if the total amount of habitat in the area falls below the needed threshold.

Although woody habitats benefit forest species, they may negatively impact many grassland species (Grant et al. 2004; Pierce et al. 2001). Thus, an important planning concern is that planting trees adjacent to grasslands may attract mammalian or avian predators or brood parasites that reduce grassland bird nesting success (Grant et al. 2004; Herkert 1994; Johnson and Temple 1990).

4.4   Grasslands

Grassland bird populations are declining faster than any other guild of North American birds (North American Bird Conservation Initiative 2009; Brennan and Kuvlesky 2005; Samson and Knopf 1994). These declines are attributed primarily to the loss and degradation of habitat (With et al. 2008; Askins et al. 2007) and conservation programs in agroecosystems are recognized as a key tool in maintaining and restoring needed grassland habitats (North American Bird Conservation Initiative 2009). Currently, many types of grassland, especially tallgrass prairies, have been severely reduced and fragmented by agriculture and other development and degraded through fire suppression and tree encroachment (Askins et al. 2007; Knopf 1994).

Vickery et al. (1999) list 48 species of North American (arctic/alpine and temperate) birds considered to be obligate grassland species and 79 considered to be facultative. The obligate species are adapted to grasslands and depend on them for their habitat needs, whereas facultative species use grasslands regularly but depend on them less and also use other habitats (Vickery et al. 1999). Examples of obligate grassland species include Northern Harrier (*Circus cyaneus*), Sprague's Pipit (*Anthus spragueii*), Grasshopper Sparrow (*Ammodramus savannarum*), Dickcissel

(*Spiza americana*), and Western Meadowlark (*Sturnella neglecta*). Examples of facultative grassland species include Northern Bobwhite, Barn Owl (*Tyto alba*), Eastern Kingbird (*Tyrannus tyrannus*), Eastern Bluebird (*Sialia sialis*), and Red-winged Blackbird (Vickery et al. 1999).

Historically, the Great Plains was the center of grassland habitats in North America, consisting of 162 million hectares of native prairie, which transitioned from tallgrass in the eastern Great Plains to mixed-grass and finally to shortgrass in western portions (Samson et al. 1998; Samson and Knopf 1994). Rangelands, which account for 60% of the terrestrial land cover in the United States, mostly in the West, are native plant communities managed for livestock production (Fuhlendorf and Engle 2001; Fleischner 1994). The type of grazing (e.g. season-long, rest-rotation), stocking rate, livestock species, and management (prescribed fire, nutrient or pesticide inputs) affect value to birds, livestock, and ecosystem services (Krausman et al. 2009; Fuhlendorf and Engle 2001; Fleischner 1994). Compared to forest systems, grasslands can vary greatly from year to year and among regions with similar vegetation (Winter et al. 2005). Thus, habitat-related research and management must remain flexible in assessment of local circumstances, habitat variation, and conservation decisions (Ribic et al. 2009b; Winter et al. 2006).

A key grassland conservation initiative in the United States is the Conservation Reserve Program (CRP), begun as part of the 1985 Food Security Act (Farm Bill) and continued in subsequent legislation. The program provides incentives to restore grasslands as set-aside in many erodible or environmentally sensitive farmland areas (Patterson and Best 1996; Johnson and Schwartz 1993b). Although originally created to reduce soil erosion and lower crop surpluses, CRP fields benefited grassland bird populations by increasing the amount of grassland habitat available (Herkert 2009; Riffell et al. 2008; Best et al. 1997; Johnson and Schwartz 1993a).

Linear grassland habitats include grassed waterways (Bryan and Best 1991, 1994), field edge buffers (Conover et al. 2007; Smith et al. 2005; Marcus et al. 2000), and grassy or herbaceous fencerows (Shalaway 1985; Best 1983), roadsides (Camp and Best 1993) and riparian zones (Renfrew et al. 2005; Tewksbury et al. 2002; Stauffer and Best 1980). Block-shaped grassland or herbaceous habitats include pastures, hayfields, small grains, and set-aside areas such as Conservation Reserve Program (CRP) fields (Ribic et al. 2009a; Best et al. 2001; Koford and Best 1996; Rodenhouse et al. 1993, 1995). Birds use linear grassland or herbaceous habitats in the breeding and non-breeding seasons for foraging, nesting, and migratory stopover (Conover et al. 2007; Smith et al. 2005; Marcus et al. 2000). Because linear habitats are relatively small and narrow, however, they often are not suitable as nesting habitats for many grassland species and may be more susceptible to predation (Koford and Best 1996; Rodenhouse et al. 1993; Gates and Gysel 1978). Wider field buffers (e.g. 30 m) appear to provide more value to birds than narrow (Conover et al. 2007; Renfrew and Ribic 2001). Except for set-aside grasslands, however, block habitats are affected by the agricultural management associated with their primary agricultural purpose (e.g. grazing). Consequently, the management practices described below may result in increased costs or benefits to bird populations depending on the bird species, season, location, and management application.

4.4.1 Hay Management

The timing and frequency of disturbances such as mowing and harvesting hay, and earlier-maturing hay varieties, affect breeding birds and have become more problematic as production efficiency has increased (Warner et al. 2005; Troy et al. 2005; Bollinger et al. 1990; Warner and Etter 1989). Some studies have evaluated delayed hay cutting as an approach to allow sufficient time for nesting birds to fledge (Nocera et al. 2005; Dale et al. 1997), but delayed cutting generally results in reduced hay quality that is economically costly (Troy et al. 2005; Frawley and Best 1991). Perlut et al. (2006) used an innovative approach in Vermont and New York that evaluated timing of four management approaches, early-, middle-, and late-hayed fields and rotational pastures, in relation to nesting Bobolinks (*Dolichonyx oryzivorus*) and Savannah Sparrows (*Passerculus sandwichensis*). They found that essentially all nests that were active at the time of haying failed in the early-hayed treatment, and that middle- and late- hayed fields were the most compatible with nesting birds. They also found that 25–40% of the grassland in the area was hayed by mid June during the nesting period before most young fledge, and recognized that balancing production and nesting needs for these species is complex. A preliminary study in Vermont found that uncut patches, which were missed or avoided because of wet soil or debris, may allow some nesting success during harvest and be compatible with production goals (Masse et al. 2008). Warner and Etter (1989) suggested a similar option of maintaining small areas of nesting cover near hayfields and noted that properly managed roadsides might serve this purpose for some species. Moreover, in contrast to privately-owned hay production fields, publicly-owned areas intended for wildlife management purposes, or set-aside acres that include wildlife goals, can plan mowing schedules within and across years to accommodate the diversity of habitat needs for grassland bird species (Warren and Anderson 2005; Dale et al. 1997; Frawley and Best 1991).

4.4.2 Grazing Effects, Managed Grasslands

In the western United States, where 70% of the land is grazed (Fleischner 2010, 1994), plant response and impacts from grazing will likely differ from those in eastern areas because western semiarid grasslands coevolved with large herbivores (Popotnik and Giuliano 2000; Milchunas et al. 1988). Thus, bird responses to various grazing management approaches may also differ. Although a review of grazing practices is beyond the scope of this review, we include some overview points and citations for further review because grazing, particularly overgrazing, affects birds and habitats. Research has, in part, attempted to better understand relationships between grazing patterns and bird nesting success and to recommend management to reduce negative impacts on birds, especially in sensitive areas (e.g. riparian zones, windbreaks, woodlots; discussed above), or in approach (e.g. stocking rates, season-long vs. rotational grazing, timing of grazing) (Krausman et al. 2009; Kauffman and Krueger 1984).

In Wisconsin studies, Temple et al. (1999) found that rotational grazing supported more birds than continuous grazing but, in contrast, Renfrew and Ribic (2001), also in Wisconsin, found no difference in these two management approaches, possibly because of location (lowland vs. upland pastures) or heavier grazing pressure in the Temple et al. (1999) study. Ungrazed control grasslands, however, had higher bird species diversity, density, nest success, and production in the Temple et al. (1999) study than did either the continuously- or rotationally-grazed pastures.

4.4.3   Landscape Perspectives, Set-Aside Grasslands

Landscapes with larger amounts of grasslands and rangelands have been found to have increasing grassland bird populations (Herkert 2009; Veech 2006). On a farm scale, more grassland birds occur in row crops (Best et al. 2001) and in grassland patches (Quinn 2010; Hanson 2007) when surrounding landscapes have higher percentages of grassland. Generally, larger patch sizes are better than small (Herkert et al. 2003), but small or remnant grassland patches also have value to birds, especially in treeless landscapes or in landscapes with a high proportion of grassland (Ribic et al. 2009a, b; Winter et al. 2006). Thus, in planning, it is generally best to locate or restore grasslands together or near other similar habitats because the increased portion of grassland in an area has greater benefit to birds (Grantham et al. 2010; Warner et al. 2005). Management for the variety of obligate grassland bird species requires a landscape mosaic with a variety of grassland structures and types, because species requirements vary (Ribic et al. 2009a; Winter et al. 2006, 2005). In agricultural areas where intensive management tends to homogenize habitat, even large grassland areas may not be sufficient to sustain grassland bird populations (With et al. 2008), so grazing and fire management to create shifting habitat mosaics is needed and appears capable of maintaining bird habitat variety and maintaining livestock production (Fuhlendorf et al. 2006; Fuhlendorf and Engle 2004, 2001).

4.5   Key Threats

Key threats to biodiversity in the United States include habitat loss or degradation, which affects 85% of 1,880 imperiled and federally listed bird species analyzed (Wilcove et al. 2000). Agriculture is the leading cause of habitat loss or degradation and, for birds, agricultural habitat impacts affected 42% of bird species assessed (Wilcove et al. 2000). Other threats include those that also face other parts of the world. Global climate change is affecting the phenology of bird migrations and nesting, raising questions about the uncoupling of birds and their food resources. Weather extremes (droughts, floods, storms) bring uncertainties for both birds and agriculture. The push for maize (corn) ethanol has resulted in habitat losses, especially removal of set-aside grasslands, which are key habitats for declining grassland bird species (Fargione et al. 2009). These global threats are briefly discussed later in this review. Although specific threats vary by location species, and situation, the expansion and intensification of agriculture has clearly affected birds and their habitats in the United States and Canada as it has in other parts of the world (Warner et al. 2005; Rodenhouse et al. 1993).

4.6 Conclusion

European settlement across the United States and Canada has brought widespread land cover change, agricultural expansion and, in the latter half of the 1900s, production intensification, resulting in major habitat losses for bird populations. Remaining field-edge and other habitat fragments, especially riparian areas and grasslands, both native and restored, are key habitats essential for bird populations in agroecosystems. Moreover, agricultural fields provide foraging sites for many species and nesting for some, with value to birds related to vegetative or crop diversity, management intensity, and practices used. Maintaining field edge habitats, riparian areas, and set-aside grasslands, and using production practices that allow for wildlife benefits, will help toward the goal of sustaining bird populations.

5   Latin America – Mexico, the Caribbean, Central and South America

Cerulean Warbler (*Dendroica cerulea*)

5.1 Historical Overview

Two of the main drivers of biodiversity loss in the tropics are forest conversion to croplands and intensification of agricultural systems (Sala et al. 2000). Although forests currently comprise approximately 47% of Central and South American land cover and account for 22% of the global forested area (FAO 2007a), land-use change continues to occur at an alarming rate. Annual net deforestation rates in Latin America from 2000 through 2005 were 0.51%, an increase from 0.46% in the 1990s (FAO 2007a).

Because 70% of animal and plant species reside in tropical forests, deforestation and conversion is one of the largest threats to global biodiversity with strong implications for avian wildlife (Sodhi et al. 2008; Donald 2004). Consequently this section covers bird use of agroecosystems and their response to fragmented landscapes. For simplicity, this section defines Latin America as incorporating the geographic regions of Mexico, the Caribbean, and Central and South America.

Six of the ten countries with the highest levels of avian biodiversity in the globe are found in Latin America (IUCN 2001). Columbia tops the list with 17.3% of the bird species followed by Peru (17.2%), Brazil (16.4%), and Ecuador (15.7%). Venezuela (13.0%) and Bolivia (12.8%) come in sixth and seventh respectively. With an estimated 624 endemic bird species (IUCN 2001), Latin America offers important habitats for avian wildlife and current land-use changes have strong potential to affect large numbers of diverse bird species.

Writing about birds and agriculture throughout Latin America is a broad and complex subject, one that is perhaps best approached by piecing together small-scale analyses to form a larger outlook on the current state of avian conservation. This section begins with a brief historical overview before focusing on birds in specific agricultural habitats, accumulating known trends in Latin America. It then scales-up to discuss birds in fragmented landscapes. Later sections of the review link these conservation concerns to the social and political pressures that act on the landscape.

5.2   Birds in Agricultural Lands

Of the agricultural habitats birds encounter in Latin America, maize is either first or second in land area throughout the region, while dry beans, sugar cane, and coffee appear on each list of the top ten agricultural products by land area (Table 1, Fig. 6). While these crop species are grown across the region, differences in age of crops; timing of blooms and fruits; distance from forest; and management practices such as pruning, pesticide use, cropping patterns, density and type of ground cover, all influence avian use of agroecosystems (Robbins et al. 1992).

Research comparing avian diversity across tropical agricultural landscapes has repeatedly found agroforestry systems to harbor greater diversity and abundance of birds than more disturbed agricultural habitats (Table 2). Agroforestry is defined as land-use practices that deliberately combine woody perennials with animals and/or crops on the same managed land (ICRAF 2010). Such systems benefit birds by providing permanence and structural diversity in the vegetative strata (MacArthur and MacArthur 1961), and are not as disturbed as annual production systems. Additionally, small increases of tree abundance in agricultural landscapes can have large benefits for migratory birds (Greenberg 1992) providing keystone structures (Tews et al. 2004). As management intensification increases, however, diversity declines (Perfecto et al. 2005).

Whether an agricultural landscape is able to provide resources for birds is undoubtedly tied to the form of agriculture practiced within the region. As management intensity increases, associated avian biodiversity declines, and forest species are

Table 1 Top ten agricultural products by regional land area (FAO 2008)

Region	Crop	Million hectares
Mexico	Maize	7.4
	Sorghum	1.8
	Dry bean	1.5
	Wheat	0.8
	Coffee, green	0.8
	Sugar cane	0.7
	Orange	0.3
	Barley	0.3
	Coconut	0.2
	Mangoes, guavas, etc.	0.2
Central America	Maize	9.2
	Dry bean	2.2
	Sorghum	2.1
	Coffee, green	1.6
	Sugar cane	1.3
	Wheat	0.8
	Orange	0.4
	Barley	0.3
	Rice, paddy	0.3
	Oil palm fruit	0.2
Caribbean	Sugar cane	0.6
	Maize	0.5
	Rice, paddy	0.3
	Coffee, green	0.3
	Dry bean	0.2
	Cacao beans	0.2
	Cassava	0.2
	Coconut	0.1
	Sweet potato	0.1
	Banana	0.1
South America	Soybean	41.8
	Maize	21.5
	Sugar cane	9.4
	Wheat	8.0
	Rice, paddy	5.0
	Dry bean	4.5
	Coffee, green	3.7
	Sunflower seed	3.0
	Cassava	2.5
	Seed cotton	1.8

replaced by woodland, scrub, and grassland species. Tscharntke et al. (2008) introduced a basic classification system dividing avian species into three categories:

- Agricultural birds – those that are associated with agricultural and grassland systems, avoiding forest cover.

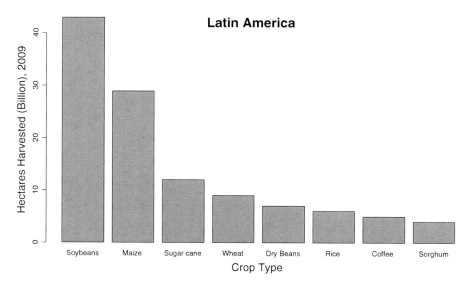

Fig. 6 Land area in the leading agricultural crops in Latin America, arranged by crop from most land area (*left*) to lesser amounts (*right*) (From: FAO (2008))

- Forest birds – those associated only with forest and woodland systems, avoiding open areas, forest specialists.
- Agriculture-forest birds – those that make use of both agricultural and forest habitat. May be common in agroforestry systems. Includes forest generalists.

Data from modeling avian species abundance in agricultural and silvopastoral habitats in Argentina support the categorization of the majority of species as agricultural birds (Filloy and Bellocq 2007). We find this delineation from Tscharntke et al. (2008) useful for Latin America and reference these groupings throughout the section.

Structurally diverse agroecosystems can provide suitable habitat for many species of birds. Forest birds though, as opposed to agriculture-forest species, tend to have stenophagous (narrow) diets and limited elevational range (Tscharntke et al. 2008; Lindell et al. 2004). Forest birds are more likely to be frugivores, nectarivores, or insectivores, and significantly less likely to be granivores, than birds in simplified agricultural habitats. Consequently, forest birds are more restricted in their habitat selection. As forest birds are lost in a landscape increasingly composed of farmland, agricultural and agriculture-forest species may be gained and the functional diversity of birds changes, potentially affecting ecosystem function (Tscharntke et al. 2008).

5.2.1 Annual Cropping Systems

Annual cropping systems are associated with low avian species diversity (Table 2) and are increasingly represented by large-scale monoculture plantings as industrial agriculture expands. Besides tilling and harvesting, these systems often receive

Table 2 Species richness declines with increasing management intensity from little to no management (*left*) to intensely managed (*right*) for all bird species and only migrant species

Land use

			Agroforest				Partially shaded			Unshaded monocultures											
Location	Forested	Acacia woodland	Cacao	Residential	Coffee	Banana	Gallery	Live fences	Fallow/shaded pasture	Milpa[a]	Citrus	Cacao	Coffee	Allspice	Mango	Cashew	Rice	Corn	Jalapeno	Banana/plantain	Unshaded pasture
All birds[b]																					
Tabasco, Mexico	141	–	–	–	110	–	–	–	–	97	–	–	–	–	–	–	–	–	–	–	–
Veracruz, Mexico	178	–	123	–	98	–	–	97	–	–	–	67	60	–	–	–	–	31	21	9	12
Panama	131	–	–	104	87	–	–	–	–	–	–	–	–	65	–	–	–	–	–	–	53
Puebla, Mexico	~120	–	–	–	–	–	–	–	93	94	–	–	–	–	–	–	–	–	–	–	–
Costa Rica	50	–	43.2	–	–	45.6	–	–	–	–	–	–	–	–	–	–	–	–	–	11.5	67
Migrant birds[c]																					
Veracruz, Mexico	46	–	44	–	44	–	–	31	–	–	37	–	–	37	–	–	–	18	18	–	7
Belize	–	–	25	–	–	–	–	–	–	–	30	–	–	–	9	9	7	6	–	–	–
Jamaica	9	–	–	–	10	–	–	–	9	–	–	–	4	–	–	–	–	–	–	–	–
Cuba	11	–	–	–	8	–	–	–	7	–	–	–	–	–	–	–	–	–	–	–	–
Dominican Republic	9	–	–	–	5	–	–	–	7	–	–	–	–	–	–	–	–	–	–	–	–
Chiapas, Mexico	–	18	–	–	11	–	13	–	–	8	–	–	–	–	–	–	–	–	–	–	5

[a] Milpa is classified as an unshaded monoculture, but due to its spatial and temporal diversity (see text) it is more diverse than other monocultures

[b] Sources: Tabasco, Mexico (Sterling and Greenberg unpublished data, reported in Greenberg et al. 2000); Veracruz, Mexico (Estrada et al. 1997); Panama (Petit et al. 1999); Puebla, Mexico (Villa-Bonilla et al. 2008); Costa Rica (Harvey and Villalobos 2007)

[c] Sources: Veracruz, Mexico (Estrada and Coates-Estrada 2005); Belize (Robbins et al. 1992); Jamaica, Cuba, Dominican Republic (Wunderle and Waide 1993); Chiapas, Mexico (Greenberg et al. 1997b)

high inputs of pesticides or fertilizers. For example, fueled by global market forces, soybean production has increased rapidly and is now the top agricultural commodity in Brazil (Fearnside 2001) and Argentina (Di Giacomo and de Casenave 2010) with a total of 41.8 million hectares planted in South America (Table 1). Such drastic changes to landscapes impact avifauna. For example in Argentina, Schrag et al. (2009) found that avian species richness was negatively correlated with pasture and annual crop cover (i.e., soybeans) and positively associated with native woodland vegetation.

Large influxes of migratory birds, in the hundreds and thousands of individuals arrive in the Neotropics during narrow time periods (Robbins et al. 1992). Fallow rice (*Oryza sativa*) fields in Belize contained low bird species richness, but of the species present, Indigo Buntings (*Passerina cyanea*) were found in such large numbers that mist netting operations were downscaled in order to process 98 buntings captured in 1 day (Robbins et al. 1992). During the same season in 1987, a previous researcher estimated bunting numbers to exceed 2000 in Belize rice fields but in 1989 the fields were overgrown and potentially not able to sustain such high numbers without the grain on which the birds feed (cited in Robbins et al. 1992).

Agricultural birds attracted to annual systems may be specialist or facultative grassland species. For example, when dividing birds into specialist and facultative grassland species, Azpiroz and Blake (2009) found that in Uruguay avian species richness was lowest in natural grassland but grassland specialists were more abundant. In contrast, species using wheat and barley crop habitats tended to be facultative grassland species.

Double-collared Seedeater (*Sporophila caerulescens*)

Not only do forest and agricultural birds respond differently to agricultural land use change, but closely-related species may as well. Filloy and Bellocq (2006) studied the response of three species of seedeaters (*Sporophila* spp.) to increasing agricultural land area in Argentina. While one species (Double-collared Seedeaters, *S. caerulescens*) did not seem to be influenced by agricultural land-use, two other species Tawny-bellied (*S. hypoxantha*) and Dark-throated (*S. ruficollis*) Seedeaters showed strong, negative associations to increasing agricultural land area and were not found in landscapes with more than 20% and 60% agricultural land use, respectively.

Incorporating plant diversity into annual fields can positively affect bird species. Similar to the value of field margins in the United States, Canada, and Europe, in Argentina, avian abundance, species richness, and diversity (H') were highest in field margins when compared with neighboring soybean and alfalfa fields (Di Giacomo and de Casenave 2010). Twenty-five of the 41 species recorded were more abundant in field margins whereas five species prefered agricultural habitats, with higher abundance in alfalfa than in soy.

Little work, however, has been done outside Argentina to characterize avian response to the increasing presence of soybean fields. For example, no studies were found comparing avian response to increasing agricultural and silvopastoral land use in South America's Pantanal, a seasonally-flooded savanna wetland that is continental-scale (147,574 km^2), ecologically diverse, and species-rich (Alho 2008). Seventeen percent of the Pantanal has been deforestated and cattle ranching, unsustainable soybean and sugarcane agriculture, and non-organized tourism are some of the factors threatening conservation efforts (Alho 2008). In the Pantanal, 665 bird species use the floodplains and upland habitats including several IUCN listed threatened species such as the Chestnut-bellied Guan (*Penelope ochrogaster*), Hyacinth Macaw (*Anodorhynchus hyacinthinus*), and Sharp-tailed Tyrant (*Culicivora caudacuta*) (Alho 2008).

In the neotropics, traditional farming systems called *milpas* can be important for avian conservation. These systems rotate annual production of maize and other crops such as beans and squash with long fallow periods to allow the soil to recover and woody flora to recolonize. Bird species richness was high in Belize milpa systems because of fallow periods and rotations (Kircher and Davis 1992). Similar results were found in Yucatán, Mexico where traditional milpa agriculture was found to benefit most migrant species, as opposed to agriculture involving mechanized plots and cattle pastures, which offered little benefit (Lynch 1992).

In addition to local effects, the diversity of the surrounding landscape affects avian use of annual fields. A comparison of the avian communities in soybean and eucalypt plantations in both forested and grassland landscapes of Argentina found that while land use was the most important factor explaining abundance of different avian species, landscape factors also were important (Filloy et al. 2010). Of the 28 species recorded in soy fields, more were found in grassland landscapes (21 species) than forested (14 species), and the opposite was true for eucalypt-affiliated species. Landscape and historical factors are important in structuring the avian community.

5.2.2 Pasture

Given the extent of current pastures and the high conversion rate, understanding avian use of pasture landscapes is critical for future conservation efforts. The expansion of pasture at the expense of forested habitat is known to have negative impacts on forest dependent avian communities. Species richness and abundance of birds is low in actively grazed pasture of introduced grasses and no trees, but Rice and Greenberg (2004) list three main ways that pastoral habitat can be improved for avian migrants and residents: (1) incorporation of fallow periods, (2) incorporation of trees and shrubs creating silvopastoral systems, and (3) maintenance of riparian corridors. Likewise, in southwestern Nicaragua, Harvey et al. (2006) found the same three factors to positively influence avifauna. Tree species richness was correlated with richness of all birds, insectivorous birds, and frugivorous birds so that habitats with lower tree cover (live fences and open pastures) contained fewer species than forested habitats. Avian abundance was highest in riparian forests and lowest in living fences and open pastures, but omnivorous bird species richness and abundance did not differ between habitats. While forest birds may still be absent in silvopastoral landscapes, increasing arboreal diversity and richness can have profound impacts on agricultural-forest birds. Below, each of the three mechanisms is further elaborated.

Fallow: The benefits of fallow periods in pasture systems benefit avian species in the same way as in annual crops such as milpa systems, namely by allowing early successional plant communities to take root, increasing resources for avifauna. Saab and Petit (1992) compared avian species richness and abundance in grazed pastures with those abandoned for 2–4 years in Belize. Migrant and resident species richness was 50% lower and avian abundance was almost 70% lower in grazed pastures. Abandoned pastures were more structurally diverse and contained four species of both nectivores and frugivores, foraging guilds that were absent in grazed pastures. While fallow periods are beneficial for many agricultural-forest species, some disturbance regimes may benefit other birds, particularly agricultural birds affiliated with grassland habitats. For example, Isacch and Martinez (2003) found that grazing by sheep in Argentian grasslands shortened grass length, which was positively associated with presence of four migrant shorebird species during the non-breeding season. Nevertheless, fallow periods and retention of woody shrubs and overstory trees are beneficial for conservation of agricultural-forest species.

Silvopastoral Systems: Incorporation of plant species increases local species diversity in pastoral landscapes, benefiting migrant and resident birds alike (Table 2) (Rice and Greenberg 2004). In the Yucatán, Mexico, Lynch (1992) found significantly higher species richness of resident and migrant birds in pasture and shrubby fields than in mature semi-evergreen forest, and Greenberg (1992) found migrants in high densities in abandoned pasture, including forest-associated species.

Some woody plants commonly found in tropical pastures are particularly beneficial for avian conservation. In Costa Rica, frugivorous birds were found to forage in fruiting *Miconia* trees both near and far from forest remnants, although avian species richness was highest in low-intensity habitats near forest (21.5 average species) and lowest in high-intensity habitats far from forest (14.1 average species; Luck and Daily 2003). The authors conclude that fruiting trees in agricultural landscapes can be assets for avian conservation but forest blocks are required to maintain forest birds. Acacia (*Acacia pennatula*) grows in disturbed sites throughout Mexico, Central America, and northern South America, at elevations from 500–2500 m (Seigler and Ebinger 1988). It has extrafloral nectaries and supports high densities of insect species, offering food resources for insectivorous birds (Greenberg and Bichier 2005; Rice and Greenberg 2004). *A. pennatula* grows in active pasture and dense single-species stands, both of which support high densities of migratory birds during the non-breeding season in Mexico (Greenberg et al. 1997a). In mixed acacia-oak woodlots of Nicaragua, migrant and resident avian abundance was significantly higher in acacia plants than oak trees (*Quercus sapotifolia*) (Greenberg and Bichier 2005). Besides the importance to aviafauna, acacia has socio-economic uses as well, and can be managed to provide timber and cattle feed (Rice and Greenberg 2004).

Riparian corridors and living fences: Tropical agricultural landscapes incorporating riparian corridors and living fences may provide critical resources to avian populations. In riparian forests along cattle pasture in Guatemala, avian capture rates in mist nets found resident and migrant species in equal proportions with abundances greatly exceeding those from secondary growth forests (Siegel and Centeno 1996). In Chiapas, Mexico, high concentrations of migrants and residents (but low abundance of forest birds) were found in riparian vegetation strips (Warkentin et al. 1995). Their data further suggest that forest width is important and the greater the land area of the riparian vegetation, the greater the amount of avian habitat. Harvey et al. (2005) found that denser and taller living fences are associated with higher species richness of birds. Likewise, in Honduras, avian species richness and abundance was significantly correlated with width of riparian corridors, particularly increasing concentrations of insectivores and nectarivores (Arcos et al. 2008).

Along with attracting bird species, living fences may facilitate avian movement through the landscape (Seaman and Schulze 2010; Estrada et al. 1997). These linear remnants along with windbreaks and wooded riparian areas may be critical to larger conservation efforts in agricultural landscapes (Hughes et al. 2002). Additionally birds may be enhancing reforestation by dispersing seeds from focal trees, providing critical ecosystem services (Cole et al. 2010; Harvey 2000). Overall it is clear that the diversification of agricultural landscapes can enhance avian species diversity and abundance in tropical pastures.

5.2.3 Perennial Systems and Agroforestry

Perennial systems benefit birds by providing permanence and structural diversity in the vegetative strata (MacArthur and MacArthur 1961), and are not as frequently or intensively disturbed as annual production systems. Mills and Rogers (1992) found high species richness and abundance of resident and migrant birds in citrus orchards of Belize. According to the scientific literature available then, the proportion and abundance of migrants was higher than any habitat within the Yucatán Peninsula. The authors propose that the high level of bird activity stemmed from high insect abundance, low human activity within the plantation, and geographical location. Indeed Belize, Mexico, and the Greater Antilles have long been recognized to contain considerably higher numbers of migrant species than Costa Rica and Venezuela (Robbins et al. 1992). Additionally, in a comparison of migrant populations during the non-breeding season in 76 sites within agricultural and forest habitats in seven Neotropical countries, Robbins et al. (1992) found that citrus (*Citrus* spp.) and cacao (*Theobroma cacao*) plantations contained extraordinary high abundance and diversity of birds. The authors conclude that while agroecosystems with greater plant diversity harbor greater avian species richness, the large exception seems to be mature citrus groves (sampled in Belize and Jamaica) which support high bird diversity and abundance.

In the scientific literature to date, two tropical agroecosystems stand out as receiving a lot of attention because of their strong ability to conserve avian species. Shade-grown coffee and cacao agroforestry systems are well-researched examples of how agricultural landscapes can be managed in concert with avian conservation goals (Greenberg et al. 2008). They are each treated below in turn.

Coffee

Coffee (*Coffea* spp.) plantations stretch throughout Mexico, the Caribbean, and Central and South America, and occur under a diversity of management intensification levels from coffee grown underneath a forest canopy (Fig. 7) to chemically-intensive monocultures of coffee grown alone (Somarriba et al. 2004; Moguel and Toledo 1999). Although the land area for cultivating shade coffee is not extensive (Table 1), the location (in mid-elevation mountain ranges that have experienced high deforestation) captures important migrant overwintering grounds and as such maintains high levels of avian species richness (Table 2) (Perfecto et al. 1996). Important factors influencing avian use of coffee systems include management intensification (level of pruning, pesticides); diversity, density, and structure of canopy trees; coffee density; understory height; and presence of epiphytes (Philpott et al. 2008).

A recent meta-analysis of six Latin American coffee biodiversity datasets found that avian species richness declined with management intensity, but that in rustic coffee systems, where coffee is grown under a tall, diverse canopy of native trees, bird richness can be higher than in nearby forests (Philpott et al. 2008). The meta-analysis combined studies from Peru, Columbia, Guatemala, Nicaragua, and four regions in Mexico. Similarly, studies in Costa Rica (Florian et al. 2008), Panama (Roberts et al. 2000; Petit et al. 1999) and the Caribbean (Johnson et al. 2006; Carlo et al. 2004; Johnson and Sherry 2001; Wunderle and Latta 1998; Wunderle and Waide 1993) continue to document the importance of shade-grown coffee systems for avian habitat, particularly benefiting migratory birds.

In coffee plantations, richness of resident birds and those that prefer to forage strictly in the canopy or understory were more negatively impacted by management intensification than were migratory birds and birds that did not exhibit strong foraging strata preferences (Philpott et al. 2008). A study comparing overwintering populations of the migratory Cerulean Warbler (*Dendroica cerulea*) in the Venezuelan Andes found that body condition increased significantly throughout the season and densities in shade coffee systems were 3–14 times higher than nearby forests, even after accounting for detectability differences (Bakermans et al. 2009). Johnson et al. (2006) found that densities of migratory birds were an appropriate (and more easily

Fig. 7 A shade-grown coffee farm in Chiapas, Mexico, viewed from the air in a bird's-eye-view (*top*) and at ground level (*bottom*) (Photos courtesy of J.A. Jedlicka)

measured) parameter for assessing habitat quality among diverse Jamaican agricultural and forested landscapes.

Cacao

Cacao (*Theobroma cacao*), like coffee, is grown under a gradient of management intensity from rustic plantations under a native forest canopy to monocultures that lack shade trees and rely on chemical inputs for crop management (Rice and Greenberg

2000). Additionally, many shade-grown cacao plantations are grown at an altitude adjacent to coffee, potentially connecting agroforestry habitat across an elevational gradient (Greenberg et al. 2000). Like rustic coffee systems, shaded cacao systems (such as under a canopy of *Erythrina*) provide habitat for many bird species matching levels of avian activity found in native broadleaf forest (Robbins et al. 1992).

There have been reports of breeding in cabrucas (agroforest systems with cacao planted under native forest trees) by tyrants, ovenbirds (Furnariidae) and two species that the IUCN lists as vulnerable: White-necked Hawk (*Leucopternis lacernulatus*) were observed singing while perching; and the endemic Pink-legged Graveteiro (*Acrobatornis fonsecai*) were found nesting in cabrucas (Faria et al. 2006). Research in Brazil found avian species richness to be nearly 90% similar in native forest and rustic cacao systems. However, some understory forest specialists such as Scaled Antbird (*Drymophila squamata*), Bahia Antwren (*Herpsilochmus pileatus*) and White-shouldered Fire-eye (*Pyriglena leucoptera*) were replaced in cabrucas by agricultural and agricultural-forest species such as seedeaters, swallows, and the nectarivorous Bananaquit (*Coereba flaveola*). Moreover, forest species were limited to cabrucas embedded within highly forested landscapes (Faria et al. 2006). Other studies support the conclusion that landscape and distance to forest are likely important factors affecting avian species richness (Van Bael et al. 2007b; Greenberg et al. 2000). In Panama, Van Bael et al. (2007b) found a strong effect of shaded cacao management intensity on avian species richness and on edge-affiliated and canopy birds (such as migratory warblers). In Costa Rica, both abandoned and managed cacao plots had higher avian species diversity and abundance than did forest, but forest birds were largely absent from cacao and were replaced by agricultural generalist species (Reitsma et al. 2001). Greenberg et al. (2000) found that cacao grown under a planted canopy in Tabasco, Mexico, provided habitat for large densities of many small, insectivorous migrant birds. However, like coffee systems, relatively few resident bird species were found, and these were mostly agricultural-forest rather than forest birds.

Clearly cacao agroforests can provide important habitat for many bird species, especially migratory birds and agricultural and agriculture-forest birds. Because of the absence of many resident forest specialists, however, cacao is by no means a substitute for Neotropical forest. Relying on agricultural lands for conservation requires the support of farmers and land managers. There is some evidence that in addition to conservation value, birds in cacao agroforests may be providing ecosystem services to farmers in the form of insect pest control. Vertebrate insectivores reduced large-arthropod density by 45% and herbivory to cacao leaves by 21% in Panama (Van Bael et al. 2007a).

5.3   Birds in a Fragmented Landscape

5.3.1   Landscape Heterogeneity

In southern Costa Rica, where several large and small patches of rainforest are distributed in an agricultural landscape composed of small-scale agricultural (usually <2 ha) and silvopastoral (5–50 ha) plots, 49% of the bird species found in the largest native rainforest reserve (Las Cruces Biological Field Station, 227 ha) also were located in the surrounding densely-populated agricultural zone (Daily et al. 2001). These authors also reported that 149 (55%) of the 272 local bird species were found only in forested habitats, 60 (22%) were found in both forest and agricultural areas and 63 (23%) were found only in open agricultural areas. Within the agricultural areas, species richness and abundance were not correlated with distance to forested patches, indicating that while distance to forest may be an important driver of avian activity at small scales (0–1,000 m from edge, Tscharntke et al. 2008), these patterns diminish at larger geographic scales. Within the forest, fragment size was positively correlated with species richness of forest birds (Daily et al. 2001). Robbins et al. (1987) also found fragment size to be important, influencing densities of many resident species (especially from the suborder Tyranni), but many migrant species were found at similar densities in isolated forest fragments and large contiguous forested areas.

Deforestation has resulted in a fragmented landscape and declines of up to 67% of tropical forest avian species richness (Sodhi et al. 2004). Conversion of forest to farmland not only affects available habitats for birds, but alters avian movement patterns as well. These changes affect both resident birds and migrant species that require appropriate land cover to navigate over great distances. In fact, in Costa Rica, persistence of insectivorous understory birds in fragmented environments was highly correlated to their ability to disperse through fragmented landscapes and not correlated with food availability (Şekercioğlu et al. 2002). Flight paths of Green Hermits (hummingbirds, *Phaethornis guy*) in Costa Rican agricultural landscapes were on average 459 m less direct than in forested landscapes and contained 36% more forested cover than the most direct path (Hadley and Betts 2009). While some hummingbirds seem to persist in fragmented landscapes (Stouffer and Bierregaard 1995a), other species are negatively impacted, although time, vegetative regeneration, avian dispersal, and landscape effects may lessen the impact on several species (Van Houtan et al. 2007; Sieving et al. 1996; Stouffer and Bierregaard 1995b). While Hadley and Betts (2009) found no difference in homing times for Green Hermits between landscapes, Belisle et al. (2001) found that two migratory forest-dependent species (Black-throated Blue Warbler (*Dendroica caerulescens*) and Ovenbird (*Seiurus aurocapilla*)) experienced longer homing times and were less likely to return to their territories when forest cover decreased in the landscape.

Other negative effects of fragmentation include increased edge effects that impact many bird species, especially in the tropics (Lindell et al. 2007). While edge effects in tropical landscapes may differ from those identified in temperate environments, and one must be careful not to over-generalize (Stratford and Robinson 2005), the composition of bird species using edge habitat in tropical regions differs from the forest interior (Laurance 2004; Restrepo and Gómez 1998). Agriculture-forest birds have been found to forage and breed in agricultural habitats in fragmented landscapes of Costa Rica, but the persistence of trees in such areas is important for many bird species (Şekercioğlu et al. 2007).

5.3.2   Breeding

Few studies have analyzed avian reproductive success in tropical agroecosystems. Lindell and Smith (2003) compared breeding success in forests, abandoned unshaded coffee plantations, and active pasture. They found that antbird (Thamnophilidae) nests were most common in forest but absent from other habitats. Two species, the White-breasted Wood-Wren (*Henicorhina leucosticta*) and Chestnut-capped Brushfinch (*Arremon brunneinucha*) were found to nest in both the forest and coffee habitats, but were absent from pasture. Overall, similar avian species nested in pasture and coffee plantations, and daily mortality rates of above-ground cup-nesting birds in both habitats were not significantly different from forest. So, forest conversion diminishes the available habitat for forest species, but agricultural and agriculture-forest birds appear to be successful in raising young in agroecosystems. More studies and species-specific data are needed, however, to better understand the mechanisms involved and to compare these findings to those from other areas.

5.3.3   Research Tools

To date, most studies comparing avian use of agricultural habitats focus on bird species richness and abundance in different production areas. Oftentimes, community similarity indices are produced to demonstrate that bird species differ in their habitat preferences. While calculations of species richness and abundance combined with similarity indices help tease apart the conservation value of various habitats, other calculations are possible. For example, Petit and Petit (2003) developed a relative conservation importance score for avian habitats that factor in vulnerability and habitat preferences for each species. These rankings are constructed by practitioners to represent whatever location-specific values and conservation concerns they deem important. Calculations such as these are relatively rare in the literature, but may prove to be important tools for assessing avian conservation value in agroecosystems.

In summary, maintaining and creating a structurally diverse agricultural landscape is beneficial for avian diversity and species richness, so the question becomes what can be done to create such a landscape. Later in this review, we analyze the social and political forces that play a large role in creating rural landscapes throughout the globe, tying in particularly useful examples from Latin America.

5.4   Key Threats

Avian conservation is threatened by deforestation, fragmentation, and agricultural intensification. Because 70% of animal and plant species reside in tropical forests, deforestation and conversion is one of the largest threats to global biodiversity with strong implications for avian wildlife (Sodhi et al. 2008; Donald 2004). Intensification of agricultural systems through the use of fertilizers and pesticides is associated with decreases in bird species richness and abundance. Other social and political drivers threaten avian conservation and these are addressed below in the "Global topics" portion of the review.

5.5   *Conclusion*

While general trends governing bird response to land-use changes are well understood, several areas of study are needed to evaluate the extent to which agroecosystems can be wildlife-friendly. Specifically, more research is needed on avian reproductive success within agroecosystems to ensure that stable breeding populations exist. More studies should focus on waterbirds and shorebirds within tropical agricultural landscapes. Further research focusing on understanding and avoiding negative effects of industrial agriculture is needed.

There seems to be a discrepancy between those agricultural systems that are well studied as habitats (such as coffee and cacao) and those that are more pervasive in terms of land area in the Neotropics (e.g. soybean). Studies of avian response to soybean and maize fields can help with forecasting scenerios of avian conservation in agricultural landscapes. While highly disturbed annual systems may be 'less interesting' from an avian behavior standpoint, they are increasingly becoming an important presence in the landscape and the exact effects of this conversion should be quantified. Finally, a focus on how birds respond to and interact with agricultural-residential landscapes such as home gardens is an important area of study that has been largely ignored but is relevant for sizable portions of Latin America (Petit et al. 1999).

Avian conservation in tropical agroecosystems is possible and birds benefit from wildlife-friendly farming practices. Forest species are generally reliant on remnant forest patches, and attempts to conserve such habitat could help maintain the high levels of species richness currently seen throughout Latin America. Forest-agricultural birds can be promoted in agricultural landscapes by increasing the densities of trees or by creating agroforestry or silvopastoral systems with field margins. Further exploration and education on social and political issues that connect bird conservation with agricultural systems can help create the momentum to act effectively and expand sustainable agroecosystem landscapes throughout the world. The Smithsonian Migratory Bird Center has been active both in researching the state of tropical avian conservation and in outreach to the public, creating checklists that explain how agroecology can help. They developed the logo Bird-Friendly® that is seen on products such as coffee to illustrate and educate how global agriculture influences migratory and resident birds across the globe. Changing agricultural production systems to work for birds and for small-scale producers will help ensure the sustainability of both people and wildlife in multi-use landscapes.

6   Global Topics

The three geographical areas examined above demonstrate the complex interplay between geography, history, and land use at global and local scales that need to be considered in sustaining the diversity of birds in agroecosystems. Other geographical areas face similar complexities. While beyond the scope of this review, readers should be aware of the similar patterns of agricultural expansion and intensification with concordant changes in patterns of avian diversity. Research efforts in

Asia (e.g. Amano 2009), Oceania (e.g. Haslem and Bennett 2008), and Africa (e.g. Brooks and Thompson 2001) are examining many of the same successful conservation strategies described above. In addition, there are shared global stressors affecting birds, including global climate change, expansion of food and biofuel production, and market uncertainties. These and other factors differ in their magnitude among different areas based on climate, history, and political and social environments. Other current topics in the scientific literature explore mechanisms or approaches to achieve both conservation and production. Examples include organic agriculture and ecosystem services that benefit production. With the enormous breadth of agricultural systems and interests, are there patterns that offer overall guidance toward a future with sustainable agriculture and birds? One point is clear – these issues cross biological, economic, social, and other science disciplines, so solutions will also require an interdisciplinary approach. In the sections below, we present a brief overview of current issues related to avian conservation in sustainable agroecosystems.

6.1   Global Climate Change, Agriculture and Birds

Global climate change is bringing uncertainty to agriculture and bird populations with sobering, research-based concerns about impacts on both (Thomson et al. 2010; Møller et al. 2008; Meza et al. 2008; Thomas et al. 2004). Recent global climate change appears to underlie altered migration and nesting schedules for some birds, and concerns about potential increases in the frequency and intensity of storms that affect migration (Faaborg et al. 2010; Jonzén et al. 2006; Jensen 2004). Of special concern is the potential uncoupling of birds and their food sources, with potential impacts on both birds and their role in suppressing insect pests (Yang and Rudolf 2010; Mac Nally et al. 2009; Visser and Both 2005; Strode 2003). Climate change is not uniform across geographic areas so responses and impacts on birds and other biota vary spatially (Primack et al. 2009). Strode (2003) found trends for earlier springs north of about 40° north latitude, but later springs south of that, so some birds were delayed in more southern locations and arrived late to more northern breeding areas. Phenology of vegetation may be affected by chilling requirements (vernalization) not being met at southern locations so timing of plants, insects, and birds may become uncoupled (Primack et al. 2009; Zhang et al. 2007b). Climate change may interact with agriculture and birds differently, resulting, for example, in timing changes for agricultural activities, which in turn can affect bird nesting and result in lower chick survival (Pearce-Higgins and Gill 2010; Kleijn et al. 2010). Global climate change is a reality (Oreskes 2004) but much remains uncertain about specifics of the impacts on agriculture and birds. Protected area networks and other conservation strategies may help birds adapt to the expected change but there is uncertainty about where impacts will be greatest (Mawdsley et al. 2009). Farms that retain wildlife-friendly components can help meet a need for habitat availability over broad areas such that suitable habitat is available where needed.

6.2 Genetically-Modified Crop Technology

The recent advent of genetically-modified (GM) crops is viewed as technology that might provide options for sustaining both agriculture and biodiversity. However, it also carries uncertainty and potential risks, direct and indirect, across multiple trophic levels (Spiroux de Vendômois et al. 2009; Pretty 2008; Groott and Dicke 2002; Hails 2002; Krebs et al. 1999). For example, some argue that GM herbicide-tolerant crops coupled with tailored herbicide application techniques could meet goals to benefit both production and birds, whereas others have raised concerns about effects of such crops on weed seed availability and potential negative impacts on granivorous bird food resources (Gibbons et al. 2006; Firbank 2005; Dewar et al. 2003). A general concern with GM crops is that potential problems would occur in landscapes already stressed by other intensification practices (Krebs et al. 1999) and there are still many unknowns and cautions expressed about potential effects, direct or indirect, on non-target organisms, including pollinators and natural enemies of crop pests (Groott and Dicke 2002; Hails 2002).

The geographical location, crop type, and local circumstances are also factors that should be considered with GM technology. For example, large amounts of GM hybrid maize imported into Mexico following free trade agreements raised questions about potential crossbreeding with native maize landraces (Canby 2010; Fitting 2006). Mexico is known as the "cradle of corn," an area where corn evolved and was domesticated, probably from the wild relative *teosinte*, and where about 59 landraces are maintained in production by small growers (Canby 2010; Fitting 2006). This maize diversity with a unique genetic reservoir of traits is of enormous value to maize crops globally, especially in the face of potential climate change issues such as floods, droughts, or diseases that could affect crops. Sustaining these unique *in situ* landraces and the associated biodiversity is important to agriculture (Canby 2010; Keleman 2010; Fitting 2006), which in turn affects bird habitats. Unknowns related to GM technology intersect agriculture, biodiversity, and other components that ultimately affect land use and bird habitats. To sustain birds and other biodiversity, GM technology must be thoroughly evaluated and controlled using appropriate safeguards, research, and monitoring (Groott and Dicke 2002; Hails 2002; Krebs et al. 1999).

6.3 The Push for Biofuels

Agricultural expansion and intensification to meet biofuel demands, especially for corn (maize) ethanol, are placing habitats at risk of conversion to biofuel crop production (Fargione et al. 2009; Pineiro et al. 2009; Searchinger et al. 2008). Already, between 2005 and 2008, the corn ethanol boom in the United States is associated with an 850,000 ha reduction in set-aside grasslands (Fargione et al. 2009). Searchinger et al. (2008) detailed how farmers respond to higher prices by converting more forests or grasslands to cropland and, by doing so, actually double greenhouse gas emissions over 30 years rather than reduce emissions as earlier predicted. An alternative is to retain set-aside acres in perennial grasses and use waste products and cellulosic ethanol production rather than using corn-based ethanol (Pineiro et al. 2009; Groom et al. 2008; Searchinger et al. 2008). Sugarcane also is a major crop used in bioethanol production; Brazil, the largest producer, accounts for about 45% of global production but there are concerns about deforestation to meet this demand (Allianz Knowledge Partnersite 2010). Ethanol production from set-aside or other grasslands has benefits of reduced soil erosion and potential habitat for grassland birds (Fargione et al. 2009; Best and Murray 2004). Switchgrass has been proposed as a perennial that could be used for cellulosic biofuels, an option that shows promise, although diverse mixtures of native prairie grasses would have higher value to grassland birds (Fargione et al. 2009; Searchinger et al. 2008; Best and Murray 2004). Algae (Fargione et al. 2009) and microalgae (Groom et al. 2008) also have promise as high-yielding bioenergy sources that would compete less for land. Research is ongoing toward finding sustainable energy options. Proposals encourage multifunctional landscapes that produce both commodities and ecological services (Jordan et al. 2007) and that include wildlife, water quality, carbon sequestration, and other ecosystem services up front in research and planning so that all components are considered in the decision process (Fargione et al. 2009; Groom et al. 2008).

6.4   Pesticide Exposure

In 1962, a turning point toward environmental awareness came with Rachel Carson's classic book, *Silent Spring*, which foretold a spring without birds and brought global attention to the negative impacts of pesticides (Aspelin 2003; Carson 1962), which can be both direct and indirect (Ehler 2006). Most of the research documenting effects of pesticide exposure to birds has been performed in the United States and Europe and is well reviewed in the literature, beyond the scope of this review. Pesticides have increased crop production but with costs of unintended environmental degradation and impacts on birds (Krebs et al. 1999; Matson et al. 1997). Currently, there is interest in finding more sustainable and environmentally-friendly approaches to pest management using, for example, biological controls, interdisciplinary approaches, safer pesticides, and integrated pest management (Bale et al. 2008; Zhang et al. 2007a; Ehler 2006).

    Tropical ecotoxicology has largely been ignored and many studies are still needed, especially focusing on industrial soybean, banana, and pineapple farming (Lacher and Goldstein 1997). Most of the ecotoxicology research focuses on the toxic effects of cholinesterase inhibitors (active ingredients of organophosphate and carbamate pesticides). These chemicals affect avian thermoregulation, reproduction, and food consumption (Grue et al. 1997). For example, in Argentina from 1995 to 1996, large amounts of organophosphate (monocrotophos) pesticides were sprayed to control caterpillar damage to crop fields. Swainson's Hawks (*Buteo swainsoni*), which are insectivorous on their non-breeding grounds, digested large amounts of the insecticide on their prey and over 5000 hawks were found dead in 19 spraying incidents (Goldstein et al. 1999). In Mexico, DDE (the chemical by-product from pesticide DDT) and other pesticides were present in all feather samples taken from eight species found dead in agricultural regions in concentrations known to negatively impact birds (Mora 1997; Mora and Anderson 1991). Residues of other organochlorine pesticides also were found in varying concentrations, dependent upon species, location, and time of year. More recently, high levels of persistent organic pollutants (such as organochlorine pesticides and DDT) from agricultural and other sources were found in soil and water samples in Sao Paulo, Brazil (Rissato et al. 2006). Additionally there is concern over high pesticide use in the expanding soybean industry and its impacts in the Pantanal in Brazil (Alho and Vieira 1997).

It can be difficult to eliminate exposure because of the migratory nature of many species. In one of the few articles focusing on migratory shorebirds, Strum et al. (2010) compared avian populations sampled in reserve wetlands (reference sites) to those found in rice agricultural fields in Uruguay, Paraguay, Argentina, and North America to determine sublethal exposure to cholinesterase-inhibiting pesticides. While five of six shorebird species did not show significant effects, the Buff-breasted Sandpiper (*Tryngites subruficollis*) in South America exhibited evidence of pesticide exposure suggesting negative impacts from agricultural activities. Previous research on cadmium levels in the Western Sandpiper (*Calidris mauri*) sampled in Panama and Canada indicated toxic exposure along their Pacific Coast migratory route (McFarland et al. 2002).

While pesticide use is still high in industrial agriculture, some good news has been uncovered. Data from Brown Booby (*Sula leucogaster*) eggs in Mexico contained only small amounts of DDE residue, apparently due to pre-1990 DDT applications (Mellink et al. 2009). Eggshell thickness in the northern Gulf of California has returned to pre-DDT levels, indicating species recovery.

6.5 Crop Damage and Birds

In some areas of the United States and Canada, where flocking blackbirds (Icteridae) or other species damage maize, sunflower, or other crops, landowners question whether management for beneficial or neutral bird species might also increase crop damage. Flocking blackbird species may perch in field-edge trees, raising concerns about whether the field-edge trees increase the likelihood of damage (Bernier-Leduc et al. 2009; Johnson and Beck 1988). Studies, however, indicate that the presence of the trees appears to have minimal relationship to likelihood or impact of bird damage to crops (Johnson and Beck 1988; Bridgeland and Caslick 1983). These flocks also perch for extended periods directly on tall corn (Johnson and Caslick 1982) or sunflower plants (Besser et al. 1979) and, in corn, consume insect pests (Dolbeer 1990; Bollinger and Caslick 1985; Bendell et al. 1981). Crop damage appears most related to proximity to large blackbird roosts (within 8–9 km) and to crop maturation date, with earlier maturing fields more likely to be damaged (Dolbeer 1990; Bridgeland and Caslick 1983; Besser et al. 1979). Birds documented using field-edge vegetation have been essentially neutral or beneficial species that consume economically-important pest species with bird damage rarely observed (Bernier-Leduc et al. 2009; Jones et al. 2005; Deschdenes et al. 2003; Jobin et al. 2001; Dix et al. 1995). Bird flocks and associated damage to crops are generally visible and, in some situations, crop losses can be substantial, especially in localized areas near large roosts. More subtle bird functions, however, such as suppression of insect pests (Jones et al. 2005; Deschdenes et al. 2003; Jobin et al. 2001; Kirk et al. 1996; Johnson et al. 1992) and wind and weather protection for crops (Brandle et al. 2009; Kort 1988) are less obvious agricultural benefits of these edge habitats.

In some areas of Latin America, birds may arrive in annual production systems before harvest and consume large amounts of grain. Such granivorous birds may be thought of as pests and growers may make attempts to control their foraging damage to crops. Basili and Temple (1995) first reported that some Venezuelan farmers used pesticides to control Dickcissel damage to rice and sorghum by poisoning water at bird drinking locations and by directly spraying entire nocturnal roosts that can contain up to three million birds. In fact, granivorous birds have a long history of being considered agricultural pests. In 1980, the United Nations Food and Agriculture Organization (FAO) estimated that agricultural losses from avian pests, mostly pigeons and doves (Columbidae) and parakeets (Psittacidae), were valued at US$6 million in Uruguay and US$36 million in Argentina (Bruggers et al. 1998). Bobolinks (*Dolichonyx oryzivorus*, the species name means 'rice eater') are considered rice pests in Bolivia where they reside during the non-breeding season and forage in flocks by the thousands not only in rice but soybean and sorghum fields as well (Renfrew and Saavedra 2007). Bird damage to crops may not be as extensive as farmers believe (Basili and Temple 1999), and research is needed to help find solutions workable for both growers and avian conservation. Pesticide exposure is an important area to consider when dealing with birds in agricultural landscapes. While the purposeful, direct contamination of birds is rare, indirect exposure can be a common effect of industrial agricultural landscapes.

Finally, bird damage impacts vary among geographical locations. For example, Red-billed Quelea (*Quelea quelea*) flocks damage food crops in >25 African countries, aggravating already short food supplies (Dalimer and Jones 2002; Bruggers et al. 1989).

Birds provide ecosystem services that are generally beneficial or neutral in agro-ecosystems but, in some situations, birds can cause economically important damage. Prevention and control of bird or other wildlife-caused damage to crops or livestock on sustainable farms merit sound research efforts to assist growers in maintaining economically-viable production.

6.6 Ecosystem Services by Birds – Insect Pest Suppression and Pollination

Besides the conservation value, birds provide ecosystem services (Jha et al., this volume) including pest suppression and pollination. Although pesticide use has contributed globally to increased yields, losses are still high and varied, but around

30–40% overall and higher in some areas, and there is increased recognition that non-chemical alternatives and integrated pest management approaches are needed (Beddington 2010; Thomas 1999). The value of birds in suppression of pest insects was once recognized in economic ornithology research, which diminished as pesticides became prevalent, but is now again gaining attention as important (Beddington 2010; Johnson et al. 2009; Whelan et al. 2008; Bianchi et al. 2006). Research to assess economic impacts of birds in pest suppression is hindered by the difficulty of linking specific bird diets with lower insect damage, which in crop fields is often highly variable in space and time and confounded by pesticide usage (Letourneau and Bothwell 2008; Tremblay et al. 2001). Yet, birds fill a clearly positive role in suppression of insect pests in many different agroecosystems (Johnson et al. 2009; Gámez-Virués et al. 2007; Jones et al. 2005; Deschdenes et al. 2003; Mols and Visser 2002; Jobin et al. 2001; Kirk et al. 1996; Johnson et al. 1992; McFarlane 1976). Moreover, habitats that benefit birds may also enhance other taxa that function to suppress agricultural pests (Tsitsilas et al. 2006; Tscharntke et al. 2005).

In North America, Downy Woodpeckers and Northern Flickers (*Colaptes auratus*) have been found to be important predators of overwintering European (*Ostrinia nubilalis*) or southwestern (*Diatraea grandiosella*) corn borers (Frye 1972; Black et al. 1970; Floyd et al. 1969; Wall and Whitcomb 1964). Similarly, in a study of winter wheat fields in Montana, McEwen et al. (1986) found that two grassland birds, Horned Larks and McCown's Longspurs (*Calcarius mccownii*), had high proportions of cutworms (mostly pale western cutworms, *Agrotis orthogonia*), grasshoppers, and other pest insects in their diets, and concluded that bird predation was a positive supplement to other controls. Jones et al. (2005) identified bird species in Florida that suppress insect pests on farms as functional insectivores and Jones and Sieving (2006) reported that intercropping sunflower (*Helianthus annuus*) strips increased beneficial birds and insect-foraging time. In apple orchards in the Netherlands, Mols and Visser (2002) found that avian predation of lepidopteran pests significantly increased apple yields by 60% compared to sites where birds were excluded from foraging. They concluded that the small initial cost of erecting nest boxes in apple orchards had value in pest reduction and may result in increased yields.

Recent studies in tropical areas have found that birds significantly reduced lepidopteran larvae on coffee plants (Perfecto et al. 2004) and lowered coffee's most significant pest (the coffee berry borer, *Hypothenemus hampei*) by 1–21%, resulting in increased quantities of saleable fruit creating an additional US$44–310 per hectare depending on annual variation and management intensity (Johnson et al. 2009; Kellermann et al. 2008). A trade-off found in some areas, however, is that vegetative complexity attractive to insect-eating birds is also associated with fungal disease symptoms on coffee, a finding that merits further study (Johnson et al. 2009). Van Bael et al. (2008) assessed data from tropical bird exclosure studies and found that insectivorous birds reduced arthropods and plant damage in both agroforestry and forest systems, and outlined questions needing further study. Koh (2008), also using exclosures, documented bird suppression of insect pests in oil palm (*Elaeis guineensis*), which translated into avoiding a potential 9–26% fruit loss.

Many food plants including fruits, vegetables, seeds, and nuts require animal pollinators such as hummingbirds, bats, and insects, yet pollinator services are being threatened by agricultural intensification, habitat losses and fragmentation, disease, and other factors (Aizen and Harder 2009; Isaacs et al. 2009; Aguilar et al. 2006; Holden 2006). Inability to recruit necessary pollinators lowers crop productivity and increases the need for cultivated area to compensate for yield losses (Aizen et al. 2009). Species richness of hummingbirds was positively correlated to size of forested remnants from 0.3 to 20 ha in Costa Rica and pollen loads varied by bird species, highlighting the importance of maintaining species diversity for maintenance of ecological services (Borgella et al. 2001). Within agricultural systems, birds are known to be important pollinators of pineapple guava (feijoa) (Stewart and Craig 1989). More research is needed to analyze the importance of avian pollinators for important food-crops globally and especially in many tropical areas (Aizen et al. 2009; Ashworth et al. 2009; Klein et al. 2007). Agricultural landscapes that retain or restore native plants can serve an important role in pollinator conservation (Isaacs et al. 2009).

6.7   The Role of Organic Agriculture

Organic agriculture is rapidly growing. Worldwide sales of organic products doubled between 2000 and 2006 from US$18 billion to US$38.6 billion, and they continue to increase by US$5 billion per year (Willer et al. 2008). In 2006, land was certified organic in over 135 countries, totaling more than 30 million hectares and 0.65% of total agricultural area (Willer et al. 2008). Almost a quarter (24%) of organic land is found in Europe, 16% in Latin America, and 7% in the United States. When compared to similar non-organic farms, organic farms most often demonstrate greater biological richness and abundance (Bengtsson et al. 2005; Hole et al. 2005) though the magnitude of difference can vary by taxa (Fuller et al. 2005b) and landscape structure (Smith et al. 2010). For birds, research has regularly demonstrated that organic farming increases richness, abundance, use and/or nesting density during breeding (Beecher et al. 2002), non-breeding (Chamberlain et al. 1999), and migration (Dänhardt et al. 2010), although there are examples where no significant difference was observed between farm types (Hole et al. 2005). The use of tillage to control weeds in organic systems has raised questions about impacts on nesting species and more data are needed (Best 1986).

At local scales, the observed benefits to birds in organic systems are attributed to a greater diversity of plants and invertebrates associated with organic management (Hole et al. 2005; Beecher et al. 2002). At larger scales, mixed farming, crop rotations, early sowing, and beneficial management of non-crop areas contribute to the difference (Kragten and De Snoo 2008; Hole et al. 2005). It was suggested that the relative benefit of organic farming is greatest in areas of intensive agricultural management (Bengtsson et al. 2005). This observation is supported by increasing evidence that surrounding landscape structure moderates observed differences in diversity between organic and conventional systems, with little to no increase observed in organic systems compared to conventional when both sites are embedded in a heterogeneous landscape (Batáry et al. 2010; Dänhardt et al. 2010; Smith et al. 2010; Piha et al. 2007; Bengtsson et al. 2005).

Consequently, comparisons between cropping systems remain a challenge because of the difficulty of equal comparisons. Moreover, organic regulations vary by country and most stipulate only what practices cannot be used. Adoption of species-friendly practices and associated land use and land cover patterns can vary greatly between and within farms (Shennan 2008; Hole et al. 2005). More recently, research efforts have begun to focus on improving organic systems, rather than on differences between farm types (Quinn 2010).

The generally positive relationship between organic farming and species conservation has prompted more frequent interactions between farming and conservation groups. These partnerships are perhaps best highlighted by the work of the IUCN and IFOAM (International Federation of Organic Agriculture Movements) to connect organic agricultural and nature conservation, building on the similarities between natural and agroecosystems.

6.8   Land Sparing and Wildlife-Friendly Farming

The need to sustain both agricultural production and biodiversity is increasingly recognized as a pressing research objective (Wilson et al. 2010; Norris 2008; Tilman et al. 2002; Krebs et al. 1999; Matson et al. 1997). Toward this goal, recent papers have presented the concepts of land sparing and wildlife-friendly farming as a structure to frame research and land planning decisions (Fischer et al. 2008; Perfecto and Vandermeer 2010; 2008; Balmford et al. 2005; Green et al. 2005).

The term 'land sparing' was used by Waggoner (1996, 1995) to argue that managing some lands more intensively for greater production would meet food needs and allow other areas to be spared for nature. Green et al. (2005) and Fischer et al. (2008) discussed this concept further and compared it with wildlife-friendly farming, which is integrating wildlife habitat components or practices into farming so that both wildlife and farming persist together in the agricultural landscape. A wildlife-friendly landscape facilitates movement of species among native habitat patches, and provides additional usable habitat for foraging and, for some species, nesting (Perfecto and Vandermeer 2010; Vandermeer and Perfecto 2007). Fischer et al. (2008) contend that land sparing and wildlife-friendly form a continuum, not an either-or. They clarify that rather than just two formats, there is a gradual range from intensified land sparing to low-intensity or wildlife-friendly. Fischer et al. (2008) argue that the suitability of a management approach will vary with the location and situation of a given landscape, depending on a variety of factors including topography, productivity of the landscape, historical land ownership patterns, socio-economic factors, and dominant paradigms and societal preferences.

One recognized weakness of the land sparing approach is lack of evidence that intensification and greater production per hectare, in reality, lead to land spared for nature, a point recognized by Green et al. (2005) and others (Rudel et al. 2009; Dorrough et al. 2007; Matson and Vitousek 2006). Moreover, land sparing has been found to be unsuitable in some landscapes, such as where land is naturally low in nutrients or is fragile and inappropriate for intensification (Dorrough et al. 2007). To be successful, land with intensified production must go beyond just the possibility of spared land, by being coupled with land actually spared for nature (Rudel et al. 2009; Matson and Vitousek 2006). To accomplish this, Hodgson et al. (2010) suggest that agri-environment policies and subsidies might be used to ensure that intensification was offset by areas set-aside for wildlife.

Research is ongoing about the merits of land sparing and wildlife-friendly approaches, especially which approach is best for birds and biodiversity. Currently, there is evidence that wildlife-friendly approaches benefit biodiversity and ecosystem services in ways compatible with sustainable agricultural production. Wildlife-friendly agroecosystems embrace not only production but also the well-being of natural habitats, the landscape of surrounding farm fields and pastures, and the local people and rural communities that support them. However, wildlife-friendly farming alone will not provide breeding habitat for all bird species, especially those requiring large grassland or forested areas. Land sparing is argued as an approach that could help retain some large habitat areas. The concept of sparing land for nature would benefit from a clear and verifiable mechanism to show that intensification does, in fact, spare land for nature. Increasing market pressures for land to produce biofuels (Koh et al 2009; Groom et al. 2008; Anderson and Fergusson 2006; Hill et al. 2006; Dias de Oliveira et al. 2005) will encourage both extensive and intensive production. Strong demand and substantial profits provide incentives and social-political pressures to intensify near-term without a comparable force to encourage land sparing. The true measure, however, is more than just one methodology versus another, but rather in trying to approach sustainable agricultural production. One such yardstick that arose in the 1940s during rapid agricultural expansion in North America was the well-known essay "The Land Ethic," (Leopold 1966: 262) that argued in part "*A thing is right when it tends to preserve the integrity, stability, and beauty of the biotic community. It is wrong when it tends otherwise.*" Thompson (1995: 119) reasoned that Leopold saw human agriculture as part of the biotic community, meaning that both are part of a holistic agroecosystem. Leopold also noted that "*A good farm must be one where the native flora and fauna have lost acreage without losing their existence*" (Dybas 2009), recognizing the value of both agriculture and biodiversity.

6.9   Social and Political Dimensions of Sustaining Agriculture and Birds

6.9.1   Decision Scales

An overriding research need cited in the introduction of this review was the challenge to ensure future food supplies while at the same time conserving biodiversity and ecosystem services (Wilson et al. 2010; Norris 2008; Krebs et al. 1999;

Matson et al. 1997). Solutions are complex and require thinking at multiple scales. Just as birds in a farm field are influenced by the habitats in the surrounding landscape, farm management decisions in that field are influenced by policies, options, and expectations at regional, national, and multinational levels. For example, in North America and Europe, there is continuing debate about goals and efficacy of current agricultural policies aimed at ensuring biodiversity and community resilience (Aviron et al. 2009; Kleijn et al. 2006). Even in biodiversity-friendly organic systems, regulations vary by country and region, resulting in a patchwork of requirements regarding biodiversity conservation. In contrast, tropical nations often lack national policies and are strongly influenced by global commodity markets and management decisions of multinational corporations. Agricultural policies and markets clearly affect farm management and bird habitats, highlighting the increasing need for collaboration between social and ecological sciences at multiple scales.

As global market systems transitioned toward a model of buyer and seller maximizing personal wealth, land and nature became commodities without safeguards for ecological systems and people (Porritt 2007; Worster 1990). As examples, Worster (1990) notes the destructive impacts on Haiti and the rain forests of Borneo as transition from traditional land use to corporate export markets affected ecological systems and traditional cultures. Polasky et al. (2004) determined that trade exports can lead to production specialization and habitat destruction, especially harmful in areas of high endemism such as tropical forests (timber exports) and Midwestern grasslands in the United States (grain exports and grassland birds). Thus, development of export markets can have especially negative effects in tropical or other areas of rich species diversity and endemism.

6.9.2   A Paradox of Poverty, Consumption and Birds

Solutions for biodiversity in agriculture are not easy. Porritt (2007) describes inherent conflicts that occur when a consumption-driven economy, a large human population, and an inability to recognize that resources have limits meet with traditional desires for improved prosperity and political realities of difficult change. Farmers already face enormous market pressures to produce more on fewer acres. Coupled with lower commodity prices relative to input costs (fuel, seed, and fertilizer), farmers are seeing their profit margins squeezed. As a result, many farmers convert more and more of their land to row crops by removing adjacent non-crop acres, bringing a slight increase in production but losing significant on-farm habitat for birds and other species of concern. Additionally, urban development takes high quality agricultural land or converts forests and grasslands to row crop agriculture.

Farm management decisions and competing land uses are intertwined with global markets, increasing human population, and inequitable use and distribution of food and resources (Godfray et al. 2010; May 2010). Serageldin (2002) has described how people live with the paradox of amazing scientific discoveries and technologies, yet in a world where 20% consume 85% of the income. He points out that the richest three individuals in the world have more wealth than the combined GDP

(Gross Domestic Product) of the 47 poorest countries and that, similarly, the richest 15 individuals have more than the combined GDP of all sub-Saharan Africa with a population of 550 million people. In addition to the increasing income disparities worldwide, Balmford et al. (2002) lament the relentless impacts on global natural habitats that erode "*overall human welfare for short-term private gain.*" Finding answers and the best routes forward is complex and will require collective efforts of many disciplines and stakeholders (Dasgupta 2010; Archer et al. 2008; Keys and McConnell 2005; Banks 2004; Lambin et al. 2001), not only for agriculture and birds, but also for sustaining the beauty and wonder of Earth. Farms can help.

6.9.3 Family Farm Perspectives

Globalization of agricultural markets, social and economic factors, and government policies intertwine and have major influences on agricultural sustainability and bird habitats worldwide (Larigauderie and Mooney 2010; Archer et al. 2008; Polasky et al. 2004). The uncertainties of competing in a global market bring new issues that affect farm-level decisions and biodiversity, in part from the need for increased scale and efficiency in order to effectively compete (Archer et al. 2008; Polasky et al. 2004). Crops grown almost exclusively for the export market, such as coffee and cacao, experience boom-and-bust price cycles that result from waves of overproduction and market fluctuation (Perfecto and Vandermeer 2010; Donald 2004; Ruf and Schroth 2004; Talbot 2004; Rice and Greenberg 2000). When prices drop, it is often small-scale producers that rely on such exports for income generation that are negatively impacted (Vandermeer 2011; Perfecto and Vandermeer 2010). So although global markets benefit shade-grown coffee and associated biodiversity when prices paid to growers are acceptable or high, it may present economic difficulties for growers and farm workers when prices fall. Moreover, increased scale and efficiency may require additional control processes that may not mesh well with dynamic natural systems or small farms (Stuart 2008). For example, food safety concerns that arise in large-scale systems can result in regulations that are difficult for small farms and that restrict management options that could benefit biodiversity (Wild Farm Alliance 2010; Stuart 2008).

As family farmers are driven out of markets, they are replaced by large-scale corporate agriculture that relies heavily on chemical inputs, favors monoculture systems, and offers significantly lower conservation value (Vandermeer 2011; Wright 2005). Additionally, small-scale farming can benefit from the increased input of local knowledge, where careful farmers observe small landscape differences and take advantage of the heterogeneity (Badgley et al. 2007; Rice and Greenberg 2000). Such heterogeneity benefits associated wildlife, offering birds not only a high-quality matrix, but foraging and breeding habitat. Small farms also conserve crop diversity such as maize landraces and associated gene reservoirs (Canby 2010; Fitting 2006). Consequently, the sustainability of small-scale farmer livelihoods is critical to the maintenance of secure agroecosystems and farming practices that benefit birds and other biodiversity.

6.9.4 Enough Food, Enough Biodiversity

Human population growth and per capita use and distribution of resources are cited as challenges related to food production and as causes for increasing competition over land, water, and energy use (Beddington 2010; Godfray et al. 2010; May 2010). Human use of natural resources is not uniform worldwide, however, with increasing consumption in some countries but growing poverty in others (United Nations 2010; Ehrlich and Holdren 1971). Humans use about 40% of Earth's primary productivity (Imhoff et al. 2004; Vitousek et al. 1986) and, over the past 150 years, both the human population and per capita use of energy have increased sevenfold (May 2010).

Neither human populations nor poverty, however, is sufficient to explain agricultural intensification in the tropics and land use and land cover change; rather much rests on factors such as global influences, economic opportunities, local and global markets, institutional constraints, and culture (Keys and McConnell 2005; Lambin et al. 2001). Focusing only on food production is not the answer to global hunger and human food security. Poverty without ability to purchase food or land, political unrest, and lack of social policies are key issues that affect both food security and biodiversity, and that require holistic solutions that sustain both people and landscapes (Perfecto and Vandermeer 2010; Alexandratos 1999). Interdisciplinary, innovative, and collaborative approaches are needed to address the complexity of these topics toward the goals of sustaining both food and biodiversity (Beddington 2010; Godfray et al. 2010; May 2010).

6.9.5 Barriers to Adoption of Sustainable Practices

Useful and relevant research findings may be ignored because of social, political, or other barriers, however, so understanding such barriers and how to accommodate the associated issues is also important (Archer et al. 2008; Pannell et al. 2006). Farmers are more likely to apply sustainable practices if they are included in the decision process and understand the value of specific practices and associated ecosystem services to their farm and situation (Roux et al. 2010; Pannell et al. 2006). Moreover, practices must be economically feasible for the farm. Crop prices are typically not set by farm production costs, and requirements for sustainability are not factored into global markets. Thus, the added on-farm costs of conserving biodiversity must be offset through incentives, compensation, or other measures that sustain working farms and rural communities, so the farms can, in turn, sustain ecosystem services broadly beneficial to society. Newton (1998) suggested a two-step approach that included top-down policy and support to provide direction for conservation as occurred with earlier promotings of intensification, and bottom-up incentive programs to aid individual farmers so that options are economically viable. Pannell et al. (2006) provide a cross-disciplinary review of adoption practices by rural landholders and suggest approaches to facilitate the decision process toward positive outcomes. In particular, they note that landholders need to understand the advantage of adopting an idea or practice and be able to easily evaluate the practice in terms of their situation. Research and education, both in classrooms and through outreach beyond universities with decision-makers, farmers, and other stakeholders, can provide needed information and clarify appropriate incentives that make application of research and information realistic.

7 The Roles of Ecotourism and Agrotourism

Nature fascinates people (Kellert and Wilson 1993), and places of rich natural diversity such as tropical rainforests or African savannahs are the compelling topics of novels and children's literature. Likewise, there seems a similar fascination with picturesque farms and agrarian landscapes, with a natural consequence being desire to experience these environments. Such experiences through ecotourism, agrotourism, and increased interaction with nature have potential to increase motivation to protect birds and associated habitats for both tourists encountering wild birds and host areas providing the experience. A key difficulty, however, is managing ecotourism income in ways that are equitable and transparent, particularly in countries where local people live in areas of rich biodiversity and beauty but lack needed skills, education, and financial resources to compete in an international ecotourism market (Honey 2008; Higham 2007).

Although ecotourism is often equated with other terms such as nature tourism, wildlife tourism, or adventure tourism, the latter are defined by tourism activities whereas ecotourism is defined by a set of principles (Honey 2008: 6–7). In 1990, The International Ecotourism Society (TIES 2010) defined ecotourism as *"Responsible travel to natural areas that conserves the environment and improves the well-being of local people."* This and other similar definitions are supported by principles designed to foster conservation and protection of natural areas and local cultures. Goals include low impact on the environment and local culture, educational opportunities for both tourists and local people, a code of conduct for travelers, empowerment of local people, and financial and other benefits for conservation and communities in the host country (TIES 2010; Honey 2008: 6f, Ceballos-Lascurain 2003). In spite of interests from both ecotourism travelers and the conservation community in meeting these principles, doing so is difficult because of competing interests that span scales from local traditional cultures to national and international markets, political circumstances, and globalized free trade (Meléndez 2010; Honey 2008; Higham 2007). Independent certification programs and increased awareness of travelers may eventually help bring ecotourism closer to sustainable outcomes envisioned in the principles (Font 2007; Ceballos-Lascurain 2003).

Ecotourism has increased markedly since the mid-1980s and is still rapidly growing, although actual total revenues stemming from ecotourism versus bird-watching tourism or just tourism or travel are not well measured (Connell 2009; Honey 2008). It encompasses a huge variety of options from visiting a local public or private protected area, outdoor-based retreat, or farm, to cross-continent travel to experience relatively undeveloped areas with few humans. For example, bird watching is encouraged as a way to increase interest in conservation and to generate tourist revenues in various parts of the United States. One example is the Nebraska Birding Trails program (http://www.nebraskabirdingtrails.com/) that generates community revenues and increases interest in conservation of birds and wildlife habitats. Birdwatching is also growing in Britain and Australia, often generating opportunities for specialist bird tour businesses and usually with limited negative impacts (Connell 2009). Australia has been noted as a leader in effective ecotourism certification programs that strive to be sustainable and to realize principles such as those above (Ceballos-Lascurain 2003). Connell (2009) points out that bird watching increased markedly since the 1990s, especially in the United States, Britain, and Australia and bird watchers, the largest group of ecotourists (Şekercioğlu 2002), tend to be well educated, upper income, and to enjoy the cerebral, competitive, and collaborative aspects of birdwatching. Birdwatching, however, can at times cause disturbance to birds or their habitats and, as with other forms of ecotourism, especially in the tropics, may come with inequities for people or conservation in how revenues are distributed (Connell 2009; Kerbiriou et al. 2009; Şekercioğlu 2002). In spite of remaining short of ideal, however, birdwatching is seen as a promising form of ecotourism because participants are typically educated and financially capable people who care about conservation and birds in natural habitats (Connell 2009; Şekercioğlu 2002). Research and collaborative efforts are needed to better understand relationships between birdwatching and ecotourism and to ensure meaningful benefits for conservation, education, local cultures, and sustainability of natural areas.

Agriculture-based tourism or agrotourism may capture interest on small or midsized farms where visitors can learn about farming techniques, see farm operations, or view birds or other wildlife (Bennett et al. 2009; Knickel et al. 2009; Brscic 2006). Many of the same challenges exist. In a mountainous area of China, Yang et al. (2009) describe rapid tourism development since the late 1990s that generated considerable income but failed to help poor farmers who needed help most and failed to stimulate adoption of sustainable conservation measures. They conclude that tourism in this area of China is an attractive source of revenue (average 34% of household income), but needs to be changed so that small farms have the needed training and opportunity to participate, and so that benefits from tourism revenues are linked to ecological agricultural practices. Agrotourism adjacent to urban areas has opportunity to sustain habitats and to provide direct access to farm products, education, and other benefits (Yang et al. 2010; Hansen and Francis 2007).

For protected areas, ecotourism can provide a source of income to help with costs of management and conservation and to benefit local people, if well-planned and managed, although benefits from tourist revenues may not necessarily translate into increased support for conservation (Bennett et al. 2009; Walpole and Goodwin 2001). Divino and McAleer (2009) argue that sustainable (low environmental and cultural impacts) international tourism has potential to offer alternatives for economic development of the Amazon without destruction of the forest.

Ecotourism appears to be gradually replacing intensive agriculture and mining in Namaqualand, South Africa (Hoffman and Rohde 2007) and in Namibia (L. Powell, personal communication, 2010), as important economic sectors providing incentives for conservation and reversing past declines in native species. Moreover, community-based conservation and tourism strategies may hold promise in some areas of East Africa (Nelson et al. 2010). Simon Seno (Narok University College, Kenya, lecture, May 2010) provided information about conservancies operated by native Maasai in Kenya that appear to be a growing ecotourism success. These collective efforts pool land resources into conservancies that offer low-impact tourism, walking safaris, and training for community youth through community-owned guiding schools. Income is used to maintain and enhance the system and to benefit landowners, resulting in pride of ownership, community empowerment, and cultural understanding and tolerance. Ongoing management and collective efforts are needed to maintain and improve the conservancies, but they hold promise to benefit both conservation and local people.

Tourism tends to be a boom-bust industry affected by outside factors that can cause rapid shifts in revenues. For example, tourism declines when economies are down, such as during the recent global recession or when there is political or social unrest, drug wars, or dangers from potential natural disasters (Meléndez 2010; Honey 2008). Where much of a local economy is based on tourism revenues, a sudden loss of those revenues can have sharp effects on the local economy, particularly in nations where there are few alternative sources of funds to buffer a downturn. A more long-term trend of importance is the reliance of tourism on energy and the increasing costs of air travel across countries or continents (Gossling 2000). Some argue that increasing fuel costs may substantially impact air travel and tourism as energy costs rise (Czúcz et al. 2010; Gossling 2000).

Ecotourism is not a panacea for nature conservation or for sustaining agroecosystems and birds. It is, however, a potential revenue source for some areas that, if effectively managed with appropriate cautions, may substantially boost local revenue opportunities and appreciation for the value of the natural resources present. It also offers options for local to international conservation education about unique resources and a potential mechanism to garner public support to help conserve and sustain farms adjacent to expanding urban areas. Ecotourism is a relatively new form of tourism with an outlook expressed in views differing from "pious hope" to "Trojan horse," potentially demonstrating an unrestrained desire for wealth and damage to ecosystems and cultures or, in contrast, the principles for sustainability envisioned as ideal (Honey 2008; Higham 2007; Butler 1990).

One new recent and promising ecotourism model combines a collaborative partnership in which volunteers provide funding and labor to carry out conservation research organized by scientists and volunteer-recruiting NGOs (Brightsmith et al. 2008). This three-way partnership appears to have promise for funding much-needed conservation research in biologically rich tropical areas. Other outcomes include education and training for local people, young biologists, and other leaders, and applied management solutions to benefit tourism sustainability and birds (Brightsmith et al. 2008).

Like many new endeavors, ecotourism has a sustainable vision, but also global complexity and a still uncertain future. Ceballos-Lascurain (2003) compares ecotourism to democracy where people strive for the ideal in spite of imperfections and failures, but also have successes and the potential to improve. Ecotourism is a strong source of revenue originating primarily from people who care about natural habitats and cultures. Realizing positive outcomes that meet the underlying principles noted above depends on how well participants collectively strive together to bring the ideal close to reality.

8   Research Needs and Conservation Applications

Hope is the thing with feathers that perches in the soul – and sings the tunes without the words – and never stops at all. — Emily Dickinson

There is a need for interdisciplinary research, policies, and infrastructure to support sustainable agriculture and to facilitate conservation of biodiversity and ecosystem services in agroecosystems. Research is needed to demonstrate methods that best optimize multiple ecosystem services from agroecosystems, including consideration of temporal and spatial scales, potential redundancy across or within trophic levels, and interactions with farm type and management practices (Whelan et al. 2008; Foley et al. 2005). The value of these services to agriculture and humanity, economic and otherwise, is truly immense (Şekercioğlu 2006; Balmford et al. 2002) but not well quantified or understood, an area where research is needed, particularly with birds (Whelan et al. 2008; Şekercioğlu 2006).

Farmer/researcher collaborations combine site-specific and experiential knowledge from farmers with experimental and science methods from researchers (Perfecto and Vandermeer 2008; Shennan 2008). Shennan (2008) argued for greater emphasis on adaptive, field-based research and interdisciplinary collaborations to address complex questions toward ecologically-based agricultural systems. Collaboration based on inclusiveness and participation between farmers and interdisciplinary research teams promotes research toward key questions and adoption of appropriate findings. The outcomes ultimately benefit society broadly through a more secure and sustainable food supply with balance among components needed for sustaining both biodiversity and agriculture, and the well-being of people that depend upon them.

Below are conservation applications and research needs that we see emerging from review of the literature and from interactions with farmers, researchers, educators, and others toward the goal of sustaining birds, agriculture, environment, and well-being of people and rural communities.

8.1   Conservation Applications to Benefit Birds in Agroecosystems

1. Align and provide policies, incentives, outreach, and education to include biodiversity maintenance, continued ecosystem services, and sustainable livelihoods

as both recognized and expected goals in agroecosystem land use decisions (Bohlen et al. 2009; Archer et al. 2008; Pretty 2008; Keys and McConnell 2005; Polasky et al. 2004; Lambin et al. 2001).

2. Enhance communication and partnerships across the science community together with organizations, agency personnel, policy makers, and farmers to apply available research and knowledge toward sustainable agricultural systems that include ecosystem services and value for rural communities (Stoate et al. 2009; Pretty 2008).

3. Ensure that farmers and farm owners are part of the process so that their knowledge of the local land is included in planning, research, and application (Pretty 2008; Shennan 2008).

4. Encourage education that teaches scientific literacy and science as inquiry so that more people engage in problem-solving discussions of conservation, agroecology, farmer livelihoods, and birds using verifiable knowledge and research-based information (Dybas 2009).

8.2 Research Questions

1. Further research is needed to monitor and understand bird populations on farms and to identify factors that cause or help reverse declines (Wilson et al. 2010; Newton 1998). How do farm management approaches, practices, and incentive programs affect bird populations and how do these differ at field, farm, and landscape scales? How will other stressors such as climate change affect the interface between bird populations and farms?

2. How can spatial landscape planning be used so that land cover and land use patterns help sustain multiple services including production, biodiversity, and other needs (Polasky et al. 2008) and how can farmlands be protected, especially prime farmlands, from urban development?

3. Native plants in agroecosystems provide shelter and food resources (seeds, invertebrates) beneficial to birds but, as weeds, can constrain crop growth (Ryan et al. 2010, 2009; Marshall et al. 2003). What is the optimum threshold for weed abundance and plant species or community composition that would best sustain both crop production and biodiversity?

4. There is need to better understand the various functions of birds in agroecosystems and how best to manage for beneficial mutualisms. How can beneficial functions be integrated with farm practices and local or regional farm goals?

    (a) Research is needed to quantify, maintain, and enhance the role of birds in suppression of insect pests (Kellermann et al. 2008; Jones et al 2005).

    (b) What are the impacts of birds consuming, dispersing, or transporting seeds? Some impacts are beneficial either to seed dispersal or regeneration in natural systems (Garcia et al. 2010; Uriarte et al. 2010; Laube et al. 2008; Holl 1998) or to suppression of weeds in crop fields (Holmes and Froud-Williams 2005); but others cause unwanted spread of early-succession woody plants in grasslands (Briggs et al. 2002; Holthuizjen and Sharik 1985).

(c) How can management practices integrate, maintain, and enhance other beneficial functions such as pollination, sustainable economic opportunities (e.g. harvest of game animals, forest and perennial plant products), and aesthetic and recreational opportunities (Brady 2006; Schulte et al. 2006)?

(d) Research is needed on methods to prevent potential bird damage to crops and to reduce other potential risks in ways that sustain or enhance farm viability and bird diversity.

5. Research is needed to develop more-inclusive perspectives of multifunctional landscapes in agroecosystems. How can disciplines be integrated toward a global unity of purpose to produce a sustainable future for agriculture, birds, and people while maintaining cooperation and value for individuals across multiple interests, disciplines, and nations (Larigauderie and Mooney 2010; May 2010)?

6. Improved decision tools and planning in agroecosystems are needed so that decisions encompass multiple ecosystem services to and from farms, including production, biodiversity, environment, and rural quality of life. There is need to develop and use transparent and verifiable measures that reflect sustainability of economic activity and that illustrate the associated impacts on nature and the well-being of people (Dasgupta 2010).

7. Mechanisms are needed to evaluate and improve the impacts of lost opportunities or reduced farm income that result from sustaining ecosystem services that benefit society more broadly. Options include providing incentives through organizations or government programs (e.g. farm bill or agri-environment schemes) or linking specific services to users who benefit (Bohlen et al. 2009; Nelson et al. 2009; FAO 2007b; Jordan et al. 2007). For example, New York City purchased development rights so that landowners could maintain forest cover to protect watersheds and water quality, a benefit to the watershed, biodiversity, landowners, and rural communities, and a savings of ≥US$5–7 billion for the City compared to constructing and maintaining a new filtration plant (Foley et al. 2005).

9 Conclusion

Writing about bird conservation becomes exceedingly more complex when paired with agriculture in an increasingly globalized world market that fails to link market decisions with impacts on biodiversity or on sustainable farming systems. Bird habitats clearly are intertwined with agricultural systems so the array of factors that influence agricultural management also affect birds. The historical expansion and intensification of agriculture have been clear factors in bird declines globally but agricultural lands produce food for people and also provide habitats used by bird populations during breeding, non-breeding, and migratory seasons. Thus, efforts to conserve birds and their habitats will benefit from understanding the relationships between bird populations and the patterns of agriculture.

Although many of the challenges facing birds and agriculture are global in context, solutions must be adaptable to local circumstances and perspectives. These challenges are complex because they involve both facts and values, are difficult to clearly define, and often lack a clear point where the issue is resolved (Roux et al. 2010; Brewer 2009; Dietz and Stern 1998). A reoccurring theme throughout this review is that different bird species require different habitat and landscape factors, so universal prescriptions are difficult. Although science is a key component, scientists alone cannot resolve the many challenges facing avian conservation in agricultural landscapes, which are transdisciplinary in nature (Roux et al. 2010). To effectively address sustainable agricultural issues, including avian wildlife, there is need for collaboration with growers, farm workers, communities, and others to enhance long-term sustainability (Roux et al. 2010; Pretty 2008; Shennan 2008). Effective outcomes will require collaboration across multiple perspectives or ideologies and verifiable information on approaches to meet local needs within a global agricultural context (Godfray et al. 2010; May 2010).

Finally, although multiple stressors now impinge on the well-being of birds and sustainable farming, there is hope for change represented in part by the flurry of scientific papers and reasoned debate about the best routes forward (Vandermeer 2011; Gliessman and Rosemeyer 2010; Perfecto and Vandermeer 2010, 2008; Fischer et al. 2008; Bohlen and House 2009; Foley et al. 2005; Green et al. 2005). There also is hope from increased global awareness of the issues and from farmers themselves – the United Nations declared 2010 to be the *International Year of Biodiversity* as "*a celebration of life on earth and of the value of biodiversity for our lives.*" As part of this celebration, the International Federation of Agricultural Producers, representing 600 million family farmers within 120 national organizations in 80 countries, committed to conserving biodiversity as part of a shared responsibility (Sorensen and Goodfellow 2010). They also recognized that the task is large and must be shared with help from scientific research appropriately disseminated and available, supportive government policies and programs, and opportunities for participation in the process with recognition of farmers' indigenous knowledge of local resource management. The future of birds is intertwined with the future of people, especially the future of rural communities and farmers who manage and care for the land (Norris 2008; Perrings et al. 2006; Jackson and Jackson 2002). The positive relationships between people, birds, and sustainable farms may be a key starting point to develop a shared conservation vision for the future.

Acknowledgements We are very grateful to Associate Editor E.C. Lefroy, D. Jackson, and an anonymous reviewer for valuable inputs on the draft manuscript. We thank M.M. Beck and J.R. Courter for reviews of draft portions of the manuscript and for logistical help, J.B. Campbell and J.D. Loomis for technical assistance, D.K. Ebbeka for graphics help in preparation of the figures, and A.I. Larson for providing artwork of birds. We also appreciate the professional efforts and contributions of the many researchers, managers, farmers, and others who have contributed to the vast literature available on birds in agroecosystems; their efforts continue to move science and knowledge forward. Last, we wish to acknowledge the financial support provided to each of us by our respective institutions and for allowing us the time to pursue this review.

References

Aguilar R, Ashworth L, Galetto L, Aizen MA (2006) Plant reproductive susceptibility to habitat fragmentation: review and synthesis through a meta-analysis. Ecol Lett 9:968–980

Aizen MA, Harder LD (2009) The global stock of domesticated honey bees is growing slower than agricultural demand for pollination. Curr Biol 19:915–918

Aizen MA, Garibaldi LA, Cunningham SA, Klein AM (2009) How much does agriculture depend on pollinators? Lessons from long-term trends in crop production. Ann Bot 103:1579–1588

Akresh ME, Dinse K, Foufopoulos J, Schubel SC, Kowalczyk T (2009) Passerine breeding and post-fledgling habitat use in riparian and upland temperate forests of the American midwest. Condor 111:756–762

Alexandratos N (1999) World food and agriculture: outlook for the medium and longer term. Proc Natl Acad Sci USA 96:5908–5914

Alho CJR (2008) Biodiversity of the Pantanal: response to seasonal flooding regime and to environmental degradation. Braz J Biol 68:957–966

Alho CJR, Vieira LM (1997) Fish and wildlife resources in the Pantanal wetlands of Brazil and potential disturbances from the release of environmental contaminants. Environ Toxicol Chem 16:71–74

Allianz Knowledge Partnersite (2010) Energy. The top five biofuel crops. http://knowledge.allianz.com/en/globalissues/energy_co2/renewable_energy/biofuels_crops.html. Accessed Oct 2010

Amano T (2009) Conserving bird species in Japanese farmland: past achievements and future challenges. Biol Conserv 142:1913–1921

Anderson GQA, Fergusson MJ (2006) Energy from biomass in the UK: sources, processes and biodiversity implications. Ibis 148:180–183

Andrén H (1994) Effects of habitat fragmentation of birds and mammals in landscapes with different proportions of suitable habitat: a review. Oikos 71:355–366

Archer DW, Dawson J, Kreuter UP, Hendrickson M, Halloran JM (2008) Social and political influences on agricultural systems. Renew Agr Food Syst 23:272–284

Arcos IT, Jiménez F, Harvey CA, Casanoves F (2008) Richness and abundance of birds in riparian forest belts of varied breadths at the Sesesmiles river microwatershed, Copan, Honduras. Rev Biol Trop 56:355–369

Argent DG, Zwier RJ (2007) Seasonal use of recently fenced agricultural riparian habitat by avifauna in Pennsylvania. Ne Nat 14:361–374

Arlt D, Forslund P, Jeppsson T, Pärt T (2008) Habitat-specific population growth of a farmland bird. PLoS ONE 3:e3006

Ashworth L, Quesada M, Casas A, Aguilar R, Oyama K (2009) Pollinator-dependent food production in Mexico. Biol Conserv 142:1050–1057

Askins RA, Chávez-Ramírez F, Dale BC, Haas CA, Herkert JR, Knopf FL, Vickery PD (2007) Conservation of grassland birds in North America: understanding ecological processes in different regions. Ornithol Monogr 64:1–46

Aspelin AL (2003) Pesticide usage in the United States: trends during the 20th century. NSF CIPM Technical Bulletin 105. Center for Integrated Pest Management, North Carolina State University, Raleigh. http://cipm.ncsu.edu/cipmpubs/index.cfm

Atkinson PW, Fuller RJ, Vickery JA (2002) Large-scale patterns of summer and winter bird distribution in relation to farmland type in England and Wales. Ecography 25:466–480

Atkinson PW, Fuller RJ, Vickery JA, Conway GJ, Tallowin JRB, Smith REN, Haysom KA, Ings TC, Asteraki EJ, Brown VK (2005) Influence of agricultural management, sward structure and food resources on grassland field use by birds in lowland England. J Appl Ecol 42:932–942

Attwood SJ, Park SE, Maron M, Collard SJ, Robinson D, Reardon-Smith KM, Cockfield G (2009) Declining birds in Australian agricultural landscapes may benefit from aspects of the European agri-environment model. Biol Conserv 142:1981–1991

Audubon (2010) State of the birds. http://stateofthebirds.audubon.org/. Accessed July 2010

Aviron S, Nitsch H, Jeanneret P, Buholzer S, Luka H, Pfiffner L, Pozzi S, Schüpbach B, Walter T, Herzog F (2009) Ecological cross compliance promotes farmland biodiversity in Switzerland. Front Ecol Environ 7:247–252

Azpiroz AB, Blake JG (2009) Avian assemblages in altered and natural grasslands in the northern campos of Uruguay. Condor 111:21–35

Badgley C, Moghtader J, Quintero E, Zakem E, Chappell MJ, Avilés-Vázquez K, Samulon A, Perfecto I (2007) Organic agriculture and the global food supply. Renew Agr Food Syst 22:86–108

Baillie JEM, Hilton-Taylor C, Stuart SN (eds) (2004) 2004 IUCN Red List of threatened species: a global species assessment. World Conservation Union

Bakermans MH, Vitz AC, Rodewald AD, Rengifo CG (2009) Migratory songbird use of shade coffee in the Venezuelan Andes with implications for conservation of Cerulean Warbler. Biol Conserv 11:2476–2483

Bale JS, van Lenteren JC, Bigler F (2008) Biological control and sustainable food production. Philos T Roy Soc B 363:761–776

Balmford A, Bruner A, Cooper P, Costanza R, Farber S, Green RE, Jenkins M, Jefferiss P, Jessamy V, Madden J, Munro K, Myers N, Naeem S, Paavola J, Rayment M, Rosendo S, Roughgarden J, Trumper K, Turner RK (2002) Economic reasons for conserving wild nature. Science 297:950–953

Balmford A, Green RE, Scharlemann JPW (2005) Sparing land for nature: exploring the potential impact of changes in agricultural yield on the area needed for crop production. Glob Change Biol 11:1594–1605

Banks JE (2004) Divided culture: integrating agriculture and conservation biology. Front Ecol Environ 2:537–545

Bas Y, Renard M, Jiguet F (2009) Nesting strategy predicts farmland bird response to agricultural intensity. Agr Ecosyst Environ 134:143–147

Basili GD, Temple SA (1995) A perilous migration. Nat Hist 104:40–46

Basili GD, Temple SA (1999) Dickcissels and crop damage in Venezuela: defining the problem with ecological models. Ecol Appl 9:732–739

Basore NS, Best LB, Wooley JB Jr (1986) Bird nesting in Iowa no-tillage and tilled cropland. J Wildl Manage 50:19–28

Batáry P, Matthiesen T, Tscharntke T (2010) Landscape-moderated importance of hedges in conserving farmland bird diversity of organic vs. conventional croplands and grasslands. Biol Conserv 143:2020–2027

Beddington J (2010) Food security: contributions from science to a new and greener revolution. Philos T Roy Soc B 365:61–71

Beecher NA, Johnson RJ, Brandle JR, Case RM, Young LJ (2002) Agroecology of birds in organic and nonorganic farmland. Conserv Biol 16:1620–1631

Bélisle M, Desrochers A, Fortin MJ (2001) Influence of forest cover on the movements of forest birds: a homing experiment. Ecology 82:1893–1904

Bendell BE, Weatherhead PJ, Stewart RK (1981) The impact of predation by red-winged blackbirds on European corn borer populations. Can J Zool 59:1535–1538

Bengtsson J, Ahnström J, Weibull A (2005) The effects of organic agriculture on biodiversity and abundance: a meta-analysis. J Appl Ecol 42:261–269

Bennett EM, Peterson GD, Gordon LJ (2009) Understanding relationships among multiple ecosystem services. Ecol Lett 12:1394–1404

Benson TJ, Dinsmore JJ, Hohman WL (2007) Responses of plants and arthropods to burning and disking of riparian habitats. J Wildl Manage 71:1949–1957

Benton TG, Vickery JA, Wilson JD (2003) Farmland biodiversity: is habitat heterogeneity the key? Trends Ecol Evol 18:182–188

Berg A (2008) Habitat selection and reproductive success of Ortolan Buntings *Emberiza hortulana* on farmland in central Sweden- the importance of habitat heterogeneity. Ibis 150:565–573

Bernier-Leduc M, Vanasse A, Olivier A, Bussiéres D, Maisonneuve C (2009) Avian fauna in windbreaks integrating shrubs that produce non-timber forest products. Agr Ecosyst Environ 131:16–24

Besser JF, Berg WJ, Knittle CE (1979) Late-summer feeding patterns of red-winged blackbirds in a sunflower-growing area of North Dakota. In; Proceedings bird control seminar, vol 8. pp 209–214

Best LB (1983) Bird use of fencerows: implications of contemporary fencerow management practices. Wildl Soc Bull 11:343–347

Best LB (1985) Conservation vs. conventional tillage: wildlife management considerations. In: D'Itri FM (ed) A systems approach to conservation tillage. Lewis Publishers, Chelsea, pp 315–326

Best LB (1986) Conservation tillage: ecological traps for nesting birds? Wildl Soc Bull 14:308–317

Best LB (2001) Temporal patterns of bird abundance in cornfield edges during the breeding season. Am Midl Nat 146:94–104

Best LB, Murray LD (2004) Bird responses to harvesting switchgrass fields for biomass. T N Am Wildl Nat Res 69:224–235

Best LB, Whitmore RC, Booth GM (1990) Use of cornfields by birds during the breeding season: the importance of edge habitat. Am Midl Nat 123:84–99

Best LB, Freemark KE, Dinsmore JJ, Camp M (1995) A review and synthesis of habitat use by breeding birds in agricultural landscapes of Iowa. Am Midl Nat 134:1–29

Best LB, Campa H, Kemp KE, Robel RJ, Ryan MR, Savidge JA, Weeks HP, Winterstein SR (1997) Bird abundance and nesting in CRP fields and cropland in the midwest: a regional approach. Wildl Soc Bull 25:864–877

Best LB, Campa H III, Kemp KE, Robel RJ, Ryan MR, Savidge JA, Weeks HP, Winterstein SR (1998) Avian abundance in CRP and crop fields during winter in the midwest. Am Midl Nat 139:311–324

Best LB, Bergin TM, Freemark KE (2001) Influence of landscape composition on bird use of rowcrop fields. J Wildl Manage 65:442–449

Bianchi FJJA, Booij CJH, Tscharntke T (2006) Sustainable pest regulation in agricultural landscapes: a review of landscape composition, biodiversity and natural pest control. Proc R Soc B 273:1715–1727

Bignal EM, McCracken DI (1996) Low-intensity farming systems in the conservation of the countryside. J Appl Ecol 33:413–424

Billeter R, Liira J, Bailey D, Bugter R, Arens P, Augenstein I, Aviron S, Baudry J, Bukacek R, Burel F, Cerny M, De Blust G, De Cock R, Diekötter T, Dietz H, Dirksen J, Dormann C, Durka W, Frenzel M, Hamersky R, Hendrickx F, Herzog F, Klotz S, Koolstra B, Lausch A, Le Coeur D, Maelfait JP, Opdam P, Roubalova M, Schermann A, Schermann N, Schmidt T, Schweiger O, Smulders MJM, Speelmans M, Simova P, Verboom J, van Wingerden WKRE, Zobel M, Edwards PJ (2008) Indicators for biodiversity in agricultural landscapes: a pan-European study. J Appl Ecol 45:141–150

BirdLife International (2010) State of the world's birds. Indicators for our changing world. http://www.birdlife.org/sowb/. Accessed July 2010

Birrer S, Spiess M, Herzog F, Jenny M, Kohli L, Lugrin B (2007) The Swiss agri-environment scheme promotes farmland birds: but only moderately. J Ornithol 148:S295–S303

Black ER, Davis FM, Henderson CA, Douglas WA (1970) The role of birds in reducing overwintering populations of the southwestern corn borer, *Diatraea grandiosella* (Lepidoptera: Crambidae), in Mississippi. Ann Entomol Soc Am 63:701–706

Boatman ND, Brickle NW, Hart JD, Milsom TP, Morris AJ, Murray AWA, Murray KA, Robertson PA (2004) Evidence for the indirect effects on pesticides on farmland birds. Ibis 146:131–143

Bohlen PJ, House G (eds) (2009) Sustainable agroecosystem management: integrating ecology, economics, and society. CRC Press/Taylor and Francis, Boca Raton

Bohlen PJ, Lynch S, Shabman L, Clark M, Shukla S, Swain H (2009) Paying for environmental services from agricultural lands: an example from the northern Everglades. Front Ecol Environ 7:46–55

Bollinger EK, Caslick JW (1985) Red-winged blackbird predation on northern corn rootworm beetles in field corn. J Appl Ecol 22:39–48

Bollinger EK, Bollinger PB, Gavin TA (1990) Effects of hay-cropping of eastern populations of the Bobolink. Wildl Soc Bull 18:142–150

Bonter DN, Gauthreaux SA Jr, Donovan TM (2009) Characteristics of important stopover locations for migrating birds: remote sensing with radar in the Great Lakes Basin. Conserv Biol 23:440–448

Borgella R Jr, Snow AA, Gavin TA (2001) Species richness and pollen loads of hummingbirds using forest fragments in southern Costa Rica. Biotropica 33:90–109

Boutin C, Freemark KE, Kirk DA (1999) Farmland birds in southern Ontario: field use, activity patterns and vulnerability to pesticide use. Agr Ecosyst Environ 72:239–254

Brady E (2006) The aesthetics of agricultural landscapes and the relationship between humans and nature. Ethics Place Environ 9:1–19

Brambilla M, Guidali F, Negri I (2008) The importance of an agricultural mosaic for Cirl Buntings *Emberiza cirlus* in Italy. Ibis 150:628–632

Brandle JR, Hodges L, Tyndall J, Sudmeyer RA (2009) Windbreak practices. In: Garrett HEG (ed) North American agroforestry, an integrated science and practice, 2nd edn. American Society of Agronomy, Madison, pp 75–104

Breitbach N, Laube I, Steffan-Dewenter I, Böhning-Gaese K (2010) Bird diversity and seed dispersal along a human land-use gradient: high seed removal in structurally simple farmland. Oecologia 162:965–976

Brennan LA, Kuvlesky WP Jr (2005) North American grassland birds: an unfolding conservation crisis? J Wildl Manage 69:1–13

Brewer GD (2009) Risk analysis, five "easy" questions. Science 325:1075–1076

Bridgeland WT, Caslick JW (1983) Relationships between cornfield characteristics and blackbird damage. J Wildl Manage 47:824–829

Briggs JM, Hoch GA, Johnson LC (2002) Assessing the rate, mechanisms, and consequences of the conversion of tallgrass prairie to *Juniperus virginiana* forest. Ecosystems 5:578–586

Brightsmith DJ, Stronza A, Holle K (2008) Ecotourism, conservation biology, and volunteer tourism: a mutually beneficial triumvirate. Biol Conserv 141:2832–2842

Brooks T, Thompson HS (2001) Current bird conservation issues in Africa. Auk 118:575–582

Brscic K (2006) The impact of agrotourism on agricultural production. J Cent Eur Agr 7:559–563

Bruggers RL, Jaeger MM, Keith JO, Hegdal PL, Bourassa JB, Latigo AA, Gillis JN (1989) Impact of fenthion on nontarget birds during *Quelea* control in Kenya. Wildl Soc Bull 17:149–160

Bruggers RL, Rodriguez E, Zaccagnini ME (1998) Planning for bird pest problem resolution: a case study. Int Biodeter Biodegr 42:173–184

Bryan GG, Best LB (1991) Bird abundance and species richness in grassed waterways in Iowa rowcrop fields. Am Midl Nat 126:90–102

Bryan GG, Best LB (1994) Avian nest density and success in grassed waterways in Iowa rowcrop fields. Wildl Soc Bull 22:583–592

Burger A (1994) The agriculture of the world. Avebury/Ashgate Publishing Company, Aldershot

Butler RW (1990) Alternative tourism: pious hope or trojan horse? J Travel Res 28:40–45

Butler SJ, Vickery JA, Norris K (2007) Farmland biodiversity and the footprint of agriculture. Science 315:381–384

Butler SJ, Boccaccio L, Gregory RD, Vorisek P, Norris K (2010) Quantifying the impact of land-use change to European farmland bird populations. Agr Ecosyst Environ 137:348–357

Camp M, Best LB (1993) Bird abundance and species richness in roadsides adjacent to Iowa rowcrop fields. Wildl Soc Bull 21:315–325

Canby P (2010) Retreat to subsistence. Nation 291:30–36

Capel SW (1988) Design of windbreaks for wildlife in the great plains of North America. Agr Ecosyst Environ 22–23:337–347

Carlo TA, Collazo JA, Groom MJ (2004) Influences of fruit diversity and abundance on bird use of two shaded coffee plantations. Biotropica 36:602–614

Carson R (1962) Silent spring. Houghton Mifflin Publishers, New York

Cassel JF, Wiehe JM (1980) Uses of shelterbelts by birds. In: Workshop proceedings, management of western forests and grasslands for nongame birds. USDA Forest Service, General Technical Report INT-86, Ogden, pp 78–87

Castrale JS (1985) Responses of wildlife to various tillage conditions. Trans N Am Wildl Nat Resour Conf 50:142–156

Ceballos-Lascurain H (2003) Preface. In: Luck M, Kirstges T (eds) Global ecotourism policies and case studies, perspectives and constraints. Channel View Publications, Sydney, pp viii–xii

Chamberlain DE, Wilson JD, Fuller RJ (1999) A comparison of bird populations on organic and conventional farm systems in southern Britain. Biol Conserv 88:307–320

Cole RJ, Holl KD, Zahawi RA (2010) Seed rain under tree islands planted to restore degraded lands in a tropical agricultural landscape. Ecol Appl 20:1255–1269

Connell J (2009) Birdwatching, twitching and tourism: towards an Australian perspective. Aust Geogr 40:203–217

Conover RR, Burger LW Jr, Linder ET (2007) Winter avian community and sparrow response to field border width. J Wildl Manage 71:1917–1923

Conservation International (2004) Hotspots revisited. http://www.biodiversityhotspots.org/xp/hotspots/resources/Pages/maps.aspx. Accessed Oct 2010

Cornulier T, Elston DA, Arcese P, Benton TG, Douglas DJT, Lambin X, Reid J, Robinson RA, Sutherland WJ (2009) Estimating the annual number of breeding attempts from breeding dates using mixture models. Ecol Lett 12:1184–1193

Czúcz B, Gathman JP, McPherson GR (2010) The impending peak and decline of petroleum production: an underestimated challenge for conservation of ecological integrity. Conserv Biol 24:948–956

Daily GC, Ehrlich PR, Sánchez-Azofeifa GA (2001) Countryside biogeography: use of human-dominated habitats by the avifauna of southern Costa Rica. Ecol Appl 11:1–13

Dale BC, Martin PA, Taylor PS (1997) Effects of hay management on grassland songbirds in Saskatchewan. Wildl Soc Bull 25:616–626

Dalimer M, Jones PJ (2002) Migration orientation behaviour of the Red-billed Quelea *Quelea quelea*. J Avian Biol 33:89–94

Dambach CA, Good EE (1940) The effect of certain land use practices on populations of breeding birds in southwestern Ohio. J Wildl Manage 4:63–76

Dänhardt J, Green M, Lindström A, Rundlöf M, Smith HG (2010) Farmland as stopover habitat for migrating birds – effects of organic farming and landscape structure. Oikos 119:1114–1125

Dasgupta P (2010) Nature's role in sustaining economic development. Philos T Roy Soc B 365:5–11

DEFRA (Department for Environment, Food, and Rural Affairs) (2010) Changes in farmland bird populations; Changes in woodland bird populations. http://www.defra.gov.uk/evidence/statistics/environment/wildlife/kf/wdkf03.htm. Accessed Oct 2010

DeFries RS, Rudel T, Uriarte M, Hansen M (2010) Deforestation driven by urban population growth and agricultural trade in the twenty-first century. Nat Geosci 3:178–181

Delgado CL (2003) Rising consumption of meat and milk in developing countries has created a new food revolution. J Nutr 133:3907S–3910S

Deschdenes M, Bélanger L, Giroux J-F (2003) Use of farmland riparian strips by declining and crop damaging birds. Agr Ecosyst Environ 95:567–577

Dewar AM, May JF, Woiwod IP, Haylock LA, Champion GT, Garner BH, Sands RJN, Qi A, Pidgeon JD (2003) A novel approach to the use of genetically modified herbicide tolerant crops for environmental benefit. Biol Sci 270:335–340

Di Giacomo AS, de Casenave JL (2010) Use and importance of crop and field-margin habitats for birds in a neotropical agricultural ecosystem. Condor 112:283–293

Dias de Oliveira ME, Vaughan BE, Rykiel EJ Jr (2005) Ethanol as fuel: energy, carbon dioxide balances, and ecological footprint. Bioscience 55:593–602

Dietz T, Stern PC (1998) Science, values, and biodiversity. Bioscience 48:441–444

Dimitri C, Effland A, Conklin N (2005) The 20th century transformation of U.S. agriculture and farm policy. USDA Economic Information Bulletin Number 3, Washington, DC. http://www.ers.usda.gov/publications/EIB3/eib3.pdf

Divino JA, McAleer M (2009) Modelling sustainable international tourism demand to the Brazilian Amazon. Environ Modell Softw 24:1411–1419

Dix ME, Johnson RJ, Harrell MO, Case RM, Wright RJ, Hodges L, Brandle JR, Schoeneberger MM, Sunderman NJ, Fitzmaurice RL, Young LF, Hubbard KG (1995) Influences of trees on abundance of natural enemies of insect pests: a review. Agroforest Syst 29:303–311

Dolbeer RA (1990) Ornithology and integrated pest management: red-winged blackbirds *Agelaius phoeniceus* and corn. Ibis 132:309–322

Donald PF (2004) Biodiversity impacts of some agricultural commodity production systems. Conserv Biol 18:17–38

Donald PF, Green RE, Heath MF (2001) Agricultural intensification and the collapse of Europe's farmland bird populations. Proc R Soc B 268:25–29

Donald P, Sanderson F, Burfield I, Vanbommel F (2006) Further evidence of continent-wide impacts of agricultural intensification on European farmland birds, 1990–2000. Agr Ecosyst Environ 116:189–196

Dorrough J, Moll J, Crosthwaite J (2007) Can intensification of temperate Australian livestock production systems save land for native biodiversity? Agr Ecosyst Environ 121:222–232

Dybas CL (2009) Report from the 2009 AIBS annual meeting: ensuring a food supply in a world that's hot, packed, and starving. Bioscience 59:640–646

Eggers J, Tröltzsch K, Falcucci A, Maiorano L, Verburg PH, Framstad E, Louette G, Maes D, Nagy S, Ozinga W, Delbaere B (2009) Is biofuel policy harming biodiversity in Europe? GCB Bioenerg 1:18–34

Ehler LE (2006) Integrated pest management (IPM): definition, historical development and implementation, and the other IPM. Pest Manage Sci 62:787–789

Ehrlich PR, Holdren JP (1971) Impact of population growth. Science 171:1212–1217

Ellis EC, Goldewijk KK, Siebert S, Lightman D, Ramankutty N (2010) Anthropogenic transformation of the biomes, 1700 to 2000. Glob Ecol Biogeogr 19:589–606

Estrada A, Coates-Estrada R (2005) Diversity of neotropical migratory landbird species assemblages in forest fragments and man-made vegetation in Los Tuxtlas, Mexico. Biodivers Conserv 14:1719–1734

Estrada A, Coates-Estrada R, Meritt DA (1997) Anthropogenic landscape changes and avian diversity at Los Tuxtlas, Mexico. Biodivers Conserv 6:19–43

Evans KL (2003) The potential for interactions between predation and habitat change to cause population declines of farmland birds. Ibis 146:1–13

Evans AD, Green RE (2007) An example of a two-tiered agri-environment scheme designed to deliver effectively the ecological requirements of both localised and widespread bird species in England. J Ornithol 148:279–286

Evenden MD (1995) The laborers of nature: economic ornithology and the role of birds as agents of biological pest control in North American agriculture, ca. 1880–1930. For Conserv Hist 39:172–183

Faaborg J, Holmes RT, Anders AD, Bildstein KL, Dugger KM, Gauthreaux SA Jr, Heglund P, Hobson KA, Jahn AE, Johnson DH, Latta SC, Levey DJ, Marra PP, Merkord CL, Nol E, Rothstein SI, Sherry TW, Sillett TS, Thompson FR III, Warnock N (2010) Conserving migratory land birds in the new world: do we know enough? Ecol Appl 20:398–418

Fargione JE, Cooper TR, Flaspohler DJ, Hill J, Lehman C, McCoy T, McLeod S, Nelson EJ, Oberhauser KS, Tilman D (2009) Bioenergy and wildlife: threats and opportunities for grassland conservation. Bioscience 59:767–777

Faria D, Laps RR, Baumgarten J, Cetra M (2006) Bat and bird assemblages from forests and shade cacao plantations in two contrasting landscapes in the Atlantic Forest of southern Bahia, Brazil. Biodivers Conserv 15:587–612

Fearnside PM (2001) Soybean cultivation as a threat to the environment in Brazil. Environ Conserv 28:23–38

Filippi-Codaccioni O, Devictor V, Bas Y, Clobert J, Julliard R (2010) Specialist response to proportion of arable land and pesticide input in agricultural landscapes. Biol Conserv 143:883–890

Filloy J, Bellocq MI (2006) Spatial variations in the abundance of *Sporophila* seedeaters in the southern neotropics: contrasting the effects of agricultural development and geographical position. Biodivers Conserv 15:3329–3340

Filloy J, Bellocq MI (2007) Patterns of bird abundance along the agricultural gradient of the pampean region. Agr Ecosyst Environ 120:291–298

Filloy J, Zurita GA, Corbelli JM, Bellocq MI (2010) On the similarity among bird communities: testing the influence of distance and land use. Acta Oecologica 36:333–338

Firbank LG (2005) Striking a new balance between agricultural production and biodiversity. Ann Appl Biol 146:163–175

Fischer J, Brosi B, Daily GC, Ehrlich PR, Goldman R, Goldstein J, Lindenmayer DB, Manning AD, Mooney HA, Pejchar L, Ranganathan J, Tallis H (2008) Should agricultural policies encourage land sparing or wildlife-friendly farming? Front Ecol Environ 6:380–385

Fitting E (2006) Importing corn, exporting labor: the neoliberal corn regime, GMOs, and the erosion of Mexican biodiversity. Agric Hum Values 23:15–26

Fitzmaurice RL (1995) Avian use of riparian corridors and adjacent cropland in east-central Nebraska. Thesis, University of Nebraska-Lincoln, Lincoln

Flannery T (2001) The eternal frontier, an ecological history of North America and its peoples. Grove Press, New York

Fleischner TL (1994) Ecological costs of livestock grazing in western North America. Conserv Biol 8:629–644

Fleischner TL (2010) Livestock grazing and wildlife conservation in the American West: historical, policy, and conservation biology perspectives. In: du Toit J, Kock R, Deutsch J (eds) Wild rangelands: conserving wildlife while maintaining livestock in semi-arid ecosystems. Zoological Society of London/ Blackwell, Oxford, pp 235–265

Florian E, Harvey CA, Finegan B, Benjamin T, Soto G (2008) The effect of structural complexity and landscape context in the avifauna of coffee agroforestry systems. Ornitologia Neotrop 19:541–548

Floyd EH, Mason L, Phillips S (1969) Survival of overwintering southwestern corn borers in corn stalks in Louisiana. J Econ Entomol 62:1016–1019

Foley JA, DeFries R, Asner GP, Barford C, Bonan G, Carpenter SR, Chapin FS, Coe MT, Daily GC, Gibbs HK, Helkowski JH, Holloway T, Howard EA, Kucharik CJ, Monfreda C, Patz JA, Prentice IC, Ramankutty N, Snyder PK (2005) Global consequences of land use. Science 309:570–574

Font X (2007) Ecotourism certification: potential and challenges. In: Higham J (ed) Critical issues in ecotourism, understanding a complex tourism phenomenon. Elsevier/Butterworth-Heinemann, Oxford, pp 386–405

Food and Agriculture Organization of the United Nations (FAO). (2007a) State of the world's forests 2007. Rome. http://www.fao.org/docrep/009/a0773e/a0773e00.htm

Food and Agriculture Organization of the United Nations (FAO) (2007b) The state of food and agriculture 2007, paying farmers for environmental services. Agricultural Development Economics Division, Food and Agriculture Organization of the United Nations. Rome. http://www.fao.org/docrep/010/a1200e/a1200e00.htm

Food and Agriculture Organization of the United Nations (FAO) (2008) FAOstat. Rome. http://faostatfaoorg. Accessed Jan 2010

Food and Agriculture Organization of the United Nations (FAO) (2009) The state of food and agriculture, livestock in the balance. Communication Division, Food and Agriculture Organization of the United Nations, Rome. http://www.fao.org/publications/sofa/en/

Foster D, Swanson F, Aber J, Burke I, Brokaw N, Tilman D, Knapp A (2003) The importance of land-use legacies to ecology and conservation. Bioscience 53:77–88

Fowler P (2002) Farming in the first millennium AD, British agriculture between Julius Caesar and William the conqueror. Cambridge University Press, Cambridge

Fox AD (2004) Has Danish agriculture maintained farmland bird populations? J Appl Ecol 41:427–439

Frawley BJ, Best LB (1991) Effects of mowing on breeding bird abundance and species composition in alfalfa fields. Wildl Soc Bull 19:135–142

Freemark KE, Kirk DA (2001) Birds on organic and conventional farms in Ontario: partitioning effects of habitat and practices on species composition and abundance. Biol Conserv 101:337–350

Freemark KE, Dunning JB, Hejl SJ, Probst JR (1995) A landscape ecology perspective for research, conservation, and management. In: Martin T, Finch D (eds) Ecology and management of neotropical migratory birds: a synthesis and review of critical issues. Oxford University Press, Cambridge, pp 381–427

Frye RD (1972) Bird predation on the European corn borer. N Dak Farm Res 29:28–39

Fuhlendorf SD, Engle DM (2001) Restoring heterogeneity on rangelands: ecosystem management based on evolutionary grazing patterns. Bioscience 51:625–632

Fuhlendorf SD, Engle DM (2004) Application of the fire-grazing interaction to restore a shifting mosaic on tallgrass prairie. J Appl Ecol 41:604–614

Fuhlendorf SD, Harrell WC, Engle DM, Hamilton RG, Davis CA, Leslie DM Jr (2006) Should heterogeneity be the basis for conservation? Grassland bird response to fire and grazing. Ecol Appl 16:1706–1716

Fuller RJ, Gregory RD, Gibbons DW, Marchant JH, Wilson JD, Baillie SR, Carter N (2005a) Population declines and range contractions among lowland farm birds in Britain. Conserv Biol 9:1425–1441

Fuller RJ, Norton LR, Feber RE, Johnson PJ, Chamberlain DE, Joys AC, Mathews F, Stuart RC, Townsend MC, Manley WJ, Wolfe MS, Macdonald DW, Firbank LG (2005b) Benefits of organic farming to biodiversity vary among taxa. Biol Lett 1:431–434

Fyfe RM, Brown AG, Rippon SJ (2003) Mid- to late-Holocene vegetation history of Greater Exmoor, UK: estimating the spatial extent of human-induced vegetation change. Veg Hist Archaeobot 12:215–232

Galle AM, Linz GM, Homan HJ, Bleier WJ (2009) Avian use of harvested crop fields in North Dakota during spring migration through North Dakota. West N Am Nat 69:491–500

Galloway JN, Aber JD, Erisman JW, Seitzinger SP, Howarth RW, Cowling EB, Cosby BJ (2003) The nitrogen cascade. Bioscience 53:341–356

Gámez-Virués S, Bonifacio RS, Gurr GM, Kinross C, Raman A, Nicol HI (2007) Arthropod prey of shelterbelt-associated birds: linking faecal samples with biological control of agricultural pests. Aust J Entomol 46:325–331

Garcia D, Zamora R, Amico GC (2010) Birds as suppliers of seed dispersal in temperate ecosystems: conservation guidelines from real-world landscapes. Conserv Biol 24:1070–1079

Gaston KJ (2010) Valuing common species. Science 327:154–155

Gates JE, Gysel LW (1978) Avian nest dispersion and fledging success in field-forest ecotones. Ecology 59:871–883

Genghini M, Gellini S, Gustin M (2006) Organic and integrated agriculture: the effects of bird communities in orchard farms in northern Italy. Biodivers Conserv 15:3077–3094

Gibbons DW, Bohan DA, Rothery P, Stuart RC, Haughton AJ, Scott RJ, Wilson JD, Perry JN, Clark SJ, Dawson JG, Firbank LG (2006) Weed seed resources for birds in fields with contrasting conventional and genetically modified herbicide-tolerant crops. Proc R Soc B 273:1921–1928

Gill F, Donsker D (eds) (2010) IOC World Bird Names (version 2.6). http://www.worldbirdnames.org/. Accessed Nov 2010

Gillings S, Fuller RJ, Sutherland WJ (2007) Winter field use and habitat by Eurasian Golden Plovers (*Pluvialis apricaria*) and Northern Lapwings (*Vanellus vanellus*) on arable farmland. Ibis 149:509–520

Gilroy JJ, Anderson GQA, Grice PV, Vickery JA, Sutherland WJ (2010) Mid-season shifts in the habitat associations of Yellow Wagtails *Motacilla flava* breeding in arable farmland. Ibis 152:90–104

Gleick PH (2000) The world's water 2000–2001: the biennial report on freshwater resources. Island Press, Washington, DC

Gliessman SR, Rosemeyer M (2010) The conversion to sustainable agriculture, principles, processes, and practices. CRC Press/Taylor & Francis, New York

Godfray HCJ, Beddington JR, Crute IR, Haddad L, Lawrence D, Muir JF, Pretty J, Robinson S, Thomas SM, Toulmin C (2010) Food security: the challenge of feeding 9 billion people. Science 327:812–818

Goldstein MI, Lacher TE, Woodbridge B, Bechard MJ, Canavelli SB, Zaccagnini ME, Cobb GP, Scollon EJ, Tribolet R, Hopper MJ (1999) Monocrotophos-induced mass mortality of Swainson's Hawks in Argentina, 1995–96. Ecotoxicology 8:201–214

Gossling S (2000) Tourism – sustainable development option? Environ Conserv 27:223–224

Grant TA, Madden E, Berkey GB (2004) Tree and shrub invasion in northern mixed-grass prairie: implications for breeding grassland birds. Wildl Soc Bull 32:807–818

Grantham HS, Bode M, McDonald-Madden E, Game ET, Knight AT, Possingham HP (2010) Effective conservation planning requires learning and adaptation. Front Ecol Environ 8:431–437

Green RE, Cornell SJ, Scharlemann JPW, Balmford A (2005) Farming and the fate of wild nature. Science 307:550–555

Greenberg R (1992) Forest migrants in non-forest habitats on the Yucatan Peninsula. In: Hagan JM III, Johnston DW (eds) Ecology and conservation of neotropical migrant landbirds. Smithsonian Institution Press, Washington, DC

Greenberg R, Bichier P (2005) Determinants of tree species preference of birds in oak-acacia woodlands of Central America. J Trop Ecol 21:57–66

Greenberg R, Bichier P, Sterling J (1997a) Acacia, cattle and migratory birds in southeastern Mexico. Biol Conserv 80:235–247

Greenberg R, Bichier P, Sterling J (1997b) Bird populations in rustic and planted shade coffee plantations of eastern Chiapas, Mexico. Biotropica 29:501–514

Greenberg R, Bichier P, Angón AC (2000) The conservation value for birds of cacao plantations with diverse planted shade in Tabasco, Mexico. Anim Conserv 3:105–112

Greenberg R, Perfecto I, Philpott SM (2008) Agroforests as model systems for tropical ecology. Ecology 89:913–914

Gregory SV, Swanson FJ, McKee WA, Cummins KW (1991) An ecosystem perspective of riparian zones. Bioscience 41:540–551

Groom MJ, Gray EM, Townsend PA (2008) Biofuels and biodiversity: principles for creating better policies for biofuel production. Conserv Biol 22:602–609

Groot AT, Dicke M (2002) Insect-resistant transgenic plants in a multi-trophic context. Plant J 31:387–406

Grue CE, Gibert PL, Seeley ME (1997) Neurophysiological and behavioral changes in non-target wildlife exposed to organophosphate and carbamate pesticides: thermoregulation, food consumption, and reproduction. Am Zool 37:369–388

Haas C (1995) Dispersal and use of corridors by birds in wooded patches on an agricultural landscape. Conserv Biol 9:845–854

Hadley AS, Betts MG (2009) Tropical deforestation alters hummingbird movement patterns. Biol Lett 5:207–210

Hafner CL, Brittingham MC (1993) Evaluation of a stream-bank fencing program in Pennsylvania. Wildl Soc Bull 21:307–315

Hagy HM, Linz GM, Bleier WJ (2007) Are sunflower fields for the birds? Proc Wildl Damage Manage Conf 12:61–71

Hails RS (2002) Assessing the risks associated with new agricultural practices. Nature 418:685–688

Hansen TM, Francis CA (2007) Peri-urban agriculture in multifunctional rural landscapes (MFRL 6). In: Hansen TM, Francis CA (eds) Multifunctional rural landscapes. University of Nebraska-Lincoln,Lincoln,pp43–48.http://cari.unl.edu/SustainableAg/pdf/MultifunctionalRural Landscapes.pdf

Hansen MC, Stehman SV, Potapov PV, Loveland TR, Townshend JRG, DeFries RS, Pittman KW, Arunarwati B, Stolle F, Steininger MK, Carroll M (2008) Humid tropical forest clearing from 2000 to 2005 quantified by using multitemporal and multiresolution remotely sensed data. Proc Natl Acad Sci USA 105:9439–9444

Hanson AV (2007) Conservation and beneficial functions of grassland birds in agroecosystems. Thesis, University of Nebraska-Lincoln, Lincoln

Harvey CA (2000) Windbreaks enhance seed dispersal into agricultural landscaped in Monteverde, Costa Rica. Ecol Appl 10:155–173

Harvey CA, Villalobos JAG (2007) Agroforestry systems conserve species-rich but modified assemblages of tropical birds and bats. Biodivers Conserv 16:2257–2292

Harvey CA, Villanueva C, Villacis J, Chacon M, Munoz D, Lopez M, Ibrahim M, Gomez R, Taylor R, Martinez J, Navas A, Saenz J, Sanchez D, Medina A, Vilchez S, Hernandez B, Perez A, Ruiz F, Lopez F, Lang I, Sinclair FL (2005) Contribution of live fences to the ecological integrity of agricultural landscapes. Agr Ecosyst Environ 111:200–230

Harvey CA, Medina A, Sanchez DM, Vilchez S, Hernández B, Saenz JC, Maes JM, Casanoves F, Sinclair FL (2006) Patterns of animal diversity in different forms of tree cover in agricultural landscapes. Ecol Appl 16:1986–1999

Haslem A, Bennett AF (2008) Countryside elements and the conservation of birds in agricultural environments. Agr Ecosyst Environ 125:191–203

Hayward MW, Kerley GIH (2009) Fencing for conservation: restriction of evolutionary potential or a riposte to threatening processes? Biol Conserv 142:1–13

Herkert JR (1994) The effects of habitat fragmentation on Midwestern grassland bird communities. Ecol Appl 4:461–471

Herkert JR (2009) Response of bird populations to farmland set-aside programs. Conserv Biol 23:1036–1040

Herkert JR, Reinking DL, Wiedenfeld DA, Winter M, Zimmerman JL, Jensen WE, Finck EJ, Koford RR, Wolfe DH, Sherrod SK, Jenkins MA, Faaborg J, Robinson SK (2003) Effects of prairie fragmentation on the nest success of breeding birds in the midcontinental United States. Conserv Biol 17:587–594

Herzon I, O'Hara RB (2007) Effects of landscape complexity on farmland birds in the Baltic States. Agr Ecosyst Environ 118:297–306

Higham J (2007) Ecotourism: which school of thought should prevail? In: Higham J (ed) Critical issues in ecotourism, understanding a complex tourism phenomenon. Elsevier/Butterworth-Heinemann, Oxford, pp 428–434

Hill J, Nelson E, Tilman D, Polasky S, Tiffany D (2006) Environmental, economic, and energetic costs and benefits of biodiesel and ethanol biofuels. Proc Natl Acad Sci USA 103:11206–11210

Hobbs PR, Sayre K, Gupta R (2008) The role of conservation agriculture in sustainable agriculture. Philos T Roy Soc B 363:543–555

Hodgson JA, Kunin WE, Thomas CD, Benton TG, Gabriel D (2010) Comparing organic farming and land sparing: optimizing yield and butterfly populations at a landscape scale. Ecol Lett 13:1358–1367

Hoffman MT, Rohde RF (2007) From pastoralism to tourism: the historical impact of changing land use practices in Namaqualand. J Arid Environ 70:641–658

Holden C (2006) Report warms of looming pollination crisis in North America. Science 314:397

Hole DG, Whittingham MJ, Bradbury RB, Anderson GQ, Lee PLM, Wilson JD, Krebs JR (2002) Widespread local house-sparrow extinctions. Nature 418:931–932

Hole DG, Perkins AJ, Wilson JD, Alexander IH, Grice PV, Evans AD (2005) Does organic farming benefit biodiversity? Biol Conserv 122:113–130

Holl KD (1998) Do bird perching structures elevate seed rain and seedling establishment in abandoned tropical pasture? Restor Ecol 6:253–261

Holmes RJ, Froud-Williams RJ (2005) Post-dispersal weed seed predation by avian and non-avian predators. Agr Ecosyst Environ 105:23–27

Holthuizjen AMA, Sharik TL (1985) Colonization of abandoned pastures by eastern red cedar (*Juniperus virginiana* L.). Can J For Res 15:1065–1068

Honey M (2008) Ecotourism and sustainable development: who owns paradise? 2nd edn. Island Press, Washington, DC

Hughes JB, Daily GC, Ehrlich PR (2002) Conservation of tropical forest birds in countryside habitats. Ecol Lett 5:121–129

ICRAF (International Center for Research in Agroforestry) (2010) Agroforestry defined. http://www.ciesin.org/IC/icraf/agrodef.html. Accessed June 2010

Imhoff ML, Bounoua L, Ricketts T, Loucks C, Harriss R, Lawrence WT (2004) Global patterns in human consumption of net primary production. Nature 429:870–873

Initiative North American Bird Conservation (2009) The state of the birds, United States of America, 2009. U.S. Department of the Interior, Washington, DC, 36 pp

Isaacs R, Tuell J, Fiedler A, Gardiner M, Landis D (2009) Maximizing arthropod-mediated ecosystem services in agricultural landscapes: the role of native plants. Front Ecol Environ 7:196–203

Isacch JP, Martinez MM (2003) Habitat use by non-breeding shorebirds in flooding pampas grasslands of Argentina. Waterbird Int J Waterbird Biol 26:494–500

IUCN (International Union for Conservation of Nature and Natural Resources) (2001) Biodiversity in development: Stratgic approach for integrating biodiversity in development cooperation. European Commission/IUCN, Brussels/Gland/Cambridge

Jackson DL, Jackson LL (2002) The farm as natural habitat, reconnecting food systems with ecosystems. Island Press, Washington, DC

Jahn AE, Levey DJ, Smith KG (2004) Reflections across hemispheres: a system-wide approach to New World bird migration. Auk 121:1005–1013

Jenkins M (2003) Prospects for biodiversity. Science 302:1175–1177

Jensen MN (2004) Climate warming shakes up species. Bioscience 54:722–729

Jobin B, Choiniére L, Bélanger L (2001) Bird use of three types of field margins in relation to intensive agriculture in Quebec, Canada. Agr Ecosyst Environ 84:131–143

Jobin B, Bélanger L, Boutin C, Maisonneuve C (2004) Conservation value of agricultural riparian strips in the Boyer River watershed, Quebec (Canada). Agr Ecosyst Environ 103:413–423

Johnson RJ (1986) Wildlife damage in conservation tillage agriculture: a new challenge. In: Proceedings of vertebrate pest conference, vol 12. pp 127–132

Johnson RJ, Beck MM (1988) Influences of shelterbelts on wildlife management and biology. Agriculture. Ecosyst Environ 22(23):301–335

Johnson RJ, Caslick JW (1982) Habitat relationships of roosting and flocking red-winged blackbirds. J Wildl Manage 46:1071–1077

Johnson DH, Schwartz MD (1993a) The Conservation Reserve Program and grassland birds. Conserv Biol 7:934–937

Johnson DH, Schwartz MD (1993b) The Conservation Reserve Program: habitat for grassland birds. Great Plains Res 3:273–295

Johnson MD, Sherry TW (2001) Effects of food availability on the distribution of migratory warblers among habitats in Jamaica. J Anim Ecol 70:546–560

Johnson RG, Temple SA (1990) Nest predation and brood parasitism of tallgrass prairie birds. J Wildl Manage 54:106–111

Johnson RJ, Brandle JR, Fitzmaurice RL, Poague KL (1992) Vertebrates for biological control of insects in agroforestry systems. In: Walla, JA, Dix ME (eds) Biological control of forest pests in the Great Plains: status and needs – a symposium. Proceedings of the 44th annual meeting, Forestry committee, Great Plains Agricultural Council, Publication No. 145, Bismark, pp 77–84

Johnson RJ, Beck MM, Brandle JR (1994) Windbreaks for people: the wildlife connection. J Soil Water Conserv 49:546–550

Johnson MD, Sherry TW, Holmes RT, Marra PP (2006) Assessing habitat quality for a migratory songbird wintering in natural and agricultural habitats. Conserv Biol 20:1433–1444

Johnson MD, Levy NJ, Kellermann JL, Robinson DE (2009) Effects of shade and bird exclusion on arthropods and leaf damage on coffee farms in Jamaica's Blue Mountains. Agroforest Syst 76:139–148

Jones GA, Sieving KE (2006) Intercropping sunflower in organic vegetables to augment bird predators of arthropods. Agr Ecosyst Environ 117:171–177

Jones GA, Sieving KE, Jacobson SK (2005) Avian diversity and functional insectivory on north-central Florida farmlands. Conserv Biol 19:1234–1245

Jonzén N, Lindén A, Ergon T, Knudsen E, Vik JO, Rubolini D, Piacentini D, Brinch C, Spina F, Karlsson L, Stervander M, Andersson A, Waldenström J, Lehikoinen A, Edvardsen E, Solvang R,

Stenseth NC (2006) Rapid advance of spring arrival dates in long-distance migratory birds. Science 312:1959–1961

Jordan N, Boody G, Broussard W, Glover JD, Keeney D, McCown BH, McIsaac G, Muller M, Murray H, Neal J, Pansing C, Turner RE, Warner K, Wyse D (2007) Sustainable development of the agricultural bio-economy. Science 316:1570–1571

Kauffman JB, Krueger WC (1984) Livestock impacts on riparian ecosystems and streamside management implications – a review. J Range Manage 37:430–438

Keleman A (2010) Institutional support and *in situ* conservation in Mexico: biases against small-scale maize farmers in post-NAFTA agricultural policy. Agr Hum Values 27:13–28

Kellermann JL, Johnson MD, Stercho AM, Hackett SC (2008) Ecological and economic services provided by birds on Jamaican blue mountain coffee farms. Conserv Biol 22:1177–1185

Kellert SR, Wilson EO (eds) (1993) The biophilia hypothesis. Shearwater Books/Island Press, Washington, DC

Kemp JC, Barrett GW (1989) Spatial patterning: impact of uncultivated corridors on arthropod populations within soybean agroecoystems. Ecology 70:114–128

Kerbiriou C, Le Viol I, Robert A, Porcher E, Gourmelon F, Julliard R (2009) Tourism in protected areas can threaten wild populations: from individual response to population viability of the chough *Pyrrhocorax pyrrhocorax*. J Appl Ecol 46:657–665

Keys E, McConnell WJ (2005) Global change and the intensification of agriculture in the tropics. Glob Environ Change 15:320–337

Kirby JS, Stattersfield AJ, Evans MI, Grimmett R, Newton I, O'Sullivan J, Tucker G (2008) Key conservation issues for migratory birds in the world's major flyways. Bird Conserv Int 18:S49–S73

Kircher JC, Davis WE (1992) Patterns of avian species richness in disturbed and undisturbed habitats in Belize. In: Hagan JM III, Johnston DW (eds) Ecology and conservation of neotropical migrant landbirds. Smithsonian Institution Press, Washington, DC

Kirk DA, Evenden MD, Mineau P (1996) Past and current attempts to evaluate the role of birds as predators of insect pests in temperate agriculture. Curr Ornithol 13:175–269

Kleijn D, Baquero RA, Clough Y, Díaz M, De Esteban JD, Fernández F, Gabriel D, Herzog F, Holzschuh A, Jöhl R, Knop E, Kruess A, Marshall EJP, Steffan-Dewenter I, Tscharntke T, Verhulst J, West TM, Yela JL (2006) Mixed biodiversity benefits of agri-environment schemes in five European countries. Ecol Lett 9:253–254

Kleijn D, Schekkerman H, Dimmers WJ, Van Kats RJM, Melman D, Teunissen WA (2010) Adverse effects of agricultural intensification and climate change on breeding habitat quality of Black-tailed Godwits *Limosa l. limosa* in the Netherlands. Ibis 152:475–486

Klein AM, Vaissiére BE, Cane JH, Steffan-Dewenter I, Cunningham SA, Kremen C, Tscharntke T (2007) Importance of pollinators in changing landscapes for world crops. Proc R Soc B 274:303–313

Knickel K, Kröger M, Bruckmeier K, Engwall Y (2009) The challenge of evaluating policies for promoting the multifunctionality of agriculture: when 'good' questions cannot be addressed quantitatively and 'quantitative answers are not that good. J Environ Policy Plann 11:347–367

Knopf FL (1994) Avian assemblages on altered grasslands. Stud Avian Biol 15:247–257

Knopf FL, Rupert JR (1999) Use of cultivated fields by breeding Mountain Plovers in Colorado. Stud Avian Biol 19:81–86

Knopf FL, Samson FB (1994) Scale perspectives on avian diversity in western riparian ecosystems. Conserv Biol 8:669–676

Knopf FL, Johnson RR, Rich T, Samson FB, Szaro RC (1988) Conservation of riparian ecosystems in the United States. Wilson Bull 100:272–284

Koford RR, Best LB (1996) Management of agricultural landscapes for the conservation of neotropical migratory birds. In: Thompson III FR (ed) Management of midwestern landscapes for the conservation of Neotropical migratory birds. USDA Forest Service, General Technical Report NC-187, North Central Forest Experiment Station, Columbia, pp 68–88

Koh LP (2008) Birds defend oil palms from herbivorous insects. Ecol Appl 18:821–825

Koh LP, Levang P, Ghazoul J (2009) Designer landscapes for sustainable biofuels. Trends Ecol Evol 24:431–438

Kort J (1988) Benefits of windbreaks to field and forage crops. Agr Ecosyst Environ 22(23): 165–190

Kragten S, de Snoo GR (2008) Field-breeding birds on organic and conventional arable farms in the Netherlands. Agr Ecosyst Environ 126:270–274

Krausman PR, Naugle DE, Frisina MR, Northrup R, Bleich VC, Block WM, Wallace MC, Wright JD (2009) Livestock grazing, wildlife habitat, and rangeland values. Rangelands 31:15–19

Krebs JR, Wilson JD, Bradbury RB, Siriwardena GM (1999) The second silent spring? Nature 400:611–612

Lacher TE, Goldstein MI (1997) Tropical ecotoxicology: status and needs. Environ Toxicol Chem 16:100–111

Laliberté E, Wells JA, DeClerck F, Metcalfe DJ, Catterall CP, Queiroz C, Aubin I, Bonser SP, Ding Y, Fraterrigo JM, McNamara S, Morgan JW, Merlos DS, Vesk PA, Mayfield MM (2010) Land-use intensification reduces functional redundancy and response diversity in plant communities. Ecol Lett 13:76–86

Lambin EF, Turner BL, Geist HJ, Agbola SB, Angelsen A, Bruce JW, Coomes OT, Dirzo R, Fischer G, Folke C, George PS, Homewood K, Imbernon J, Leemans R, Li X, Morano EF, Mortimore M, Ramakrishnan PS, Richards JF, Skanes H, Steffen W, Stone GD, Svedin U, Veldkamp TA, Vogel C, Xu J (2001) The causes of land-use and land-cover change: moving beyond the myths. Glob Environ Change 11:261–269

Larigauderie A, Mooney HA (2010) The intergovernmental science-policy platform on biodiversity and ecosystem services: moving a step closer to an IPCC-like mechanism for biodiversity. Curr Opin Environ Sustain 2:9–14

Larson G, Albarella U, Dobney K, Rowley-Conwy P, Schibler J, Tresset A, Vigne J-D, Edwards CJ, Schlumbaum A, Dinu A, Balaçsescu A, Dolman G, Tagliacozzo A, Manaseryan N, Miracle P, Lo Wijngaarden-Bakker LV, Masseti M, Bradley DG, Cooper A (2007) Ancient DNA, pig domestication, and the spread of the Neolithic into Europe. Proc Natl Acad Sci USA 104:15276–15281

Laube I, Breitbach N, Böhning-Gaese K (2008) Avian diversity in a Kenyan agroecosystem: effects of habitat structure and proximity to forest. J Ornithol 149:181–191

Laurance SGW (2004) Responses of understory rain forest birds to road edges in central Amazonia. Ecol Appl 14:1344–1357

Leopold A (1966) A Sand County almanac with essays on conservation from Round River. Oxford University Press/Ballantine Books, New York

Letourneau DK, Bothwell SG (2008) Comparisons of organic and conventional farms: challenging ecologists to make biodiversity functional. Front Ecol Environ 6:430–438

Lima SL, Zollner PA (1996) Towards a behavioral ecology of ecological landscapes. Trends Ecol Evol 11:131–135

Lindell C, Smith M (2003) Nesting bird species in sun coffee, pasture, and understory forest in southern Costa Rica. Biodivers Conserv 12:423–440

Lindell CA, Chomentowski WH, Zook JR (2004) Characteristics of bird species using forest and agricultural land covers in southern Costa Rica. Biodivers Conserv 13:2419–2441

Lindell CA, Riffell SK, Kaiser SA, Battin AL, Smith ML, Sisk TD (2007) Edge responses of tropical and temperate birds. Wilson J Ornithol 119:205–220

Litvaitis JA (1993) Response of early successional vertebrates to historic changes in land use. Conserv Biol 7:866–873

Luck GW, Daily GC (2003) Tropical countryside bird assemblages: richness, composition, and foraging differ by landscape context. Ecol Appl 13:235–247

Lynch JE (1992) Distribution of overwintering nearartic migrants in the Yucatan Peninsula, ii: Use of native and human-modified vegetation. In: Hagan JM III, Johnston DW (eds) Ecology and conservation of neotropical migrant landbirds. Smithsonian Institution Press, Washington, DC

MacNally R, Bennett AF, Thomson JR, Radford JQ, Unmack G, Horrocks G, Vesk PA (2009) Collapse of an avifauna: climate change appears to exacerbate habitat loss and degradation. Divers Distrib 15:720–730

MacArthur R, MacArthur JW (1961) On bird species-diversity. Ecology 42:594–598

Magana M, Alonso JC, Martin CA, Bautista LM, Martin B (2010) Nest-site selection by Great Bustards *Otis tarda* suggests a trade-off between concealment and visibility. Ibis 152:77–89

Marcus JF, Palmer WE, Bromley PT (2000) The effects of farm field borders on overwintering sparrow densities. Wilson Bull 112:517–523

Marshall EJP, Brown VK, Boatman ND, Lutman PJW, Squire GR, Ward LK (2003) The role of weeds in supporting biological diversity within crop fields. Weed Res 43:77–89

Martin TE (1980) Diversity and abundance of spring migratory birds using habitat islands in the Great Plains. Condor 82:430–439

Martin TE (1981) Limitation in small habitat islands: chance or competition? Auk 98:715–734

Martin TE, Finch DM (1995) Ecology and management of neotropical migratory birds, a synthesis and review of the issues. Oxford University Press, New York

Martin TE, Vohs PA (1978) Configuration of shelterbelts for optimum utilization by birds. In: Trees, a valuable Great Plains multiple-use resource. Proceedings of the 30th annual meeting, Forestry Committee, Great Plains Agricultural Council, Tulsa, Publication No. 87, pp 79–88

Masse RJ, Strong AM, Perlut NG (2008) The potential of uncut patches to increase the nesting success of grassland songbirds in intensively managed hayfields: a preliminary study from the Champlain Valley of Vermont. Northeast Nat 15:445–452

Matson PA, Vitousek PM (2006) Agricultural intensification: will land spared from farming be land spared for nature? Conserv Biol 20:709–710

Matson PA, Parton WJ, Power AG, Smith MJ (1997) Agricultural intensification and ecosystem properties. Science 277:504–509

Maul JD, Smiley PC Jr, Cooper CM (2005) Patterns of avian nest predators and a brood parasite among restored riparian habitats in agricultural watersheds. Environ Monit Assess 180:133–150

Mawdsley JR, O'Malley R, Ojima DS (2009) A review of climate-change adaptation strategies for wildlife management and biodiversity conservation. Conserv Biol 23:1080–1089

May RM (2010) Ecological science and tomorrow's world. Philos T Roy Soc B 365:41–47

McComb WC, Noble RE (1981) Nest-box and natural-cavity use in three mid-south forest habitats. J Wildl Manage 45:93–101

McDonald RI, Fargione J, Kiesecker J, Miller WM, Powell J (2009) Energy sprawl or energy efficiency: climate policy impacts on natural habitat for the United States of America. PLoS ONE 4:e6802

McEwen LC, DeWeese LR, Schladweiler P (1986) Bird predation on cutworms (Lepidoptera: Noctuidae) in wheat fields and chlorpyrifos effects on brain cholinesterase activity. Environ Entomol 15:147–151

McFarland CN, Bendell-Young LI, Guglielmo C, Williams TD (2002) Kidney, liver and bone cadmium content in the Western Sandpiper in relation to migration. J Environ Monit 4:791–795

McFarlane RW (1976) Birds as agents of biological control. Biologist 58:123–140

Meléndez LM (2010) NAFTA, tourism, and environment in Mexico. Int Environ Agreements 10:107–131

Mellink E, Riojas-López ME, Luevano-Esparza J (2009) Organchlorine content and shell thickness in Brown Booby (*Sula leucogaster*) eggs in the Gulf of California and the southern Pacific Coast of Mexico. Environ Pollut 157:2184–2188

Meyer WB, Turner BL II (1992) Human population growth and global land-use/cover change. Annu Rev Ecol Syst 23:39–61

Meza FJ, Hansen JW, Osgood D (2008) Economic value of seasonal climate forecasts for agriculture: review of *ex-ante* assessments and recommendations for future research. J Appl Meteorol Climatol 47:1269–1286

Milchunas DG, Sala OE, Lauenroth WK (1988) A generalized model of the effects of grazing by large herbivores on grassland community structure. Am Nat 132:87–106

Mills ED, Rogers DT (1992) Ratios of neotropical migrant and neotropical resident birds in winter in a citrus plantation in central Belize. J Field Ornithol 63:109–116

Mineau P, McLaughlin A (1996) Conservation of biodiversity within Canadian agricultural landscapes: integrating habitat for wildlife. J Agr Environ Ethics 9:93–113

Mize CW, Brandle JR, Schoeneberger MM, Bentrup G (2008) Ecological development and function of shelterbelts in temperate North America. In: Jose S, Gorden AM (eds) Agroforestry design: an ecological approach. Springer, New York, pp 27–54

Moguel P, Toledo VM (1999) Biodiversity conservation in traditional coffee systems of Mexico. Conserv Biol 13:11–21

Møller AP, Rubolini D, Lehikoinen E (2008) Populations of migratory bird species that did not show a phenological response to climate change are declining. Proc Natl Acad Sci USA 105:16195–16200

Mols CMM, Visser ME (2002) Great tits can reduce caterpillar damage in apple orchards. J Appl Ecol 39:888–899

Mora MA (1997) Transboundary pollution: persistent organochlorine pesticides in migrant birds of the southwestern United States and Mexico. Environ Toxicol Chem 16:3–11

Mora MA, Anderson DW (1991) Seasonal and geographical variation of organochlorine residues in birds from northwest Mexico. Arch Environ Contam Toxicol 21:541–548

Morris AJ, Holland JM, Smith B, Jones NE (2004) Sustainable arable farming for an improved environment (SAFFIE): managing winter wheat structure for Skylarks *Alauda arvensis*. Ibis 146:155–162

Naiman RJ, Decamps H, Pollock M (1993) The role of riparian corridors in maintaining regional biodiversity. Ecol Appl 3:209–212

Naiman RJ, Decamps H, McClain ME (2005) Riparia: ecology, conservation, and management of streamside communities. Academic, San Diego

Nelson E, Mendoza G, Regetz J, Polasky S, Tallis H, Cameron DR, Chan KMA, Daily GC, Goldstein J, Kareiva PM, Lonsdorf E, Naidoo R, Ricketts TH, Shaw MR (2009) Modeling multiple ecosystem services, biodiversity conservation, commodity production, and tradeoffs at landscape scales. Front Ecol Environ 7:4–11

Nelson F, Foley C, Foley LS, Leposo A, Loure E, Peterson D, Peterson M, Peterson T, Sachedina H, Williams A (2010) Payments for ecosystem services as a framework for community-based conservation in northern Tanzania. Conserv Biol 24:78–85

Newton I (1998) Bird conservation problems resulting from agricultural intensification in Europe. In: Marzluff JM, Sallabanks R (eds) Avian conservation: research and management. Island Press, Washington, DC

Newton I (2008) The migration ecology of birds. Elsevier, London

Nocera JJ, Parsons GJ, Milton GR, Fredeen AH (2005) Compatibility of delayed cutting regime with bird breeding and hay nutritional quality. Agr Ecosyst Environ 107:245–253

Norment C (2002) On grassland bird conservation in the Northeast. Auk 119:271–279

Norris K (2008) Agriculture and biodiversity conservation: opportunity knocks. Conserv Lett 1:2–11

NRC (National Research Council) (2002) Riparian areas: functions and strategies for management. National Academy Press, Washington, DC. http://www.nap.edu/catalog.php?record_id=10327#toc

O'Conner RJ, Shrubb M (1986) Farming and birds. Cambridge University Press, New York

Olsen P (Compiler) (2008) The state of Australia's birds. Supplement to Wingspan, vol 18, pp 1–40. http://www.birdsaustralia.com.au/wingspan/supplements.html

Oreskes N (2004) The scientific consensus on climate change. Science 306:1686

Packett DL, Dunning JB Jr (2009) Stopover habitat selection by migrating landbirds in a fragmented forest-agricultural landscape. Auk 126:579–589

Palmer TS (1900) A review of economic ornithology in the United States. In: United States Department of Agriculture, Yearbook, 1899. Washington, DC, pp 259–292. http://naldr.nal.usda.gov/

Pannell DJ, Marshall GRN, Barr N, Curtis A, Vanclay F, Wilkinson R (2006) Understanding and promoting adoption of conservation practices by rural landholders. Aust J Exp Agr 46:1407–1424

Patterson MP, Best LB (1996) Bird abundance and nesting success in Iowa CRP fields: the importance of vegetation structure and composition. Am Midl Nat 135:153–167

Peak RG, Thompson FR III (2006) Factors affecting avian species richness and density in riparian areas. J Wildl Manage 70:173–179

Peak RG, Thompson FR III, Shaffer TL (2004) Factors affecting songbird nest survival in riparian forests in a Midwestern agricultural landscape. Auk 121:726–737

Pearce-Higgins JW, Gill JA (2010) Unravelling the mechanisms linking climate change, agriculture and avian population declines. Ibis 152:439–442

Perfecto I, Vandermeer J (2008) Biodiversity conservation in tropical agroecosystems – a new conservation paradigm. In: Schlesinger WH, Ostfeld RS (eds) Year in ecology and conservation biology 2008, vol 1134. Annals of the New York Academy of Sciences, Oxford, pp 173–200

Perfecto I, Vandermeer J (2010) The agroecological matrix as alternative to the land-sparing/agriculture intensification model. Proc Natl Acad Sci USA 107:5786–5791

Perfecto I, Rice RA, Greenberg R, VanderVoort ME (1996) Shade coffee: a disappearing refuge for biodiversity. Bioscience 46:598–608

Perfecto I, Vandermeer JH, Bautista GL, Nunez GI, Greenberg R, Bichier P, Langridge S (2004) Greater predation in shaded coffee farms: the role of resident neotropical birds. Ecology 85:2677–2681

Perfecto I, Vandermeer J, Mas A, Pinto LS (2005) Biodiversity, yield, and shade coffee certification. Ecol Econ 54(4):435–446

Perkins MW, Johnson RJ, Blankenship EE (2003) Response of riparian avifauna to percentage and pattern of woody cover in an agricultural landscape. Wildl Soc Bull 31:642–660

Perlut NG, Strong AM, Donovan TM, Buckley NJ (2006) Grassland songbirds in a dynamic management landscape: behavioral responses and management strategies. Ecol Appl 16:2235–2247

Perrings C, Jackson L, Bawa K, Brussaard L, Brush S, Gavin T, Papa R, Pascual U, De Ruiter P (2006) Biodiversity in agricultural landscapes: saving natural capital without losing interest. Conserv Biol 20:263–264

Petit LJ, Petit DR (2003) Evaluating the importance of human-modified lands for neotropical bird conservation. Conserv Biol 17:687–694

Petit LJ, Petit DR, Christian DG, Powell HDW (1999) Bird communities of natural and modified habitats in Panama. Ecography 22:292–304

Philpott SM, Arendt WJ, Armbrecht I, Bichier P, Diestch TV, Gordon C, Greenberg R, Perfecto I, Reynoso-Santos R, Soto-Pinto L, Tejeda-Cruz C, Williams-Linera G, Valenzuela J, Zolotoff JM (2008) Biodiversity loss in Latin American coffee landscapes: review of the evidence on ants, birds, and trees. Conserv Biol 22:1093–1105

Pierce RA II, Farrand DT, Kurtz WB (2001) Projecting the bird community response resulting from the adoption of shelterbelt agroforestry practices in Eastern Nebraska. Agroforest Syst 53:333–350

Piha M, Tiainen J, Holopainen J, Vepsäläinen V (2007) Effects of land-use and landscape characteristics on avian diversity and abundance in a boreal agricultural landscape with organic and conventional farms. Biol Conserv 140:50–61

Pineiro G, Jobbagy EG, Baker J, Murray BC, Jackson RB (2009) Set-asides can be better climate investment than corn ethanol. Ecol Appl 19:277–282

Pino J, Rodá F, Ribas J, Pons X (2000) Landscape structure and bird species richness: implications and conservation in rural areas between natural parks. Landscape Urban Plann 49:35–48

Piperno DR, Ranere AJ, Holst I, Iriarte J, Dickau R (2009) Starch grain and phytolith evidence for early ninth millennium BP maize from the Central Balsas River Valley, Mexico. Proc Natl Acad Sci USA 106:5019–5024

Podoll EB (1979) Utilization of windbreaks by wildlife. In: Windbreak management. Great Plains Agricultural Council Publication No. 92, Lincoln, p 132

Polasky S, Costello C, McAusland C (2004) On trade, land-use, and biodiversity. J Environ Econ Manage 48:911–925

Polasky S, Nelson E, Camm J, Csuti B, Fackler P, Lonsdorf E, Montgomery C, White D, Arthur J, Barber-Yonts B, Haight R, Kagan J, Starfield A, Tobalske C (2008) Where to put things? Spatial land management to sustain biodiversity and economic returns. Biol Conserv 141:1505–1524

Popotnik GJ, Giuliano WM (2000) Response of birds to grazing of riparian zones. J Wildl Manage 64:976–982

Porritt J (2007) Capitalism as if the world matters, revised edn. Earthscan, London

Postel SL, Daily GC, Ehrlich PR (1996) Human appropriation of renewable fresh water. Science 271:785–788

Pretty J (2008) Agricultural sustainability: concepts, principles and evidence. Philos T Roy Soc B 363:447–465

Price TD (2009) Ancient farming in eastern North America. Proc Natl Acad Sci USA 106:6427–6428

Primack RB, Ibáñez I, Higuchi H, Lee SD, Miller-Rushing AJ, Wilson AM, Silander JA Jr (2009) Spatial and interspecific variability in phenological responses to warming temperatures. Biol Conserv 142:2569–2577

Pringle H (1998) The slow birth of agriculture. Science 282:1446–1450

Puckett HL, Brandle JR, Johnson RJ, Blankenship EE (2009) Avian foraging patterns in crop field edges adjacent to woody habitat. Agr Ecosyst Environ 131:9–15

Quinn JE (2010) Assessment and communication of the value of biodiversity. PhD Dissertation, University of Nebraska-Lincoln, Lincoln

Ramankutty N, Foley JA (1999) Estimating historical changes in global land cover: croplands from 1700 to 1992. Glob Biogeochem Cycles 13:997–1027

Ranere AJ, Piperno DR, Holst I, Dickaua R, Iriarte J (2009) The cultural and chronological context of early Holocene maize and squash domestication in the Central Balsas River Valley, Mexico. Proc Natl Acad Sci USA 106:5014–5018

Reif J, Voříšek P, Šastný K, Bejček V, Petr J (2008) Agricultural intensification and farmland birds: new insights from a central European country. Ibis 150:596–605

Reitsma R, Parrish JD, McLarney W (2001) The role of cacao plantations in maintaining forest avian diversity in southeastern Costa Rica. Agroforest Syst 53:185–193

Renfrew RB, Ribic CA (2001) Grassland birds associated with agricultural riparian practices in southwestern Wisconsin. J Range Manage 54:546–552

Renfrew RB, Saavedra AM (2007) Ecology and conservation of bobolinks (*Dolichonyx oryzivorus*) in rice production regions of Bolivia. Ornitologia Neotrop 18:61–73

Renfrew RB, Ribic CA, Nack JL (2005) Edge avoidance by nesting grassland birds: a futile strategy in a fragmented landscape. Auk 122:618–636

Restrepo C, Gómez N (1998) Responses of understory birds to anthropogenic edges in a neotropical montane forest. Ecol Appl 8:170–183

Ribic CA, Guzy MJ, Sample DW (2009a) Grassland bird use of remnant prairie and Conservation Reserve Program fields in an agricultural landscape in Wisconsin. Am Midl Nat 161:110–122

Ribic CA, Koford RR, Herkert JR, Johnson DH, Nieumuth ND, Naugle DE, Bakker KK, Sample DW, Renfrew RB (2009b) Area sensitivity in North American grassland birds: patterns and processes. Auk 126:233–244

Rice RA, Greenberg R (2000) Cacao cultivation and the conservation of biological diversity. Ambio 29:167–173

Rice RA, Greenberg R (2004) Silvopastoral systems: ecological and socioeconomic benefits and migratory bird conservation. In: Schroth G, Fonseca GABd, Harvey CA, Gascon C, Vasconcelos HL, Izac A-MN (eds) Agroforestry and biodiversity conservation in tropical landscapes. Island Press, Washington, DC

Richardson JS, Naiman RJ, Swanson FJ, Hibbs DE (2005) Riparian communities associated with Pacific Northwest headwater streams: assemblages, processes, and uniqueness. J Am Water Resour Assoc 41:935–947

Riffell S, Scognamillo D, Burger LW (2008) Effects of the Conservation Reserve Program on northern bobwhite and grassland birds. Environ Monit Assess 146:309–323

Rissato SR, Galhiane MS, Ximenes VF, de Andrade RMB, Talamoni JLB, Libanio M, de Almeida MV, Apon BM, Cavalari AA (2006) Organochlorine pesticides and polychlorinated biphenyls in soil and water samples in the northeastern part of Sao Paulo State, Brazil. Chemosphere 65:1949–1958

Robbins CS (1979) Effects of forest fragmentation on bird populations. In: DeGraaf RM, Evans KE (eds) Management of north central and northeastern forests for nongame birds. USDA Forest Service, General Technical Report NC-51, North Central Experiment Station, St. Paul, pp 198–212

Robbins CS, Dowell BA, Dawson DK, Colón J, Espinoza F, Rodriguez J, Sutton R, Vargas T (1987) Comparison of neotropical winter bird populations in isolated patches versus extensive forest. Acta Oecologica 8:285–292

Robbins CS, Sauer JR, Greenberg RS, Droege S (1989) Population declines in North American birds that migrate to the neotropics. Proc Natl Acad Sci USA 86:7658–7662

Robbins CS, Dowell BA, Dawson DK, Colón JA, Estrada R, Sutton A, Sutton R, Weyer D (1992) Comparison of neotropical migrant landbird populations wintering in tropical forest, isolated forest fragments, and agricultural habitats. In: Hagan JM III, Johnston DW (eds) Ecology and conservation of neotropical migrant landbirds. Smithsonian Institution Press, Washington, DC

Roberts DL, Cooper RJ, Petit LJ (2000) Flock characteristics of ant-following birds in premontane moist forest and coffee agroecosystems. Ecol Appl 10:1414–1425

Robinson RA, Wilson JD, Crick HQP (2001) The importance of arable habitat for farmland birds in grassland landscapes. J Appl Ecol 38:1059–1069

Rodenhouse NL, Best LB, O'Connor RJ, Bollinger EK (1993) Effects of temperate agriculture on Neotropical migrant landbirds. In: Finch DM, Stangel PW (eds) Status and management of neotropical migratory birds, 1992. USDA Forest Service, General Technical Report RM-229. Rocky Mountain Forest and Range Experiment Station, Fort Collins, pp 280–295

Rodenhouse NL, Best LB, O'Connor RJ, Bollinger EK (1995) Effects of agricultural practices and farmland structures. In: Martin TE, Finch DM (eds) Ecology and management of neotropical migratory birds. Oxford University Press, New York, pp 269–293

Rodewald PG, Brittingham MC (2004) Stopover habitat of landbirds during fall: use of edge-dominated and early-successional forests. Auk 121:1040–1055

Rodgers RD, Wooley JB (1983) Conservation tillage impacts on wildlife. J Soil Water Conserv 38:212–213

Roux DJ, Stirzaker RJ, Breen CM, Lefroy EC, Cresswell HP (2010) Framework for participative reflection on the accomplishment of transdisciplinary research programs. Environ Sci Policy 13(8):733–741

Rowley-Conwy P (2009) Human prehistory: hunting for the earliest farmers. Curr Biol 19:R948–R949

Rudel TK, Schneider L, Uriarte M, Turner BL II, DeFries R, Lawrence D, Geoghegan J, Hecht S, Ickowitz A, Lambin EF, Birdenholtz T, Baptista S, Grau R (2009) Agricultural intensification and changes in cultivated areas, 1970–2005. Proc Natl Acad Sci USA 106:20675–20680

Ruf F, Schroth G (2004) Chocolate forests and monocultures: a historical review of cocoa growing and its conflicting role in tropical deforestation and forest conservation. In: Schroth G, Fonseca GABd, Harvey CA, Gascon C, Vasconcelos HL, Izac A-MN (eds) Agroforestry and biodiversity conservation in tropical landscapes. Island Press, Washington, DC

Ryan MR, Smith RG, Mortensen DA, Teasdale JR, Curran WS, Seidel R, Shumway DL (2009) Weed–crop competition relationships differ between organic and conventional cropping systems. Weed Res 49:572–580

Ryan MR, Mortensen DA, Bastiaans L, Teasdale JR, Mirsky SB, Curran WS, Seidel R, Wilson DO, Hepperly PR (2010) Elucidating the apparent maize tolerance to weed competition in long-term organically managed systems. Weed Res 50:25–36

Saab VA, Petit DR (1992) Impact of pasture development on winter bird communities in Belize, Central America. Condor 94:66–71

Sala OE, Chapin FS, Armesto JJ, Berlow E, Bloomfield J, Dirzo R, Huber-Sanwald E, Huenneke LF, Jackson RB, Kinzig A, Leemans R, Lodge DM, Mooney HA, Oesterheld M, Poff NL, Sykes MT, Walker BH, Walker M, Wall DH (2000) Biodiversity – global biodiversity scenarios for the year 2100. Science 287(5459):1770–1774

Samson F, Knopf F (1994) Prairie conservation in North America. Bioscience 44:418–421

Samson FB, Knopf FL, Ostlie WR (1998) Grasslands. In: Mac MJ, Opler PA, Haecker CEP, Doran PD (eds) Status and trends of the Nation's biological resources, vol 2. U.S. Geological Survey, Reston, pp 437–472

Sanderson FJ, Kloch A, Sachanowicz K, Donald PF (2008) Predicting the effects of agricultural change on farmland bird populations in Poland. Agr Ecosyst Environ 129:37–42

Santelmann M, Freemark K, Sifneos J, White D (2006) Assessing effects of alternative agricultural practices on wildlife habitat in Iowa, USA. Agr Ecosyst Environ 113:243–253

Scharlemann JPW, Green RE, Balmford A (2004) Land-use trends in endemic bird areas: global expansion of agriculture in areas of high conservation value. Glob Change Biol 10:2046–2051

Schrag AM, Zaccagnini ME, Calamari N, Canavelli S (2009) Climate and land-use influences on avifauna in central Argentina: broad-scale patterns and implications of agricultural conversion for biodiversity. Agr Ecosyst Environ 132:135–142

Schroeder RL, Cable TT, Haire SL (1992) Wildlife species richness in shelterbelts: test of a habitat model. Wildl Soc Bull 20:264–273

Schulte LA, Liebman M, Asbjornsen H, Crow TR (2006) Agroecosystem restoration through strategic integration of perennials. J Soil Water Conserv 61:164A–169A

Scott VE, Evans KE, Patton DR, Stone CP (1977) Cavity-nesting birds of North American forests, vol 511, Agricultural Handbook. USDA Forest Service, Washington, DC, 112 pp

Seaman BS, Schulze CH (2010) The importance of gallery forests in the tropical lowlands of Costa Rica for understory forest birds. Biol Conserv 143:391–398

Searchinger T, Heimlich R, Houghton RA, Dong F, Elobeid A, Fabiosa J, Tokgoz S, Hayes D, Yu T-H (2008) Use of U.S. croplands for biofuels increases greenhouse gases through emissions from land-use change. Science 319:1238–1240

Seavy NE, Viers JH, Wood JK (2009) Riparian bird response to vegetation structure: a multiscale analysis using LiDAR measurements of canopy height. Ecol Appl 19:1848–1857

Sedgwick JA, Knopf FL (1987) Breeding bird response to cattle grazing of a cottonwood bottomland. J Wildl Manage 51:230–237

Sedgwick JA, Knopf FL (1991) Prescribed grazing as a secondary impact in a western riparian floodplain. J Range Manage 44:369–373

Seigler DS, Ebinger JE (1988) *Acacia macracantha, A. pennatula,* and *A. cochliacantha* (Fabaceae: Mimosoideae) species complexes in Mexico. Syst Bot 13:7–15

Şekercioğlu CH (2002) Impacts of birdwatching on human and avian communities. Environ Conserv 29:282–289

Şekercioğlu CH (2006) Increasing awareness of avian ecological function. Trends Ecol Evol 21:464–471

Şekercioğlu CH, Ehrlich PR, Daily GC, Aygen D, Goehring D, Sandi RF (2002) Disappearance of insectivorous birds from tropical forest fragments. Proc Natl Acad Sci USA 99:263–267

Şekercioğlu CH, Loarie SR, Brenes FO, Ehrlich PR, Daily GC (2007) Persistence of forest birds in the Costa Rican agricultural countryside. Conserv Biol 21:482–494

Serageldin I (2002) World poverty and hunger – the challenge for science. Science 296(54–55):57–58

Shalaway SD (1985) Fencerow management for nesting birds in Michigan. Wildl Soc Bull 13:302–306

Shennan C (2008) Biotic interactions, ecological knowledge and agriculture. Philos Trans R Soc B 363:717–739

Shiklomanov IA (2000) Appraisal and assessment of world water resources. Water Int 25:11–32

Siegel RB, Centeno MV (1996) Neotropical migrants in marginal habitats on a Guatemalan cattle ranch. Wilson Bull 108:166–170

Sieving KE, Willson MF, DeSanto TL (1996) Habitat barriers to movement of understory birds in fragmented south-temperate rainforest. Auk 113:944–949

Siriwardena GM, Crick HQP, Baillie SR, Wilson JD (2000) Agricultural land-use and the spatial distribution of granivorous lowland farmland birds. Ecography 23:702–719

Skagen SK, Melcher CP, How WH, Knopf FL (1998) Comparative use of riparian corridors and oases by migrating birds in southeast Arizona. Conserv Biol 12:896–909

Skagen SK, Kelly JF, van Riper C, Hutto RL, Finch DM, Krueper DJ, Melcher CP (2005) Geography of spring landbird migration through riparian habitats in southwestern North America. Condor 107:212–227

Smiley PC Jr, Maul JD, Cooper CM (2007) Avian community structure among restored riparian habitats in northwestern Mississippi. Agr Ecosyst Environ 122:149–156

Smith BD, Yarnell RA (2009) Initial formation of an indigenous crop complex in eastern North America at 3800 BP. Proc Natl Acad Sci USA 106:6561–6566

Smith MD, Barbour PJ, Burger W Jr, Dinsmore SJ (2005) Density and diversity of overwintering birds in managed field borders in Mississippi. Wilson Bull 117:258–269

Smith TA, Osmond DL, Moorman CE, Stucky JM, Gilliam JW (2008) Effect of vegetation management on bird habitat in riparian buffer zones. Southeast Nat 7:277–288

Smith HG, Dänhardt J, Lindström A, Rundlöf M (2010) Consequences of organic farming and landscape heterogeneity for species richness and abundance of farmland birds. Oecologia 162:1071–1079

Söderström B, Pärt T (2000) Influence of landscape scale on farmland birds breeding in semi-natural pastures. Conserv Biol 14:522–533

Sodhi NS, Liow LH, Bazzaz FA (2004) Avian extinctions from tropical and subtropical forests. Annu Rev Ecol Evol Syst 35:323–345

Sodhi NS, Posa MRC, Lee TM, Warkentin IG (2008) Effects of disturbance or loss of tropical rainforest on birds. Auk 125:511–519

Somarriba E, Harvey CA, Samper M, Anthony F, González J, Staver C, Rice RA (2004) Biodiversity conservation in neotropical coffee (*Coffea arabica*) plantations. In: Schroth G, Fonseca GABd, Harvey CA, Gascon C, Vasconcelos HL, Izac A-MN (eds) Agroforestry and biodiversity conservation in tropical landscapes. Island Press, Washington, DC

Sorensen N, Goodfellow J (2010) International year of biodiversity: farmers say that conserving biodiversity is a shared responsibility. International Federation of Agricultural Producers. http://www.cbd.int/iyb/doc/partners/iyb-ifap-press-en.pdf. Accessed Jan 2010

Spangenberg JE, Matuschik I, Jacomet S, Schiblerc J (2008) Direct evidence for the existence of dairying farms in prehistoric Central Europe (4th millennium BC). Isot Environ Health Stud 44:189–200

Spiroux de Vendômois JF, Roullier F, Cellier D, Séralini GD (2009) A comparison of the effects of three GM corn varieties on mammalian health. Int J Biol Sci 5:706–726

Stanley TR, Knopf FL (2002) Avian responses to late-season grazing in a shrub-willow floodplain. Conserv Biol 16:225–231

Stauffer DF, Best LB (1980) Habitat selection by birds of riparian communities: evaluating effects of habitat alterations. J Wildl Manage 44:1–15

Stauffer DF, Best LB (1986) Nest-site characteristics of open-nesting birds in riparian habitats in Iowa. Wilson Bull 98:231–242

Steenhof K, Peterson BE (2009) American Kestrel reproduction in southwestern Idaho: annual variation and long-term trends. J Raptor Res 43:283–290

Stevens DK, Bradbury RB (2006) Effects of the Arable Stewardship Pilot Scheme on breeding birds at field and farm-scales. Agr Ecosyst Environ 112:283–290

Stewart AM, Craig JL (1989) Factors affecting pollinator effectiveness in Feijoa-Sellowiana. New Zeal J Crop Hort 17:145–154

Stoate C, Báldi A, Beja P, Boatman ND, Herzon I, van Doorn A, de Snoo GR, Rakosy L, Ramwell C (2009) Ecological impacts of early 21st century agricultural change in Europe- a review. J Environ Manage 91:22–46

Stouffer PC, Bierregaard RO (1995a) Effects of forest fragmentation on understory hummingbirds in Amazonian Brazil. Conserv Biol 9:1085–1094

Stouffer PC, Bierregaard RO (1995b) Use of Amazonian forest fragments by understory insectivorous birds. Ecology 76:2429–2445

Stratford JA, Robinson WD (2005) Gulliver travels to the fragmented tropics: geographic variation in mechanisms of avian extinction. Front Ecol Environ 3:85–92

Strode PK (2003) Implications of climate change for North American wood warblers (Parulidae). Glob Change Biol 9:1137–1144

Strum KM, Hooper MJ, Johnson KA, Lanctot RB, Zaccagnini ME, Sandercock BK (2010) Exposure of nonbreeding migratory shorebirds to cholinesterase-inhibiting contaminants in the western hemisphere. Condor 112:15–28

Stuart D (2008) The illusion of control: industrialized agriculture, nature, and food safety. Agr Hum Values 25:177–181

Sugden LG, Beyersbergen GW (1984) Farming intensity on waterfowl breeding grounds in Saskatchewan Parklands. Wildl Soc Bull 12:22–26

Szaro RC (1980) Factors influencing bird populations in southwestern riparian forests. In: Workshop proceedings, management of western forests and grasslands for nongame birds. USDA Forest Service, General Technical Report IN-86, Salt Lake City, pp 403–418

Talbot JM (2004) Grounds for agreement: the political economy of the coffee commodity chain. Rowman & Littlefield, New York

Taylor MW, Wolfe CW, Baxter WL (1978) Land-use change and Ring-Necked Pheasants in Nebraska. Wildl Soc Bull 6:226–230

Temple SA, Fevold BM, Paine LK, Undersander DJ, Sample DW (1999) Nesting birds and grazing cattle: accommodating both on Midwestern pastures. Stud Avian Biol 19:196–202

Tewksbury JJ, Black AE, Nur N, Saab VA, Logan BD, Dobkin DS (2002) Effects of anthropogenic fragmentation and livestock grazing on western riparian bird communities. Stud Avian Biol 25:158–202

Tews J, Brose U, Grimm V, Tielbörger K, Wichmann MC, Schwager M, Jeltsch F (2004) Animal species diversity driven by habitat heterogeneity/diversity: the importance of keystone structures. J Biogeogr 31:79–92

Thomas MB (1999) Ecological approaches and the development of "truly integrated" pest management. Proc Natl Acad Sci USA 96:5944–5951

Thomas JW, Anderson RG, Maser C, Bull EL (1979) Snags. In: Thomas JW (ed) Wildlife habitats in managed forests, vol 553, Agriculture handbook. USDA Forest Service, Washington, DC, pp 60–77

Thomas CD, Cameron A, Green RE, Bakkenes M, Beaumont LJ, Collingham YC, Erasmus BFN, de Siqueira MF, Grainger A, Hannah L, Hughes L, Huntley B, van Jaarsveld AS, Midgley GF, Miles L, Ortega-Huerta MA, Peterson AT, Phillips OL, Williams SE (2004) Extinction risk from climate change. Nature 427:145–148

Thompson PB (1995) The spirit of the soil, agriculture and environmental ethics. Routledge, London

Thomson LJ, Macfadyen S, Hoffmann AA (2010) Predicting the effects of climate change on natural enemies of agricultural pests. Biol Control 52:296–306

Tichit M, Doyen L, Lemel JY, Renault O, Durant D (2007) A co-viability model of grazing and bird community management in farmland. Ecol Modell 206:277–293

TIES (The International Ecotourism Society) (2010) Learning center – What is ecotourism? http://www.ecotourism.org/. Accessed Aug 2010

Tilman D (1999) Global environmental impacts of agricultural expansion: the need for sustainable and efficient practices. Proc Natl Acad Sci USA 96:5995–6000

Tilman D, Cassman KG, Matson PA, Naylor R, Polasky S (2002) Agricultural sustainability and intensive production practices. Nature 418:671–677

Tremblay A, Mineau P, Stewart RK (2001) Effects of bird predation on some pest insect populations in corn. Agr Ecosyst Environ 83:143–152

Troy AR, Strong AM, Bosworth SC, Donovan TM, Buckley NJ, Wilson JL (2005) Attitudes of Vermont dairy farmers regarding adoption of management practices for grassland songbirds. Wildl Soc Bull 33:528–538

Tscharntke T, Klein AM, Kruess A, Steffan-Dewenter I, Thies C (2005) Landscape perspectives on agricultural intensification and biodiversity – ecosystem service management. Ecol Lett 8:857–874

Tscharntke T, Sekercioglu CH, Dietsch TV, Sodhi NS, Hoehn P, Tylianakis JM (2008) Landscape constraints on functional diversity of birds and insects in tropical agroecosystems. Ecology 89:944–951

Tsitsilas A, Stuckey S, Hoffmann AA, Weeks AR, Thomson LF (2006) Shelterbelts in agricultural landscapes suppress invertebrate pests. Aust J Exp Agr 46:1379–1388

Tubbs AA (1980) Riparian bird communities of the Great Plains. In: Workshop proceedings, management of western forests and grasslands for nongame birds. USDA Forest Service, General Technical Report IN-86, Salt Lake City, pp 419–433

Tucker GM, Evans MI (1997) Habitats for birds in Europe: a conservation strategy for the wider environment, vol 6, BirdLife conservation. BirdLife International, Cambridge

Turner MG (2005) Landscape ecology: what is the state of the science? Annu Rev Ecol Evol Syst 36:319–344

United Nations (2010) Rethinking poverty: report on the world social situation 2010. United Nations, Department of Economic and Social Affairs. United Nations, New York, http://www. Un.Org/esa/socdev/rwss/index-print.html. Accessed Oct 2010

Uriarte M, Bruna EM, Rubim EM, Rubim P, Anciães M, Jonckheere I (2010) Effects of forest fragmentation on the seedling recruitment of a tropical herb: assessing seed vs. safe-site limitation. Ecology 91:1317–1328

Van Bael SA, Bichier P, Greenberg R (2007a) Bird predation on insects reduces damage to the foliage of cocoa trees (*Theobroma cacao*) in western Panama. J Trop Ecol 23:715–719

Van Bael SA, Bichier P, Ochoa I, Greenberg R (2007b) Bird diversity in cacao farms and forest fragments of western Panama. Biodivers Conserv 16:2245–2256

Van Bael SA, Philpott SM, Greenberg R, Bichier P, Barber NA, Mooney KA, Gruner DS (2008) Birds as predators in tropical agroforestry systems. Ecology 89:928–934

Van Houtan KS, Pimm SL, Halley JM, Bierregaard RO, Lovejoy TE (2007) Dispersal of Amazonian birds in continuous and fragmented forest. Ecol Lett 10:219–229

Vance DR (1976) Changes in land use and wildlife populations in southeastern Illinois. Wildl Soc Bull 4:11–15

Vandermeer JH (2011) The ecology of agroecosystems. Jones and Bartlett, Sudbury

Vandermeer J, Perfecto I (2007) The agricultural matrix and a future paradigm for conservation. Conserv Biol 21:274–277

Veech JA (2006) A comparison of landscapes occupied by increasing and decreasing populations of grassland birds. Conserv Biol 20:1422–1432

Vickery PD, Hunter ML Jr, Melvin SM (1994) Effects of habitat area on the distribution of grassland birds in Maine. Conserv Biol 8:1087–1097

Vickery PD, Tubaro PL, Da Silva JMC, Peterjohn BG, Herkert JR, Cavalcanti RB (1999) Conservation of grassland birds in the western hemisphere. Stud Avian Biol 19:2–26

Vickery J, Carter N, Fuller RJ (2002) The potential value of managed cereal field margins as foraging habitats for farmland birds in the UK. Agr Ecosyst Environ 89:41–52

Villa-Bonilla B, Rojas-Soto OR, Colodner-Chamudis AG, Tejeda-Cruz C (2008) Municipal-level inventories of the avifauna and its application to conservation: the case of Zacapoaxtla, Puebla, Mexico. Ornitologia Neotrop 19:531–551

Visser ME, Both C (2005) Shifts in phenology due to global climate change: the need for a yardstick. Proc R Soc B 272:2561–2569

Vitousek PM, Ehrlich PR, Ehrlich AH, Matson PA (1986) Human appropriation of the products of photosynthesis. Bioscience 36:368–373

Waggoner PE (1995) How much land can ten billion people spare for nature? Does technology make a difference? Technol Soc 17:17–34

Waggoner PE (1996) How much land can ten billion people spare for nature? Daedalus 125:73–93

Wall ML, Whitcomb WH (1964) The effect of bird predators on winter survival of the southwestern and European corn borers in Arkansas. J Kansas Entomol Soc 37:187–192

Walpole M, Goodwin H (2001) Local attitudes towards conservation and tourism around Komodo National Park, Indonesia. Environ Conserv 28:160–166

Warburton DB, Klimstra WD (1984) Wildlife use of no-till and conventionally tilled corn fields. J Soil Water Conserv 39:327–330

Warkentin IG, Greenberg R, Ortiz JS (1995) Songbird use of gallery woodlands in recently cleared and older settled landscapes of the Selva-Lacandona, Chiapas, Mexico. Conserv Biol 9:1095–1106

Warner RE (1994) Agricultural land use and grassland habitat in Illinois: future shock for midwestern birds? Conserv Biol 8:147–156

Warner RE, Etter SL (1989) Hay cutting and the survival of pheasants: a long-term perspective. J Wildl Manage 53:455–461

Warner RE, Walk JW, Hoffman CL (2005) Managing farmlands for wildlife. In: Braun CE (ed) Techniques for wildlife investigations and management, 6th edn. The Wildlife Society, Bethesda, pp 861–872

Warren KA, Anderson JT (2005) Grassland songbird nest-site selection and response to mowing in West Virginia. Wildl Soc Bull 33:285–292

Whelan CJ, Wenny DG, Marquis RJ (2008) Ecosystem services provided by birds. Ann NY Acad Sci 1134:25–60

Whittingham MJ, Evans KL (2004) The effects of habitat structure on predation risk of birds in agricultural landscapes. Ibis 146:210–220

Whittingham MJ, Swetnam RD, Wilson JD, Chamberlain DE, Freckleton RP (2005) Habitat selection by yellowhammers *Emberiza citrinella* on lowland farmland at two spatial scales: implications for conservation management. J Appl Ecol 42:270–280

Wilcove DS, Rothstein D, Dubow J, Phillips A, Losos E (2000) Leading threats to biodiversity: what's imperiling U.S. species? In: Stein BA, Kutner LS, Adams JS (eds) Our precious heritage: the status of biodiversity in the United States. The nature conservancy and association for biodiversity information. Oxford University Press, New York, pp 239–254

Wild Farm Alliance (2010) Food safety and conservation conflicts. http://www.wildfarmalliance.org/. Accessed Nov 2010

Willer H, Yussefi-Menzler M, Sorensen N (eds) (2008) The world of organic agriculture – statistics and emerging trends 2008. International Federation of Organic Agriculture Movements (IFOAM)/Research Institute of Organic Agriculture (FiBL), Frick/Bonn

Williams M (2003) Deforesting the earth: from prehistory to global crisis. University of Chicago Press, Chicago

Willner GR, Gates JE, Devlin WJ (1983) Nest box use by cavity-nesting birds. Am Midl Nat 109:194–201

Wilson JD, Whittingham MJ, Bradbury RB (2005) The management of crop structure: a general approach to reversing the impacts of agricultural intensification on birds? Ibis 147:453–463

Wilson JD, Evans AD, Grice PV (2009) Bird conservation and agriculture. Cambridge University Press, Cambridge

Wilson JD, Evans AD, Grice P (2010) Bird conservation and agriculture: a pivotal moment? Ibis 152:176–179

Winter M, Johnson DH, Shaffer JA (2005) Variability in vegetation effects on density and nesting success of grassland birds. J Wildl Manage 69:185–197

Winter M, Johnson DH, Shaffer JA (2006) Does body size affect a bird's sensitivity to patch size and landscape structure? Condor 108:808–816

With KA, King AW, Jensen WE (2008) Remaining large grasslands may not be sufficient to prevent grassland bird declines. Biol Conserv 141:3152–3167

Worster D (1990) Transformations of the earth: toward an agroecological perspective in history. J Am Hist 76:1087–1106

Wretenberg J, Lindström A, Svensson S, Thierfelder T, Pärt T (2006) Population trends of farmland birds in Sweden and England: similar trends but different patterns of agricultural intensification. J Appl Ecol 43:1110–1120

Wretenberg J, Pärt T, Berg A (2010) Changes in local species richness of farmland birds in relation to land-use changes and landscape structure. Biol Conserv 143:375–381

Wright AL (2005) The death of Ramon Gonzalez: the modern agricultural dilemma, 2nd edn. University of Texas Press, Austin

Wunderle JM, Latta SC (1998) Avian resource use in Dominican shade coffee plantations. Wilson Bull 110:271–281

Wunderle JM, Waide RB (1993) Distribution of overwintering nearctic migrants in the Bahamas and Greater-Antilles. Condor 95:904–933

Yahner RH (1981) Avian winter abundance patterns in farmstead shelterbelts: weather and temporal effects. J Field Ornithol 52:50–56

Yahner RH (1982a) Avian nest densities and nest-site selection in farmstead shelterbelts. Wilson Bull 94:156–175

Yahner RH (1982b) Avian use of vertical strata and plantings in farmstead shelterbelts. J Wildl Manage 46:50–60

Yahner RH (1983) Seasonal dynamics, habitat relationships, and management of avifauna in farm-stead shelterbelts. J Wildl Manage 47:85–104

Yang LH, Rudolf VHW (2010) Phenology, ontogeny and the effects of climate change on the timing of species interactions. Ecol Lett 13:1–10

Yang M, Hens L, Ou X, De Wulf R (2009) Tourism: an alternative to development? Reconsidering farming, tourism and conservation incentives in northwest Yunnan mountain communities. Mt Res Dev 29:75–81

Yang Z, Cai J, Sliuzas R (2010) Agro-tourism enterprises as a form of multi-functional urban agriculture for peri-urban development in China. Habitat Int 34:374–385

Zeder MA (2008) Domestication and early agriculture in the Mediterranean Basin: origins, diffusion, and impact. Proc Natl Acad Sci USA 105:11597–11604

Zhang W, Ricketts TH, Kremen C, Carney K, Swinton SM (2007a) Ecosystem services and dis-services to agriculture. Ecol Appl 64:253–260

Zhang X, Tarpley D, Sullivan JT (2007b) Diverse responses of vegetation phenology to a warming climate. Geophys Res Lett 34:L19405

A Review of Ecosystem Services, Farmer Livelihoods, and Value Chains in Shade Coffee Agroecosystems

Shalene Jha, Christopher M. Bacon, Stacy M. Philpott, Robert A. Rice, V. Ernesto Méndez, and Peter Läderach

Abstract Cultivation, processing, and consumption of coffee are dynamic processes that connect coffee farmers and agro-ecosystems with coffee drinkers spanning the globe. As a cash crop, coffee cultivation gained popularity in the Old and then the New world, and flourished under colonial regimes of the nineteenth and twentieth century. Coffee production patterns and management styles have changed drastically

S. Jha (✉)
Department of Environmental Science, Policy and Management, University of California,
54 Mulford Hall, Berkeley, CA, USA 94720
e-mail: s.jha@berkeley.edu

C.M. Bacon
Environmental Studies Institute, Santa Clara University, 500 El Camino Real, Santa Clara,
CA, USA 95050-4901
e-mail: CBacon@scu.edu

S.M. Philpott
Department of Environmental Sciences, University of Toledo, 2801 W. Bancroft St,
MS 604, Toledo, OH, USA 43606
e-mail: stacy.philpott@utoledo.edu

R.A. Rice
Migratory Bird Center, Smithsonian Conservation Biology Institute, National Zoological Park,
Washington, DC, USA
e-mail: ricer@si.edu

V.E. Méndez
Environmental Program and Plant and Soil Science Department, University of Vermont,
The Bittersweet- 153 South Prospect St, Burlington, VT, USA 05401
e-mail: emendez@uvm.edu

P. Läderach
Centro Internacional de Agricultura Tropical (CIAT), Km 17 Recta Cali-Palmira,
Cali, Colombia
e-mail: p.laderach@cgiar.org

W.B. Campbell and S. López Ortíz (eds.), *Integrating Agriculture, Conservation and Ecotourism: Examples from the Field*, Issues in Agroecology – Present Status and Future Prospectus 1, DOI 10.1007/978-94-007-1309-3_4,
© Springer Science+Business Media B.V. 2011

in the past two centuries and continue to shift, with the greatest recent expansions in East Asia. Traditionally, coffee is cultivated under a canopy of shade trees, a practice that ensures the longevity of the farm, supports biodiversity, and provides communities with a broad array of ecosystem services. However, many modern management schemes abandon shade practices. On the other hand, specialty coffee markets, like certified organic, certified shade (Bird Friendly), Fair Trade, and other certified coffees have gained recent popularity, though they still represent a small fraction of the global coffee economy. The global coffee economy is comprised of a wide array of coffee value chains that connect farmers with consumers, and thus impact farmer livelihoods at multiple spatial scales. Key players in the coffee value chain include local cooperatives, national government agencies, and global certification agencies. Similarly, ecosystem services provided by shade coffee occur at local, regional, and global scales, including pollination, erosion-control, and carbon sequestration, respectively. While the ecological and socio-economic costs and benefits associated with shade coffee are clear, this review reveals that there are many challenges to bridging sustainable coffee management with livelihood security. Furthermore, in this review we identify existing gaps in the literature and a number of promising research directions concerning the ecological and socio-economic impacts of coffee production.

1 Introduction

In this review, we synthesize the history and current standing of coffee production and the state of science on ecosystem services and farmer livelihoods associated with coffee production. We use a multi-scalar approach to organize ecological and social interactions taking place at local, regional, and global scales. Specifically, we address the following questions: (1) What is the history of coffee? (2) How is coffee produced, and by whom? (3) What are the ecological costs and benefits associated with coffee? (4) What are the socio-economic costs and benefits associated with coffee? Ultimately, as a synthesis of these topics, we ask (5) What future directions can research take in order to address current gaps in our understanding of the ecological and socio-economic aspects of coffee production?

Across the globe, over 400 billion cups of coffee are served per year (Illy 2002). While coffee is consumed around the world, few people recognize the extensive journey taken by the beverage. From seed to cup, this journey employs more than 25 million people, from farmers and laborers to roasters and distributors (Donald 2004). The first step in the coffee life cycle begins on coffee farms (Fig. 1), which in 2008, covered over 9.7 million ha of land worldwide (FAO 2008). Within these farms, coffee is cultivated under a wide range of vegetation management types that provide varying levels of shade (e.g. Philpott et al. 2008a; Moguel and Toledo 1999). For example, coffee management can span from 'rustic' coffee, where coffee shrubs are grown under a dense canopy of tropical trees (approximately 90% cover), to 'sun' coffee, where coffee shrubs are grown in the absence of shade trees and in direct sunlight (0% cover) (Fig. 2). Coffee bushes need 4–6 years before they begin producing the ripe cherries that farmers and workers harvest. After harvesting, the

Fig. 1 Forest fragments and coffee plantations in Chiapas, Mexico

Fig. 2 Sun coffee in Costa Rica (*left*), and shade coffee in Nicaragua (*right*)

cherries are processed to separate the fruit and hull from the beans or seeds. The beans are then dried, sorted multiple times, roasted, shipped for sale and distribution, brewed and consumed (Prendergast 1999). If stored properly, coffee beans can last for more than 8 months and maintain much of their flavor. This makes the coffee value chain more flexible than most other tropical agricultural products, such as bananas and oranges (Talbot 2004).

The simplicity of the coffee production process, however, masks the complexity and diversity of networks that are involved in organizing coffee landscapes, coffee farm owners, and coffee farm workers across local, regional, and global scales. At a local scale, human effort combines with ecological processes through different farming practices to produce coffee beans. Thus, if managed appropriately, coffee farms can dually produce coffee and support biodiversity (reviewed in Perfecto and Vandermeer 2008a; Perfecto et al. 1996). At a local and landscape scale, biological diversity maintained within coffee farms offers a range of provisioning and regulating ecosystem services, such as water storage, coffee flower pollination, and pest control (e.g. Lin 2007; Perfecto et al. 2004; Klein et al. 2003c). At a global scale, coffee vegetation management affects a grower's ability to qualify for premium-providing 'organic' or 'bird-friendly' certification (e.g. Philpott et al. 2007) or potentially earn carbon credits (e.g. Dossa et al. 2008). Thus, coffee management impacts ecological systems and socio-economic livelihoods, rendering these two aspects of coffee cultivation inextricably linked at local, regional, and global scales.

Despite this interconnectedness, few reviews have moved beyond the case study approach to attempt a global synthesis of ecological and socioeconomic costs and benefits of shade coffee production. In this review, we will examine the ecosystem service and farmer livelihood issues associated with coffee production. Specifically, we will review (1) the history, ecology and geography of shade coffee, (2) coffee production patterns, (3) the ecological costs and benefits associated with coffee, (4) the socio-economic costs and benefits associated with coffee, and (5) the current gaps in the literature concerning the ecosystem science and livelihood security involved in coffee production.

2   Ecology, History, and Geography of Shade Coffee

2.1   Crop Characteristics

Coffee belongs to the genus *Coffea*, which includes more than 103 species (Davis et al. 2006). Only two species are commercially viable: Arabica coffee (*Coffea arabica* L.) and Robusta coffee (*Coffea canephora* Pierre ex Froehner). Arabica grows in mid-elevation (600–1,500 m) regions and yields a smooth, slightly acidic beverage after roasting, whereas the lower-elevation (0–800 m) Robusta is more tolerant to growth in full sun (Wilson 1999) and produces a relatively harsher cup of coffee with higher caffeine content (Charrier et al. 2009). Because the Arabica species produces higher quality coffee, it generates more economic value; in contrast, Robusta generates higher yields per plant than Arabica, but produces beans that specialty markets generally consider of lower quality and economic value (Bacon 2005a). A third species, *C. liberica* Bull ex Hiern., is regionally important within Africa and Asia but is not sold globally (Charrier et al. 2009). Of the 48 coffee exporting countries listed by the International Coffee Organization (ICO), 27% export Robusta exclusively, 29% export both Arabica and Robusta, and 44% export only Arabica (ICO 2010).

Coffee growth, photosynthesis, and production require specific ecological and physical environmental characteristics, limiting the specific regions in which coffee is grown. For example, coffee is dependent on seasonal rainfall in the tropics both for production of flower buds (following a drought) and flowering (following a dry-season rain) (Carr 2001; Cannell 1983; Magalhaes and Angelocci 1976). Water availability, as well as small changes in temperatures, can affect coffee photosynthesis (Cannell 1976; Nunes et al. 1968). Because coffee is not frost resistant (DaMatta 2004), the upper elevations and latitudes at which coffee can be cultivated are limited. Likely due to its evolution in the understory of tropical forests, the maximum photosynthetic rate of Arabica plants are at moderate temperatures and under moderate levels of shade (Lin et al. 2008; Nutman 1937) and thus it has traditionally been cultivated as an understory crop. Understory crops are trees, shrubs, vines, or other plants that thrive in the environment under the canopy of taller trees, are often grown within orchards, and may also be cultivated in natural forests or conservation areas (Elevitch and Wilkinson 2000).

While coffee's genetic center of origin and its early beginnings as a product lie in Ethiopia, concerted plantation production has its roots in the Near East, amid the terraced slopes of what is now Yemen. The beans moved around the world with Arab traders, religious leaders, many undocumented social networks, and later with European colonial powers seeking to disengage from dependency upon the Near East traders for the bean. Spanish traders introduced the beverage to Western Europe in 1528, and upon reaching Italy, coffee caused such a stir as to be targeted by a number of priests as "Satan's Drink". Its aroma and taste, however, moved Pope Clement III to bestow baptismal status on it shortly thereafter, securing coffee a place in Christendom as an acceptable beverage (Ukers 1922).

Once coffee gained a foothold in Western Europe, its spread throughout the colonial world was all but certain. While the French, British, and Dutch took coffee to the tropical regions of the Old World, it was the French who first brought it to the New World tropics where, as an introduced crop, it was free of most of its natural enemies (insect pests and fungal diseases) and thrived. Like many tropical agricultural commodities pursued by the Colonial governments, coffee's early history also was intertwined with that of slavery (Clarence-Smith 2003). Something of a novelty at first, coffee formed the backbone of newly found economic freedom in Latin America after the Spanish started to relinquish their colonial hold in the 1820s. Coffee became closely allied with the Liberal movement in Central America, for instance, as the crop that would replace faltering dyestuffs like indigo and cochineal, which had fallen in economic value (Biderman 1982).

The latter half of the 1800s saw coffee emerge as one of Latin America's principal cash crops, rising to prominence as an important generator of foreign exchange. Labor was cheap as slavery and forced labor were common on larger coffee plantations, and land, often following displacements of indigenous peoples, was plentiful. With aid from governments using repressive policies to secure both labor and land, coffee flourished throughout the American tropics (McCreery 1995; Williams 1994). By 1900, coffee's physical and social landscapes were well on their ways to changing the region. As mentioned, coffee's spread in the Old World pre-dated its expansion in the Americas, due largely to the efforts of the Arabs, the first to discover and cultivate coffee for large scale production, and the Dutch (Ukers 1922). As coffee spread to more and more countries in order to satisfy European and later North American demand, the management of shade within the coffee farms became a hotly debated subject.

The United States Department of Agriculture, as a consequence of the "…recent acquirement of tropical territory by the United States…", as well as "…the much controverted question of the shading of the coffee tree", tasked Special Agent for Tropical Agriculture Mr. Orator Fuller Cook to examine the shade issue for the USDA. Through personal observation and a literature review, he produced the authoritative report in 1901 on the subject which is still cited today by coffee researchers. Cook (1901) stated in his report on global shade coffee trends, that Brazil and parts of the East Indies favored a reduced shade or open-to-the-sun management style, a condition possibly due more to the natural land cover and climate at least in Brazil, than other factors. His assessment of the degree to which shade is needed in coffee plantations hinged on production, but always with an eye toward the health of the plant and some of shade's indirect effects. He especially identified the role of the canopy in protecting against drought and erosion, as well as the beneficial effects of nitrogen fixing by leguminous shade trees.

2.2   Modernization or 'Technification' of the Coffee Sector: From Shade to Sun

Unlike most of the basic grains and certain other food crops, coffee escaped the early pressures of the Green Revolution and the intensification of production that was the hallmark of that transformative process. Yet, different situations and forces converged to alter the production practices of coffee in a number of countries. In Central America and parts of South America, for instance, the arrival of the coffee leaf rust, *Hemileia vastatrix* Berk., created a virtual panic among producers and national level institutes responsible for production. With the assistance of the United States Agency for International Development (USAID) of $81 million and eight multi-year (and some multi-country) projects, a modernization or renovation wave swept the countries of Nicaragua, Guatemala and Honduras, among others, in efforts to head off the effects of the rust (Rice and McLean 1999). The efforts, spearheaded by a USAID-funded regional office called Promecafe, promoted the introduction of new high-yielding varieties, the removal of shade and an increase in the planting density of coffee bushes. The rationale behind the widespread modification was both commercial and agronomic. An 'open-to-the-sun' environment would diminish any dampness, which is conducive to the rust's development, and the planting changes would increase yields, provided the appropriate kinds and levels of inputs were used. The regional transformation represented an intensification of coffee that had been proven in Costa Rica, where yields of 1,500–2,000 kg/ha had been reported for a number of years. However, countries like Nicaragua and El Salvador did not experience such dramatic technological change in their coffee farming, due – at least in part – to decreased investments on coffee plantations and social marginalization as associated with armed conflict in the 1970s and 1980s (Bacon et al. 2008b).

Conversion from shade to sun coffee in Colombia was more abrupt, with nearly a century of shade coffee production before intensification efforts began. In 1895, shade management in Colombia was displayed as an example to emulate for Jamaica, the British colony's prize coffee producer at the time. A Mr. Thompson of the British Foreign Office applauded and attributed the success of Colombia's yields (1,022 kg/ha) to its use of shade trees, even to the point of identifying the elevation ranges in which genera like *Cassia*, *Erythrina*, and *Inga* were used. So impressed was Thompson by Colombia's yields that he stated "…were the Jamaica plantations yielding to the same extent as those of Colombia, the value of the output would be increased …to double…yearly" (Cook 1901). He concluded, moreover, that the quality of Colombian coffee was far superior to that of Jamaica.

The variability in shade management styles across the global coffee landscape today in many ways relates back to the time of Cook's assessment. Environmental factors, such as altitude, climate, and local disease problems combined with social processes and structures, to produce a range or gradient of shade management across the globe. The Western Ghats region of India also has a history of shade mainly for reasons of protection from the coffee leaf rust. In these systems, the native forest was retained but trees were thinned because of the perceived detriment to coffee plants due to heightened local competition (Cook 1901). In parts of Indonesia such as Java, shade trees were maintained as a windbreak and foil against the spread of fungal diseases (Cook 1901). Cook's conclusion about shade is one of geographic conditionality: farmers should develop site and subject-dependent plans based upon local conditions and growers' attitudes about how best to deal with the vagaries of nature while cultivating this perennial cash crop.

Transformation of the coffee landscape from shade to sun coffee is extensive but uneven across the globe. Approximately 40% of Latin American shade coffee farms have been converted to low shade systems (Rice and Ward 1996). Today, we find Latin American farming systems largely unchanged since the 1996 survey. Colombia is still dominated by a relatively intensively managed coffee sector that was modified beginning in the 1970s to control disease and increase yields (Guhl 2004). Sun coffee still characterizes Brazil's sector, with a very few producers in places like northern Saô Paulo or Pernambuco maintaining diverse canopies over their coffee (R. Rice, personal communication with Marco Croce).

Differences in shade management are evident within countries as well (Table 1). Guatemala's Huehuetenango region tends to have a diverse shade cover dominated by native *Inga* spp., whereas the region around the city of Antigua (which suffers periodic near-frost temperatures from cold air masses from the north) has a monoculture canopy of *Grevillea robusta* A. Cunningham ex R. Br., an exotic Australian native that can withstand low temperatures. In the Guatemalan cloud forest regions of San Marcos or Coban, by contrast, farmers manage little to no shade because of daily cloud cover. When shade trees are planted, such as *Erythrina* spp. and *Gliricidia* spp., they are pollarded into low-stature cover. Guatemala's national

Table 1 Percent coffee area managed beneath different technological/shade levels

Country	% Area in diverse shade/traditional management	% Area in monoculture shade/medium technology	% Area in sun coffee/intensified management
Peru	90	8	2
Haiti	100	–	–
Vietnam	5	20	75
Kenya	15	←-------------------85*------------------→	
Honduras	35	45	20
Indonesia	25	35	40
Brazil	←-------------------5*------------------→		95
Guatemala	40	58	2
El Salvador	24	75	1
Colombia	30	←-------------------70*------------------→	

Asterisk denotes no differentiation between categories. (Sources: Interviews and mail correspondence with the following individuals and/or institutions: Peru: Jessica Rojas, Junta Nacional del Café, 2010, and agronomist Gerardo Medina of Rainforest Alliance; Vietnam, Truong Hong, Vice Director of Vietnam's Coffee Research Centre, 2010; Colombia, SICA/AFIC, 2009; Haiti, Centre National de l'Information Geo-Spatiale, 1998; Mexico, SIAP and Rene Avila Nieto, staff statistician at AMCAFE 2010; Honduras, Edgar Ibarra and Filiberto Olloa, at the Instituto Hondureño del Café, 2010; Indonesia, Dr. Misnawi, researcher at the Indonesian Cocoa and Coffee Research Institute, 2010; Kenya, Isabella Nkonge at the Coffee Board of Kenya and Juliana Jaramillo at the International Centre of Insect Physiology and Ecology (Nairobi), 2010; Guatemala, Anacafe Director of Research Dr. Francisco Anzueto, 2010; Colombia, intensified management can include scant, monoculture shade cover, Guhl 2004).

coffee association, Anacafe, reports that some 98% of the country's coffee grows beneath a shade cover, dominated by *Inga* spp. trees, with some 15 species accounting for 47% of the shade trees found in coffee (Anacafe 2008). Coffee defined as 'shade-grown' in these cases has a fairly low-diversity tree cover composed mainly or completely of the native genus *Inga* or the exotic *Grevillea robusta*. The percent area of coffee grown beneath a diverse shade cover in Guatemala is estimated to be only 40% (R. Rice, personal communication with F. Anzueto).

In Colombia, much of the coffee area underwent intensification (i.e., shade tree removal) since the 1970s, due principally to the fear of coffee rust's imminent arrival. Yet, the Santander region in the southeastern part of the country maintained a diverse shade cover of towering trees, many of which were once part of the original forest. The cultural identities and values of coffee farmers deeply influence the types of shade coffee maintained (Moguel and Toledo 1999). Observers in both Mexico and Colombia have described patterns of more diverse shade and trends towards organic coffee production in communities with stronger indigenous identities (Moguel and Toledo 1999).

In Vietnam, recent decades have seen the coffee area expand in the northern highland region. The species *C. canephora* is tended in irrigated systems in the open sun. A quick look at the United Nations Food and Agriculture Organization's data for coffee production in Vietnam since 1965 show this phenomenal rise, mainly in

Table 2   Vietnam's coffee transformation, 1965–2008 (FAO 2010)

Year	Area harvested (ha)	Yield (kg/ha)	Production (tons)
1965	22,800	329	7,500
1970	18,600	392	7,300
1975	11,400	596	6,800
1980	10,820	776	8,400
1985	14,060	875	12,300
1990	61,857	1,487	92,000
1995	155,000	1,406	218,000
2000	476,900	1,683	802,500
2005	497,400	1,512	752,100
2008	530,900	1,989	1,055,800

the 1990s, that positioned that country as one of the top two or three in production (Table 2). Between 1965 and 2008, area increased by 2,200%, yields by 83% and production by 13,900%. Nearly all increases were for Robusta coffee, produced beneath little or no shade cover and aided by irrigation and chemical inputs, the results of which have led to large scale environmental and socioeconomic decline in the highlands region of that country (D'Haeze et al. 2005; Kotecha et al. 2003).

Although few debate the social and ecological importance of shade coffee, there is a lack of independent empirical research documenting the extent of shade grown coffee and landscape changes in ecologically important coffee growing territories. The most comprehensive review of these issues focused on Latin America and was conducted nearly 15 years ago (Rice and Ward 1996). In many countries, such as Nicaragua and El Salvador, 95% of the coffee is managed under a diversified shade canopy (Rice and Ward 1996). Studies conducted since 2000 have documented high levels of shade tree diversity in smallholder farmers, with more than 100 species found on 34 farms in Nicaragua and over 120 species on 54 plots in El Salvador (Méndez et al. 2010b). In contrast, shade tree diversity has declined in some larger coffee farms. Furthermore, as a result of government incentives and desires to increase yields, farmers have gradually removed or reduced shade cover assuming that higher light and more dense cropping patterns lead to higher yields (Staver et al. 2001). Although more research is needed to fully understand the multiple drivers of change in coffee landscapes, it is clear that the changing structure of global coffee value chains will continue to exert a substantial influence upon these processes (Topik et al. 2010; Jaffee 2007; Perfecto et al. 1996).

3   Conventional and Alternative Coffee Value Chains

Here, we summarize global coffee production, trade statistics and trends, which are relevant to describing the coffee value chain. Our review of coffee value chains considers the dominant trends in global markets, the emergence of specialty and sustainably certified value chains and the key stakeholders that participate at local, regional, and global scales.

3.1 Global Production and Trade Statistics

Coffee is one of the most valuable legally traded commodities from the developing world (FAO 2010), bearing relevance to many national economies (O'Brien and Kinnaird 2003). Between 14 and 25 million families are actively involved in coffee production, and millions more depend on coffee for their livelihoods (Lewin et al. 2004; Oxfam 2001). The vast majority of producers, estimated at more than 70%, are smallholders farming less than 10 hectares (ha) (Lewin et al. 2004; Oxfam 2001). These producers make a significant, though currently uncalculated, contribution to the 8.2 million metric tons produced in 2008 (FAO 2010).

Although global production statistics tend toward smallholders, there are large differences in the relative fraction of smallholder vs. estate farms among the top 20 producing countries (Table 3). During 2008, coffee was produced in more than 70 countries, located throughout the tropics. The top producers are Brazil, Vietnam, Colombia, and Indonesia, with each country generating more than 68,000 metric tons of green coffee in 2008 (FAO 2010, Table 3), and Brazil providing more than twice that of second-place Vietnam (2.7 vs. 1.1 million tons, respectively). Yields among global producers vary substantially, with the highest yields recorded coming from Martinique (25,000 kg/ha) and the lowest from Suriname (190 kg/ha) (FAO 2009). Specialty coffee (e.g. organic, fair trade, and shade-grown coffee) accounts for approximately 9–12% of all coffee production (Raynolds et al. 2007; Van der Vossen 2005) of which Mexico, Central America, Columbia, and Peru are the market leaders (Lewin et al. 2004). Likewise, coffee area varies greatly between countries (e.g. from >2 million ha in Brazil to 10 ha in Tonga), with around 10 million ha a constant feature in tropical landscapes globally since at least 1965. Worldwide, land in coffee production in developing countries is significant, with several of the top producers controlling more than 5% of agricultural land area in coffee production (FAO 2010). Data for Table 3 were gathered directly from the Embassies and agricultural ministries among the world's top 20 coffee producing countries. However, we complemented this data with a review of the published and grey literature and consultations with FAO databases.

A simple farm-sized based typology of coffee producers provides important background for our subsequent analysis seeking to understand coffee commodity chains and the drivers of conservation practices in shade and sun coffee landscapes. Coffee smallholders represent most coffee farmers, yet they may not represent the majority of all coffee produced. Furthermore, only a limited number have formed smallholder cooperatives that enable them to have a direct stake in coffee exports and further downstream in the coffee commodity chain (Rice 2000). Although often more evenly distributed than other agricultural and ranching landscapes, land ownership patterns in many coffee growing communities and countries remains highly concentrated. Large coffee estates, including those with more than 50 ha of coffee production, often control exports and purchase coffee from small and micro producers.

The majority of producers worldwide are coffee smallholders managing less than 10 ha of coffee (Table 3). This is an important global figure, but it should be

Table 3 Coffee production statistics for 2008[a]

Country	Coffee yield (hg/ha)	Coffee production (tons)	Area in production (ha)	No. of producers (<10 ha)	No. of producers (10–100 ha)	No. of producers (>100 ha)
Brazil	12,594	2,790,858	2,216,014	220,554	30,900	1,656
Vietnam	19,886	1,055,800	530,900	500,000	200[b]	
Colombia	9,399	688,680	732,656	520,069	7,540	0
Indonesia[c]	6,987	682,938	1,313,309	1,259,656	53,653[b]	
Ethiopia	6,715	273,400	407,147			
Mexico	3,516	265,817	755,843	507,377	3,166	–
India	7,660	262,000	342,000			
Guatemala[d]	10,400	254,800	245,000	50,000	9,092	3,557
Peru	6,973	225,992	324,062	136,000	23,900	100
Honduras	9,476	217,951	230,000	81,680	5,026	51
Uganda	7,989	211,726	265,000			
Costa Rica[e]	11,102	107,341	96,681	46,705	3,120	802
Haiti	3,863	35,000	90,600	800	–	–
El Salvador	6,345	97,727	154,000	17,869	2,881	282
Philippines	7,903	97,428	123,269	263,836	11,845	–
Côte d'Ivoire	3,478	80,000	230,000			
Kenya	2,709	42,000	155,000	700,000	4,000[b]	
Papua New Guinea	10,771	75,400	70,000			

(continued)

Table 3 (continued)

Country	Coffee yield (hg/ha)	Coffee production (tons)	Area in production (ha)	No. of producers (<10 ha)	No. of producers (10–100 ha)	No. of producers (>100 ha)
Nicaragua	6,275	72,727	115,883	29,000		167 (more than 50 ha)
Venezuela	3,810	70,311	184,536			
Madagascar	5,360	67,000	125,000			
Thailand	8,111	50,442	62,186			

[a] Source: FAO statistics (http://faostat.fao.org) for most yield, production and area data

[b] No. of producers >10 ha

[c] Indonesia data given in ha by Dr. Misnawi and the Cocoa and Coffee Research Institute of Indonesia (2008), with no differentiation above the 10 ha category; Kenya, Coffee Board of Kenya and Dr. Juliana Jaramillo, 2010; El Salvador, Ana Elena Escalante of the Consejo del Café de El Salvador, 2010

[d] Guatemala (size categories are <10 ha, 10–45 ha, and >45 ha) data from Anacafe's Dr. Francisco Anzueto, Director of Research, 2010, and CEPAL, nd; Vietnam, Mr. Truong Hong of Vietnam's Coffee Research Centre, 2010; Colombia, SICA/AFIC, 2009; Haiti, Centre National de l'Information Geo-Spatiale, 1998; Mexico, SIAP and Rene Avila Nieto, staff statistician at AMCAFE, 2010; Honduras, Edgar Ibarra and Filiberto Olloa, at the Instituto Hondureño del Café, 2010; Brazil, Instituto Brasileiro de Geografia e Estatística, Censo Agriopecuario, 2006; Philippines, Census of Agriculture, 2002. Peru, farm size classes are <5 ha, 5–100 ha, and >100 ha, and obtained from Peru's Junta Nacional del Café staff member Jessica Rojas and agronomist Gerardo Medina of Rainforest Alliance

[e] Costa Rica size categories based on production data provided by Sr. Deryhan Muñoz Barquero at the Instituto del Café de Costa Rica, 2010

used with caution. A closer look demonstrates that size ranges mask several important trends observable within specific countries and regions when the data are further segmented. In Mesoamerica (including Mexico and Central America), most coffee producers are substantially smaller than the 10 ha standard used for delineating a producer as a smallholder. A 2002 report published by a regional office of the United Nations (CEPAL 2002) found that more than 68% or 394,716 of the 573,000 plus farmers in this region are micro-producers managing less than 2 ha of coffee. In other coffee producing countries, such as Rwanda, the majority of farms are so small that they are measured in the number of coffee trees (about 300, as compared to many Mesoamerican smallholder farms that generally have from 1,500 to 2,500 coffee bushes per ha). In Central America, smallholders represent 85% of coffee producers and control 18% of coffee production lands, while the largest producers and industrial operations managing farms larger than 50 ha represent fewer than 3.5% of all coffee farmers and control about 49% the area in coffee production (CEPAL 2002). However, it should be noted that the trends in Latin America, especially after the 1999 coffee crisis, show a decrease in the number of large estates and an increase in the number of smallholder and micro-producers (Topik et al. 2010). These trends in the size of coffee producer operations are strongly influenced by the changing structures and incentives within the coffee value chain.

3.2   Coffee Value Chains and Global Markets: An Introduction

In its journey from tree to cup, coffee passes through the hands – directly or indirectly – of several players in the commodity chain. This value chain runs thread-like through a number of sequential steps, supported tangentially by production networks like machine manufacturers and transport services (see Sturgeon 2000), all of which are essential to getting the finished product to its destination. Growers, processors, exporters, importers, roasters, distributors and retailers form the normal categories of those involved, with repetitive handler groups (except for producers and roasters) being inserted in the chain in some cases (Fig. 3).

While the division of surplus (profits) has bounced back and forth over time, with growers usually getting a smaller share, recent years have seen those in producing countries – growers and national governments alike – receiving a smaller fraction of the profits (Fridell 2006; Oxfam 2002). Low international prices are one of the problems that are accentuated when the crisis of oversupply recurs (which is cyclical for most commodities). The early 2000s, for instance, saw coffee prices fall to levels that resulted in the value of coffee itself representing only 18% of the retail price – compared to 64% in the mid-1980s (Oxfam 2002). This reflected not only the general deterioration of terms of trade for producing countries over the last several decades, but the sharp collapse in coffee prices due to the breakup of the ICA and neoliberal policies spawned by multi-lateral institutions like the IMF and the World Bank. Neoliberalism is a political-economic theory,

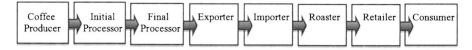

Fig. 3 The coffee commodity chain

class-based project, and regulatory practice (Harvey 2005). The central proposition is that by forcefully liberating individual entrepreneurial freedoms, through robust private property rights, free trade, and the power of free markets, well-being of all in society will be maximized (Watts 2007). According to this proposition, neither the state nor civil society should influence market factors, such as prices or costs of production.

Profits in coffee value chains are uneven and often dependent on the costs of production. They are not particularly high compared to other industries, but in certain cases, especially in times of crisis for growers, profits elsewhere in the value chain can be enormous (Oxfam 2002). In 2000–2001, Ugandan farmers received $0.14 for a kilo of unprocessed coffee that at retail would fetch more than $26.00 as instant coffee in the United Kingdom (Oxfam 2002). Accounting for weight loss during the processing and roasting of the coffee, that represents a 7,000% price increase in the journey from farm to shopping cart. For a roasted and ground package of the same coffee in the US, the increase would be around 4,000% (Oxfam 2002). Seen another way, if we assume that 5 pounds of the Ugandan farmer's fresh cherries are needed to make a pound of roasted beans which makes 40 servings of coffee that retail for $2.00 a cup, the $0.70 received by the grower fetches $80.00 at retail, which is an 11,000% increase. A recent and systematic comparison conducted during a period of low green coffee prices (in 1999 and 2000) and selling coffee from Tanzania to Italy by the pound (not the cup), found that 8.7% of final retail value of low quality Robusta coffees stayed on the farm, in comparison to only 3.9% of the high end 100% Arabica coffees (Daviron and Ponte 2005). However, green coffee prices have increased substantially from their depths in the coffee crisis of 2001 (Bacon et al. 2008a). From 2006 through the end of 2008, prices for green Arabica coffee increased by 24% and they were 60% above 2006 levels through September 2010 (FAO 2010). This has resulted in slightly higher percentages of retail price for bulk roasted conventional coffee accruing to exporters and growers. Recent data from the FAO show that global average coffee prices paid to growers increased 25% from January of 2006 through the end of 2008 (FAO 2010). Global data is not readily available to estimate changes in retail prices. However, during the same time period the average price of bulk conventional roasted coffee in the US cities increased by only 13% (US Department of Labor 2010). These numbers suggest that in the case of conventional coffees sold to supermarkets, and not specialty coffee sold by the cup, exporters and growers have recently captured 15–20% of the total retail value, a situation that was similar to those in the 1970s and early 1980s when the international coffee agreement sought to control supply to maintain more stable prices to producers (Talbot 2004).

3.3   Specialty and Certified Coffees

The specialty coffee market seeks to differentiate its coffees from the bulk commercial coffees in the mainstream markets (i.e., those purchased from supermarket shelves in large cans of Folgers® and Maxwell House®) based on sensorial attributes expressed in the cup (Läderach et al. 2006) and, to a lesser extent, sustainability. The closer attention to the qualities of coffee and the relationships with coffee producing communities and exporters initially led to the creation of many distinct global coffee value chains organized around coffee qualities and in some cases (especially those associated with early fair trade and organic coffees) notions of fairness, livelihoods, and ecology (Goodman 2008). Several countries are taking advantage of their promising production conditions for specialty coffee to develop Denominations of Origin (DO) such as Antigua in Guatemala, Marcela in Honduras, Veracruz in Mexico and several denominations in Colombia, among others (Daviron and Ponte 2005). DO are based on unique quality growing conditions expressed in a unique sensorial quality (Läderach et al. 2009).

Although the specialty coffee market segment was pioneered by small-scale artisanal roasting companies active since the early 1970s (Bacon 2005a; Dicum and Luttinger 1999), during the past decade several large coffee companies have diversified their rent capturing strategies into this market. In the past, most profits were sought via an 'economies of scale' approach. However, the recent emphasis on the qualities of coffee and the coffee drinking experience could be broadly categorized as an emergence of a more 'flexible' value chain, where an array of coffee products (i.e., espressos, lattes, and now frappuccinos) targeting specific consumer categories and niche marketing opportunities have emerged. Many small-scale roasters and cafes have also used the qualities and more direct relationships with coffee producing communities as an effective business strategy to expand their market share. Most of these businesses are organized within the specialty coffee market segment.

During the past two decades the specialty segment has gained a considerable following, sustaining annual retail market value growth rates that generally topped 10% since the mid 1980s (Giovannucci et al. 2008). The decline of the International Coffee Agreement (ICA) and withdrawal of national coffee marketing boards and rural assistance programs also contributed to the rise of the specialty coffee market sector (Bacon et al. 2008a). The Specialty Coffee Association of America (SCAA), one of the few industry associations with a relatively progressive track record, also provided fertile ground for launching several grower and civil society-based sustainability certification programs. The shade coffee category, along with organic and fair-trade coffees, may well represent a challenge for the established markets, conceptually if not economically.

Sustainable coffee certification is an umbrella term encompassing several types of certifications, and combinations of certifications. While Fair Trade focuses on the trade relationships, organic certification standards regulate the production process and require a separate chain of custody throughout different processing stages in the value chain. The overall organic market, which extends well beyond coffee, is significantly larger than Fair Trade markets. This is in part because the organic certification system has existed for a longer period of time and also developed a very diverse and often contested decentralized regulatory system. Most organic standards include the need for 'ecological' management of farms, including soil conservation practices which permit very little or no use of synthetic fertilizers and pesticides, prohibit genetically modified crops, and require intensive on-farm record keeping, among many other criteria (Van der Vossen 2005). Farms are certified organic by third party inspectors who follow an international code for each crop. Mexico exported the first certified organic coffee in 1967 (S. Philpott, personal interview with Walter Peters), and as of 2007, North American coffee drinkers had spent over one billion dollars on organic coffee (Giovannucci et al. 2008). Today, the leading certified organic coffee exporting countries include Ethiopia, Peru, and Mexico. Nicaragua is also among the top exporters with close to 10% of its coffee farmers certified as organic.

Table 4 offers a comparative analysis that considers the largest third-party sustainability certifications in the coffee industry. While the five certification programs listed in the table below have initially targeted the rapidly expanding specialty coffee market segment, both the Rainforest Alliance and Utz Certified have started to sell large volumes of certified products to the conventional coffee industry. Smithsonian's Bird Friendly certification program has the highest agro-environmental standards, requiring more than ten different species of diverse shade trees and certified organic production as well as general guidelines to conserve soil and water (Bacon et al. 2008a). Rainforest Alliance, Utz Certified, and Fair Trade all have several agro-environmental standards restricting the use of many of the most toxic pesticides and herbicides (generally based on an expanded version of the 'dirty dozen' list initially popularized by the Pesticide Action Network) and the expectation that all national laws will be implemented, but synthetic fertilizers and most pesticides, fungicides, and herbicides are permitted. A discussion of the enforcement of these standards is beyond the scope of this review. However, it is important to note that some of these standards are basic requirements that must be attained prior to certification, while others are goals towards which farms, farmers, and local organizations are expected to move over several years of annual inspections. The social standards, often based on non-discriminatory conventions from the International Labor Organization, are also summarized in the table below. The final column in Table 4 shows that Fair Trade, organic, and the Smithsonian's Bird Friendly certification programs have first sought to partner with small-scale farmers and their collective organizations, while Rainforest Alliance and Utz Certified started by certifying large-scale coffee plantations (Ponte 2008).

Given the number of stakeholders involved in the coffee value chain, it is not surprising that they operate at multiple, and often overlapping scales. These scales are at once spatial and temporal, and the boundaries characterizing them are not easily defined. This social science approach to scale defines the term as emerging

Table 4 A comparison of sustainable coffee certification standards

Certification	Coffee market segment	Agro-environmental standards	Social criteria	Size and geography of producers
Fair Trade (FT)	Specialty and some conventional markets	Highly toxic agrochemicals use restricted (standards go beyond national laws), water conservation buffer zones around water bodies No genetically modified organisms (GMOs)[a]	Prioritizes smallholder producer cooperatives (co-ops receive minimum coffee prices plus premium for social development), standards restrict child labor, guarantee freedom of association and right (rt) to collective bargaining, buyers encouraged to sign long term contracts directly with smallholder co-ops and provisions access to credit	Started w/ indigenous smallholder cooperative in Latin America 800,000 plus individual smallholders affiliated with 250 producer cooperatives in Latin America, expanded coverage to Asia and Africa post 2000
Organic	Specialty	Prohibit the use of synthetic fertilizers and agrochemicals, encourage integral soil management, no GMOs	Freedom of association and right to collective bargaining, working conditions, equal treatment, etc.	Started w/ larger farms and indigenous smallholder cooperative in Latin America (Mexico), now prevalent in Latin America, Ethiopia and elsewhere

(continued)

Table 4 (continued)

Certification	Coffee market segment	Agro-environmental standards	Social criteria	Size and geography of producers
Utz certified	Specialty and conventional market segments	Focus on enforcing adherence to national laws and avoiding use of illegal agrochemicals, several standards to reduce contaminations	Originally were a direct adaption of the Global Good Agricultural Practices (GAP) criteria to coffee, standards for record keeping, better and documented use of agrochemicals, labour rights and access to health care and education for employees and their families	Started w/ larger farms in Guatemala partnering with large retail outlet
Rainforest alliance	Specialty and increasingly conventional	Restricted agrochemical use, encourages social and water conservation, shade trees standards: canopy cover of mixed native trees	Freedom of association, safe and clean working environment, the national legal minimum wage, dignified housing, medical care, free education, health, training	Started w/ larger farms in Central America, post 2004–2006 has adjusted standards for smallholder organizations, many operations in Brazil, Vietnam
Smithsonian migratory center's Bird-Friendly (BF)	Specialty coffee markets	Requires organic certification plus at least 10 shade tree woody species on farm and presence of larger trees, stream buffer zones, and secondary plant diversity	Social criteria are same as organic, much of BF coffee is also FT certified	Started with smallholder cooperatives in southern Mexico and also piloted on some larger farms

Sources: Modified and adopted Bacon et al.(2008a): 348–149; based on data from Ponte (2008), Raynolds et al. (2007)

Sources: [a]according to TransFair USA's website

http://www.transfairusa.org/content/about/overview.php, but we have yet to see this clearly elaborated within the international regulations

http://nationalzoo.si.edu/ConservationAndScience/MigratoryBirds/Coffee/quick_reference_guide.cfm

http://www.rainforest-alliance.org/agriculture.cfm?id=standards_farms

Fig. 4 Representation of local, national/regional, and global stakeholders (•) and ecosystem services provided within shade coffee farms (+)

out of economic, cultural, and social interactions and thus representing a social construction difficult to divorce from human interactions or activities (Sayre 2005; Brenner 2001; Marston 2000). However, those interested in interdisciplinary research and the connections between social and ecological approaches to scale must seek the commonality in definitions. As Sayre (2005) states: "It is obvious that social and ecological phenomena are intimately linked across scales; it follows that the problems of one cannot be resolved in isolation from those of the other". From the standpoint of the producer involved with shade coffee and its associated benefits, controversies and nuances, we find three scales at which stakeholders operate and/or interact: (1) local/community, (2) national/regional, and (3) global (Fig. 4). Despite the economic connections between scales, the stakeholders themselves rarely understand the depth or scope of issues facing the others within the commodity chain.

3.4   Local and Community Scale Coffee Value Chains

The local or community scale of the coffee value chain includes farmers, farm workers and others falling within a farm's sphere of influence at the level of production. At the most fine-grained scale, the farm itself is the unit, with decisions by the operator affecting ecological processes as well as his or her own socioeconomic rewards. While concerned with the management of the farm, a very local, site-based operation, small landholders often are members of cooperative organizations, placing them into a distinct scale of activities, commitments and benefits. Faced with phenomena at the global scale, a small producer subjected to international price fluctuations might respond by deciding to seek organic, fair trade, or shade certification. The challenge in this case could be of temporal scales, defined on the one hand by niche market fashions that shift rapidly, and on the other by the perennial nature of coffee and shade trees, and the time it can take to be certified (the transition period imposed by certification standards). A single grower practicing environmentally beneficial land stewardship (i.e., maintaining a biologically diverse shade, low-input coffee system, and protection of water sources) because of necessity (little income for inputs, managing a diverse system for the array of products it provides, etc.) can be catapulted into community, national, and global scale arenas once he or she decides to connect with a local agency that certifies coffees according to international standards. Unlike growers laboring within the anonymous collective of producers supplying beans to meet the global demand for industrially produced coffees, farmers involved in certified coffee production aim at meeting codified standards and satisfying specific interests of consumers. The documentation associated with certifications creates an audit trail as the certified coffee passes from player to player along the chain. Paperwork leads back to the individual farm, and documents not only all stakeholders handling the coffee, but obliterates the anonymity in which non-certified producers exist. The documentation, identification, and recognition of certified producers create relationships and scalar interactions arguably unrealized prior to certification.

3.5   National and Regional Scale Value Chains

The national or state scale is one of cooperative unions, social movements, non-governmental organizations and government ministries, that along with other organizations and within legal institutional frameworks, create the web of social connections that enable and influence the journey of the coffee seed from plant to the point of export (Bair 2009). An individual farmer and cooperative member operating at this scale is often and usually brokered by the cooperative leadership or professional staff. In other cases, like Indonesia, this is done by private exporters who prepare and fund activities related to certification – and reap some of the associated rents. Growers contend with national tax laws often attached to coffee exports as a way of funding marketing and/or research institutes as in Colombia or Costa Rica, or marketing boards that control nearly all exports as has been the case for Kenya and Ethiopia (Akiyama et al. 2003). These same national scale entities often provide technical advice through extension workers, and sometimes support a degree of social development (e.g. the Colombian Coffee Federation) the aim of which is to increase quantity and quality exports and thus subsequently capture more revenues.

The regional scale may well encompass more than a single country, as with the mountain ranges of Central or South America where much of the world's coffee is produced. Growers at this scale, while distinct in terms of nationality, live and manage farms in indistinguishable locales ecologically and culturally, as with growers living on either side of the border between Chiapas, Mexico, and Guatemala. As ecosystems are often defined by their watersheds and topographic structure, growing practices are often similar within regional watersheds, even though political boundaries may divide them. Price differentials and coffee origins are shaped by international perceptions and the quality, consistency, communications, and marketing of coffee exporters and thus growers on one side of a national boundary may suffer price punishment due to origin, even though the ecological, climatic and processing conditions of the two origins are the same (Daviron and Ponte 2005). Climatic phenomena like frost, drought or hurricanes can also affect entire regions. The result of such extreme events can devastate production across national borders, affecting local farmers adversely while growers in untouched regions can benefit from the higher prices caused by scarcity in supply.

3.6 Global Coffee Value Chains

Globally, there are trade organizations, certifiers, and governmental bodies accrediting certifiers, roasters, and consumers. From a basic, traditional commodity chain relationship, growers enter into global relationships, directly or via mediators, in questions of quality and quantity. Increasingly, however, the "latte revolution" (Ponte 2002) has pushed many growers toward specialty coffees defined by high-quality processing, fortunate origin location, certification, or some blend of these features. The growth of specialty coffee has created a consumer who is more aware of where, how, and by whom the coffee is produced, and what its impact on the environment, the grower, biodiversity and even climate change might be. To the extent that demand for specialty coffees with some characteristics addressing consumers' concerns increases, the local farmer (in this case a member of the Global South) will be influenced to produce in specific ways governed by the interests of northern roasters, retailers and consumers.

A definite re-orientation of scale related to organic certification came into play within the last decade when the global organic community shifted from a relatively self-monitored organizational structure channeled through the International Federation of Organic Agricultural Movements (IFOAM) to the more formalized regulations of the USDA's National Organic Program, as well as the Japanese analogue (JAS) and the European Union's supra-national control over organic products. Even though the activities under the IFOAM period were global in nature, the state and supra-state power brought to bear on farmers and certification agencies when these markets of the global North moved to oversee certification introduced new geographies of political regulation into the system. The audits and quality control hurdles that agencies currently face are not simply bureaucratically tangled; they are costly in terms of personnel workloads and the payment of fees for accreditation (Mutersbaugh 2005).

3.7 Neoliberal Reforms and the Post 1999 Coffee Crisis

The changing management of coffee systems is also influenced by the evolving structures of the coffee value chain and the prices paid for this global commodity. In 1999, prices paid to producers for the green beans they sold through the international coffee commodities market plunged causing a humanitarian and in some cases an ecological crisis in many coffee growing regions (Oxfam 2002). However, the 1999 coffee crisis, also known as the 'global coffee crisis' provided researchers with insight into the mechanism of change in coffee landscapes (Bacon et al. 2008b; Rice 2003; Varangis et al. 2003). Consensus has it that the withdrawal of the United States from the International Coffee Agreement (ICA), the established pact between producing and consuming countries that controlled global inventory and prices, resulted in the dumping of warehoused stocks into the market and causing prices to plummet in the early 1990s (Eakin et al. 2006; Varangis et al. 2003). This, combined with increased consolidation in the roasting and trading phases of the value chain, rapid roll-back of direct state involvement in coffee production and marketing, and with existing farmer vulnerabilities created the most recent coffee crisis (Goodman 2008). Corrected for inflation, the "30-year" low price levels were actually 100-year lows, well below the price of production (Varangis et al. 2003). A buyer's market undoubtedly helped to keep prices at basement levels, resulting in a scramble to sell coffee with little leverage for growers. But growers were not the only ones to suffer.

The low coffee prices resulted in a crisis due to the persistent vulnerabilities among many coffee producers, conditions exacerbated by a broader, deeper crisis related to the systematic exclusion of farmers and agricultural workers, global economic woes, low commodity prices generally, and extreme weather events like hurricanes (Bacon et al. 2008b). The generalized low prices translated into stress within the banks and government coffers, which in turn meant that capital usually flowing from coffee revenues was not to be found, adding to national anxieties and frustration. Low prices, weakened financial linkages, and diminished government revenues also resulted in disruption of commerce, transportation and other socio-economically linked activities. Producers resorted to strategies like planting alternative crops, migrating to the US to find work, neglecting and/or outright abandoning their farms in order to cope with imperiled livelihoods. The act of curtailing all cultural practices in order to save on production costs was a strategy that obviously cut into rural wages from day labor, a common source of income for rural families in coffee regions. Some strategies were more severe; reports from Anacafe staff in Guatemala included farmer suicides.

While the low prices were devastating for producers and others with economic links to the coffee production sector, it did not necessarily translate into lower prices for consumers. Roasters seemed to have maintained or only slightly lowered prices at the retail level, turning what was a crisis for growers into a golden opportunity for their own bottom lines – at least for the large coffee companies. One report from an industry trader stated that the 15% return seen by roasters in normal times climbed to 110% during this crisis period (Rice 2003). An extended case study reveals how several of these processes interplayed.

The trajectory and institutional linkages related to Mexico's coffee sector over the past several decades showcase the efforts, aims, and consequences of commodity production in a dynamic global environment buffeted by economic and sociopolitical winds. As a country, Mexico is representative of many coffee producers in that its 95,000 producers tending 400,000 ha of coffee in 1985 were dominated by smallholders with an average of 3 ha of coffee, accounting for 84% of the coffee area (Nolasco 1985). An expanding global economy and the concomitant increasing demand for coffee since WWII helped, coupled with state led development (and electoral patronage) models prevalent at the time combined to establish the National Mexican Coffee Institute (INMECAFE) in 1958, the charge of which was to oversee production, processing and marketing of coffee (Jaffee 2007). State-based coffee marketing and support agencies, like INMECAFE, were fundamentally important in retaining and storing coffee exports as part of the international coffee agreements established in an effort to maintain prices that could sustain a degree of positive development outcomes. Coffee exporters and important countries negotiated important economic clauses (including quotas for production and imports) through the International Coffee Agreements (ICA). The ICA was put into force in the 1960s not only for price stability, but also as a geopolitical strategy to help stem social unrest and the threat of communism so feared at the time, and providing dependable (if perhaps not totally adequate) prices to growers (Dicum and Luttinger 1999).

During this expansion period for coffee, INMECAFE promoted the intensification of coffee production via experimental stations and a network of offices providing technical assistance. A monoculture, shade-less coffee system was advocated, even though the yields resulting from INMECAFE's technical assistance did not match those associated with other sources of technical assistance (Nolasco 1985). However, many of the state-led efforts to convince smallholders to eliminate shade trees failed. Furthermore, INMECAFE had greater influence in certain areas of Mexico, such as Veracruz, but much less among the more marginalized states of Oaxaca and Chiapas. The indigenous populations represent a larger proportion of the inhabitants in both states, and these states would also emerge as global pioneers in organic, shade and Fair Trade coffee production (Nigh 1997). The strong networks of smallholder coffee cooperatives, indigenous community level management or ejidos and community associations provided the social capital for partnering with northern certifiers, scientist and coffee roasters that led to the early pilot testing and eventual development of the major certification programs for organic, shade, and fair trade coffee (Bacon et al. 2008a). By the early 1990s, coffee cultivation area had nearly doubled and the number of growers nearly tripled (Calo and Wise 2005). INMECAFE targeted small and medium sized farmers with the goal of introducing and spreading a technological package involving the coffee monoculture mentioned above. The widespread adoption of neoliberal political and economic reforms as evidenced by the passage of international trade agreements, such as the North American Free Trade Agreement, and the privatization of state based industry and activities accelerated in the 1990s profoundly influenced the coffee sector (Topik et al. 2010; McCarthy 2004). Under the Neoliberal model, free markets are expected to optimize benefits to society. Damages to the environment and or social wellbeing are often characterized as an 'externalities' (e.g. pollution). In these cases, most, though not all promoters of this approach suggest that the state play an important role creating new property rights and establishing a new market that proponents claim will enable private profit seeking to spur innovative solutions (Kay et al. 1997). Critics of Neoliberalism highlight the often violent means that governments and corporations employ to maintain this system (Harvey 2005) and show the negative empirical consequences as measured by uneven development patterns (Watts 2007), persistent economic poverty, and usurped rights of many local and indigenous communities. Researchers have also questioned the efficacy of Neoliberal approaches to solving pressing environmental problems (Marsden et al. 1996; McCarthy 2004), suggesting they are not up to addressing the root social causes and long term drivers of climate change, pollution, and biodiversity loss at global scales (Peet and Watts 2004).

The deregulation of the international coffee markets following the collapse of the international coffee agreement in 1989, the rollback of state investments in coffee marketing, technical assistance and exports, and the fraying rural social safety net are all evidence of Neoliberal trends in the coffee sector (Topik et al. 2010; Bacon et al. 2008b). In Mexico, the national government began to pull support from social programs (although it maintained more than many other governments) and state supported coffee marketing and technical assistance institutions like INMECAFE. With the collapse of the ICA in 1989 and the subsequent dismantling of INMECAFE in 1993, growers were left virtually on their own to face the shocking reality of trade liberalization. Price supports that had given them $1.00–$1.40 per pound for their coffee gave way to below-production cost prices of only $0.50 per pound (Calo and Wise 2005). It is worth noting that the social unrest seen as a threat in the 1960s when the ICA was formed actually blossomed 1 year after INMECAFE's breakup and 5 years after the collapse of the ICA (ICAFE 1989). In 1994, during the Zapatista uprising in Chiapas, 36% and 30% of the coffee area and producers, respectively, protested the signing of the North American Free Trade Agreement (NAFTA) and made headlines throughout the world (AMECAFE 2010).

4   Ecological Processes and Ecosystem Services

4.1   Ecological Processes in Coffee Landscapes

Ecosystem services are ecological functions that sustain and improve human life (Daily 1997). Globally, ecosystem services such as pollination, pest control, erosion control, watershed management, and carbon sequestration, provide an estimated economic value of $18 trillion annually (Costanza et al. 1997). According to the Millenium Ecosystem Assessment (MEA 2005), an international and comprehensive study of global resources, an ecosystem service is defined as any benefit that humans obtain from an ecosystem. The MEA divided ecosystem services into four categories, (1) 'provisioning services', such as water, food, and forest products, (2) 'regulating services', such as the regulation of climate, waste, and floods, (3) 'cultural services', such as aesthetic, spiritual, or recreational benefits, and (4) 'supporting services', such as nutrient cycling and photosynthesis. Thus, the key 'provisioning services' within shade coffee farms are the coffee yields themselves, along with the fruits and forest products often gathered within these systems. A 'regulating service' value of pollination would be the increase in production of coffee within a farm, while the 'supporting service' value of pollination would be the reproduction of native non-crop plants that benefit other ecosystem services, such as the provision of erosion control by a native tree that also grew as a result of pollination (i.e., Kremen et al. 2007).

Shaded coffee plantations are increasingly valued for their contributions to biodiversity conservation and the provisioning of ecosystem services. Within shade coffee farms, as in other landscapes, ecosystem services function at different spatial scales (local, regional, and global), thus the ecological and economic benefits garnered from these services depend on the stakeholder composition at multiple spatial scales. Additionally, ecosystem services interact with one another in complex ways (Bennett et al. 2009), making it important to examine how these interactions play out within coffee plantations. In the sections below, we review the ecosystem services provided by shaded coffee plantations at local, regional, and global scales (Fig. 4). While there is overlap between services provided across spatial scales, we believe that it is beneficial to highlight the scales at which specific ecosystem services have the greatest impact on stakeholders.

4.2   Coffee Management Paradigms

Coffee plantations were traditionally cultivated under the canopy of a native forest, but coffee management systems practiced today follow a strong gradient from rustic to sun plantations. These different management systems have drastically different names depending on the farmers, researchers, or conservationists asked, but have many common features. Typically, more traditional practices include growing coffee under the canopy of a native forest ('rustic', or 'home garden' in Ethiopia where Arabica coffee evolved). As shade management is 'intensified', the resulting planta-tions have lower canopy cover, fewer shade trees, fewer shade tree species, fewer epiphytes, and more weeds (Philpott et al. 2008b; Moguel and Toledo 1999). Generally, although not always, shade management intensification is accompanied by increases in the use of synthetic agrochemicals (e.g. pesticides, fungicides, herbicides, and fertilizers). Many previous authors have outlined the details of coffee management gradients specific to Mexico (Moguel and Toledo 1999), Latin America (Philpott et al. 2008b), and parts of Asia (Craswell et al. 1997). Here, we summarize common characteristics of different management systems and the ecosystem services they provide (Table 5).

4.3   Local Scale Ecological Processes and Ecosystem Services

Biodiverse shade coffee plantations can support substantial native biodiversity, much of which contributes to provisioning, regulating, and supporting ecosystem services, including the supply of firewood, pollination and pest control services, erosion control, and nitrogen fixation. Dozens of studies have documented and sum-marized that shade coffee intensification, defined as the reduction in shade tree canopy richness and complexity (Moguel and Toledo 1999), generally leads to sig-nificant losses of diversity for trees, epiphytes, birds, bats, arthropods, small mam-mals, and amphibians (Perfecto et al. 1996, 2007; Greenberg et al. 1997a; Gallina et al. 1996). Reductions in tree diversity, removal of epiphytes, or other changes in the vertical structure of the vegetation can lead to further losses of animal diversity within agroforestry systems (Jha and Vandermeer 2010; Cruz-Angon et al. 2008; Philpott et al. 2008b; Gillison et al. 2004).

Specifically, biodiversity losses due to this type of intensification lead to signifi-cant losses of diversity of natural enemies (e.g. ants, birds, parasitoid wasps) with important implications for pest control services, a key regulating service provided by the shade coffee landscape (Philpott et al. 2008a; Perfecto et al. 1996, 2007). For example, ants and spiders reduce damage to coffee plants caused by the coffee berry borer, *Hypothenemus hampei* Ferrari (Larsen and Philpott 2010; Armbrecht and Gallego 2007; Perfecto and Vandermeer 2006; Vélez et al. 2001) and the coffee leaf miner, *Leucoptera coffeella* Guer. (De la Mora et al. 2008; Lomeli-Flores 2007). These studies report up to a 74–99% removal of the borers from occupied coffee berries (Armbrecht and Gallego 2007), suggesting that at a field scale, farmers could substantially benefit from reduced coffee berry losses if their farms provide sufficient ant habitat. Both birds (Kellermann et al. 2008) and bats (S. Philpott, personal communication with K. Williams-Guillen) also prey on the borers, and these services can save farmers from costly coffee losses due to borer damage (Kellermann et al. 2008). More generally, birds are important predators of arthro-pods in shaded coffee plantations (Borkhataria et al. 2006; Greenberg et al. 2000;

Table 5 Characteristics of coffee management systems[a]

Management style	Tree composition	Tree richness (No.)	Canopy cover (%) and height (m)	Shade strata	Coffee density	Canopy management techniques	Additional management / certification	Potential ecosystem services provided	Reviewed in
Rustic	Native forest canopy	25	>90% >15 m	3	Low – medium	Minimal canopy intervention	Bird-friendly, Rainforest Alliance or organic typically possible	Pollination, pest control, biodiversity, natural disaster protection, climate regulation, nutrient maintenance	Jha and Vandermeer (2010) and Philpott et al. (2008a)
Traditional polyculture	Some forest trees and some planted timber and fruit trees	10–20	60–90% 15 m	3	Low – medium	No or little pruning of the shade canopy	Bird-friendly, Rainforest Alliance or organic typically possible, usually with compost	Alternative food/ timber sources, pollination, pest control, biodiversity, natural disaster protection, climate regulation	Jha and Vandermeer (2010), Philpott et al. (2008a), and Méndez et al. (2007)

(continued)

Table 5 (continued)

Management style	Tree composition	Tree richness (No.)	Canopy cover (%) and height (m)	Shade strata	Coffee density	Canopy management techniques	Additional management / certification	Potential ecosystem services provided	Reviewed in
Commercial polyculture	Mostly planted canopy trees (timber and fruit trees) and N-fixing legumes. Few very abundant genera	5–10	30–60% 12–15 m	2	Medium – high	Regular pruning of canopy, removal of epiphytes	Rainforest alliance or organic typically possible, mixture of compost and agrochemical control	Alternative food/ timber sources, pollination, pest control, biodiversity	Jha and Vandermeer (2010) and Philpott et al. (2008a)
Shade monoculture	Canopy dominated by one species or genus of tree (i.e. *Inga* spp.)	1–5	<30% 10 m	1	High	Regular pruning of canopy, removal of epiphytes	Usually with agrochemical inputs	Minimal soil erosion control	Lin (2007)
Sun coffee	With rare isolated trees or without tree canopy	0	0% NA	0	High	Na	Usually with agrochemical inputs	Minimal soil erosion control and organic matter incorporation from coffee leaf litter	Philpott et al. (2008a)

[a]Based on Philpott et al. (2008a)

Johnson 2000) and may be able to quickly respond to pest outbreaks (Perfecto et al. 2004). Ants are more important predators in shaded coffee farms than in sun farms (Armbrecht and Gallego 2007) and functional or behavioral diversity of predatory species within coffee agroecosystems may enhance ecosystem services in general (Philpott et al. 2009; Van Bael et al. 2008). Thus, not only is the loss of predators significant for conservation purposes, but it will likely limit the regulating service of pest predation.

Furthermore, biodiversity within shaded coffee plantations may also perform important pollination services for crops (Klein et al. 2008), another key regulating and supporting ecosystem service. Both commercial species of coffee (*C. arabica* and *C. canephora*) benefit from pollinator visits (Klein et al. 2003a) and studies have shown that coffee pollinator species may be lost with agroforestry management intensification (Jha and Vandermeer 2010; Klein et al. 2003c). Large numbers of visits by honeybees (*Apis mellifera* L.), for example, correlate with higher coffee fruit set and fruit weight (Manrique and Thimann 2002; Roubik 2002; Raw and Free 1977). Native bees (including both social and solitary bees) augment pollination services to coffee, especially where diverse assemblages visit coffee plants (Klein et al. 2003b). Increased fruit set due to enhanced insect pollination at a per-bush level, can contribute to increased yields and farmer income at a farm scale, often worth tens of thousands of dollars (Ricketts et al. 2004).

In addition to pest control and pollination services, shaded coffee plantations provide a variety of other regulating services at the local level. Moderate levels of shade can hinder fungal diseases, such as the coffee leaf rust, which can have major impacts on coffee foliage and yields (Beer et al. 1998). Namely, trees create windbreaks, slowing the horizontal spread of spores of the coffee leaf rust (Soto-Pinto et al. 2002; Schroth et al. 2000), though incidence of other fungal diseases (e.g. coffee leaf spot, *Mycena citricolor* Cke.) may increase with vegetation complexity at local and regional scales (Johnson et al. 2009). Vegetation complexity at the canopy level can also provide weed reduction. In plantations with at least 40% canopy cover, many weeds, including grasses, can be completely eliminated (Beer et al. 1998; Muschler 1997). Furthermore, many common shade trees used in coffee agroforests (i.e., *Inga* spp.) provide the regulating service of fixing nitrogen and augmenting the nutrient content of soils (Beer et al. 1998), saving farmers the cost of expensive nitrogen inputs. Thus shaded plantations offer a number of potential ecosystem services at the local scale.

The shade component also generates important provisioning services in the form of direct products that provide socioeconomic benefits to coffee farming communities. Understandably, the array of tree species providing shade can also yield useful products in the form of fuelwood, building materials, fruits and ornamental or ceremonial plants (Rice 2008; Escalante 1995; Escalante et al. 1987; Lagemann and Heuveldop 1983), showing how non-coffee products can supply income to the farm household – especially during months when coffee income is depleted. In El Salvador, the shade tree canopy provides firewood for smallholder households for an equivalent value of 1 month of income generated by all the members of the household (Bacon et al. 2008a). The trees and plants within some shade systems

Table 6 Shade levels and impact on quality reviewed in studies from Latin America

Reference	Country	Positive impact	Negative impact	Observation
Läderach et al. (2009)	Colombia	>50% shade	<50% shade	Optimal growing zone
Vaast et al. (2006)	Costa Rica	45% shade	0% shade	Optimal growing zone
Muschler (2001)	Costa Rica	High shade level	Low shade level	Sub-optimal growing zone
Lara-Estrada (2005)	Nicaragua	46–63%	≤45%	Optimal growing zone
Decazy et al. (2003)	Honduras	Not evaluated	<44%	Optimal growing zone
Guyot et al. (1996)	Guatemala	High shade level	Low shade level	Optimal growing zone

also provide cultural services for coffee growers, as in the Peruvian cases of ritual plants from the farm being taken and given as offerings to the earth/mountain – "Pacha Mama" – in some of the indigenous communities on the Apurimac/Ene River region (R. Rice, personal communication with growers 2000, Peru).

Another local scale provisioning service provided by shade is the potential to improve the quality and flavor, since quality is a characteristic of production, and adds economic value to the product. While it has long been agreed that shade is the main factor enhancing coffee plantation sustainability in sub-optimal coffee zones (Beer et al. 1998), recent studies have also revealed that shade cover is beneficial as a means to improve coffee quality (e.g. taste, texture, pH), though the amount of shade needed for optimal quality varies for each bioregion. For example, a study in Colombia found that higher shade levels yield better quality than lower shade levels (Läderach et al. 2009). In Costa Rica, zero shade has a negative impact and 45% shade has a positive impact on coffee quality (Vaast et al. 2006), while in Nicaragua 45% or less had a negative effect and 46–63% had a positive effect (Lara-Estrada 2005). In Honduras, coffee with less than 45% shade was of inferior quality (Decazy et al. 2003), and in Guatemala high shade levels benefited coffee quality. The optimal shade level for the 0–20°N latitude is therefore probably somewhere between 45% and 70%, though the actual numbers are site specific and related to the overall production system and environment (Läderach et al. 2009) (Table 6).

4.4   Regional Scale Ecological Processes and Ecosystem Services

At a regional level, shade coffee plantations contribute to the regulation of services such as water conservation, watershed management, soil conservation, and landslide prevention. Coffee is grown throughout the tropics, but is susceptible to changes in local weather patterns (Carr 2001), with yield declines in years with lower precipitation (DaMatta et al. 2003; Salinas-Zavala et al. 2002). Furthermore, there is also a narrow temperature range under which coffee growth rates and yields are highest (Cannell 1976; Alegre 1959). Maintenance of these temperature and humidity conditions can benefit coffee producers with greater yields, but climate extremes (including regional dry and wet periods) may put coffee producers at risk (Lin 2010, 2007). A study conducted in the coffee growing Soconusco region of Chiapas, Mexico, examined daily and seasonal temperature and humidity conditions in the soil under three coffee management systems ranging in shade intensification (traditional polyculture, commercial polyculture, and shade monoculture) (Lin 2007). Lin (2007) found much greater daily fluctuations in temperature and relative humidity in the low shade sites compared with the higher shade sites – fluctuations that put coffee outside of the ideal temperature range for the region. There was also greater water loss from the soils in the low shade sites. Thus, shaded and diversified shade coffee farms provided greater climate regulating services, with potential impacts on coffee berry development and overall per bush yield.

Because coffee is grown in wet tropical climates, often on sloping mountainous regions, coffee landscapes are highly at risk of natural disasters including landslides associated with hurricanes, and will likely experience more frequent disturbances as climates continue to change. In 2005, a hurricane passed through the Soconusco region of Chiapas, Mexico, and caused extensive damage to the coffee harvest and to the landscape (Philpott et al. 2008c). Philpott and colleagues (2008c) examined economic damage to coffee farms (e.g. fruits lost to heavy rainfall) and the number of roadside landslides in a full range of coffee shade management systems. They found no differences in terms of economic damage depending on shade management system; however, they found that farms with more complex vegetation (i.e., less intensive farms) experienced significantly fewer landslides as a result of the hurricane. Additionally, this factor was more important than the amount of forest nearby and a number of topographic features (distance to rivers, elevation, and slope). The climate-regulating protection provided by increased vegetation complexity also has been recognized by coffee cooperative leaders in Guatemala and Mexico.

Furthermore, impacts from Hurricane Stan and land use changes following the hurricane in the Siltepec municipality of Chiapas have been examined; an area previously dominated by coffee production (G. Cruz-Bello, personal communication and unpublished data). Riparian areas suffered more total soil loss from the hurricane than non-riparian areas, and farmers were more keenly aware of the risks of growing coffee near rivers. Given concern about erosion and potential crop loss, many farmers were changing their land use practices. While some chose to grow maize in order to improve food security, many continued growing coffee. In these coffee growing areas, the majority of farmers with coffee left standing after the storm were changing their practices in order to increase the number of shade trees within their fields, with the knowledge that this may help buffer future climate-related disasters. Thus, across a number of regions, coffee growers have come to know that shade coffee can at least partially mitigate some climate-related natural disasters, saving potential crop loss and providing a key regulating ecosystem service.

Shade coffee farms also have gained recent attention for their role in serving as a corridor for organisms, such as pollinators and pest predators, moving between forest fragments within the region. Specifically, migratory birds, which are often pest predators, often utilize shade coffee farms while making their long-distance journey between temperate and tropical regions (e.g. Bakermans et al. 2009; Greenberg et al. 1997b). Molecular-based and mark-recapture studies have shown that key pollinators are able to migrate through shade coffee farms, between forest fragments. These include organisms such as butterflies (Muriel and Kattan 2009), and native bees (Jha and Dick 2010). Because shade coffee farms facilitate pollen and seed dispersing animals, native trees dependent on these dispersers are able to maintain reproduction and key gene flow processes across shade coffee systems (Jha and Dick 2008, 2010). These trees provide regulating services in the form of erosion control (Jha and Dick 2008) and also support native pollinators that are essential during the coffee bloom (Jha and Dick 2010). Thus, unlike sun coffee systems, which are often less permeable to dispersing organisms (e.g. Muriel and Kattan 2009), shade coffee farms can serve as habitat corridors for ecosystem service providing organisms moving regionally between forest fragments. Shaded coffee may also provide regional scale ecosystem services related to biodiversity conservation by enhancing the ecological quality of buffer zones near protected areas.

In order to take a closer look at the global spatial relationship between coffee cultivation and protected areas (PAs), we used the World Data Base on Protected Areas (WDPA consortium, 2005) and the Spatial Production Allocation Model (SPAM) database on crop production (You 2005). The WDPA was initiated by a United Nations (UN) General Assembly resolution in 1962 to record the status of the world's PAs, known as the UN List. There have been 13 editions of the UN List between 1962 and 2003, produced collaboratively by IUCN and the United Nations Environment Program -World Conservation Monitoring Centre (UNEP-WCMC). We used the latest digital version of 2005. SPAM relies on a collection of relevant spatially explicit input data, including crop production statistics, land cover and land use data, biophysical crop "suitability" assessments as well as any prior knowledge about the spatial distribution of specific crops or crop systems. Additionally SPAM uses crop production data at the national level reported by Food and Agriculture Organization of United Nations (FAO) and similar data within subnational boundaries compiled through a network of organizations.

The coffee institutions included in the analysis quantify their coffee areas using Geographical Information Science (GIS), either through delimitation of the areas by GIS, by remote sensing or by expert knowledge. Depending on the method, the information is more or less precise. As well, in the course of farmers shifting to other crops or renovation programs, the estimated coffee areas change constantly. On a global scale, we combined the SPAM and WDPA data to quantify the protected areas and areas under coffee production (Fig. 5). The output of the SPAM database on crop production generates maps with a 10 by 10 km resolution; the different shading of the pixels indicates the amount of harvested area per 100 km^2. According to the SPAM data, the global extension of coffee is approximately 1,008,600 km^2 and the extension of protected area registered under the WDPA database is 2,515,600 km^2. To assess the national coffee and protected areas in Mesoamerica we used information from coffee areas obtained from national coffee institutions and the WDPA data used for the global assessment (Fig. 6).

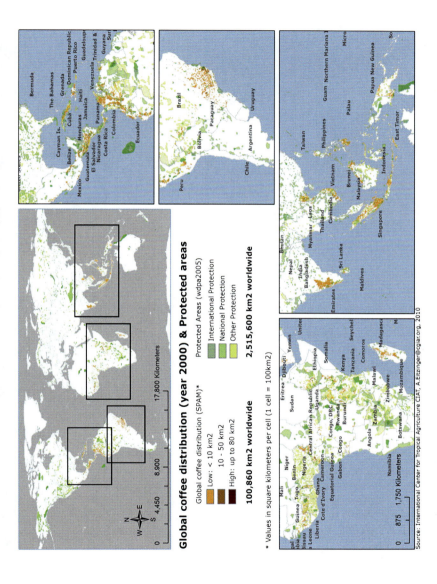

Fig. 5 Spatial distribution of global coffee cultivation and protected areas (Source: International Center for Tropical Agriculture, CIAT, A.Eitzinger@CGIAR. ORG, 2010)

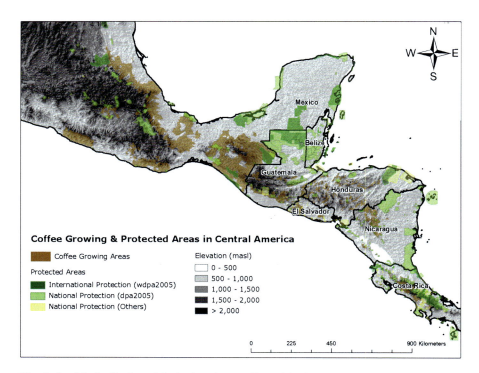

Fig. 6 Spatial distribution of Latin American coffee cultivation and protected areas. The data sources for examining correspondence between coffee producing regions and protected areas varied by country. For Mexico we used the reedited data of the *El Colegio del La Frontera Sur* GIS lab, based on Nolasco (1985), for Guatemala we used the digitized Coffee Atlas 2006/2007, for Honduras we used the GIS data of the Honduran Coffee Institute (IHCAFE), for El Salvador we used GIS data of the Salvadorian coffee institute (PROCAFE), for Nicaragua we used census data of the Nicaragua ministry of agriculture and forestry (MAGFOR), and for Costa Rica we used GIS data of the Costa Rican coffee institute (ICAFE) (Source: International Center for Tropical Agriculture, CIAT, A.Eitzinger@CGIAR.ORG, 2010)

Many protected areas are located in mountain chains, where they house important natural resources such as biodiversity, water, carbon, etc. The areas just below the protected mountainous areas are often designated for coffee, and if grown with shade, these areas serve as natural buffers around the protected areas. The map and table display the fact that coffee and protected areas jointly form important biological corridors (Table 7). We chose to examine the percent of protected area within 10 km and 50 km distances from coffee area, since organisms like birds, bats, and bees in tropical habitats disperse across short and long distances (Dick et al. 2008). In El Salvador, 72% of the protected areas are within a 10 km radius of all coffee growing areas, whereas in Costa Rica it is 32%, and in other Mesoamerican countries less than 15%. In El Salvador, 100% of the protected areas are within a 50 km radius of all coffee growing areas, in Costa Rica 84%, and in remaining countries less than 40%.

Table 7 Calculated protected and coffee areas by country

Country	Coffee areas (ha)	Protected areas (ha)	Coffee and protected area overlapping (ha)	Protected areas within 10 km distance (ha)	Percent of protected areas within 10 km of coffee area (%)	Protected areas within 50 km distance (ha)	Percent of protected areas within 50 km of coffee area (%)
Costa Rica	674,960	1,495,944	58,037	477,326	32	1,253,415	84
Guatemala	1,015,706	3,236,582	95,976	289,280	9	1,162,055	36
Honduras	1,304,765	1,272,725	32,678	146,450	12	485,793	38
Mexico	14,638,625	15,538,540	1,199,191	1,790,808	12	2,816,498	18
Nicaragua	777,004	2,202,118	70,930	145,384	7	495,950	23
El Salvador	312,689	58,650	12,188	42,261	72	58,650	100

4.5   Global Scale Ecological Processes and Ecosystem Services

At the global level, shaded agroforestry systems may be large contributors to the regulating services of carbon sequestration and climate change mitigation. Tropical deforestation and the use of fire in agricultural areas are leading contributors to increases in atmospheric CO_2 concentrations (Canadell and Raupach 2008; IPCC 2007). However, agroforestry systems, such as shaded coffee, have received attention for their potential to store and sequester relatively high levels of carbon (Canadell and Raupach 2008; Roncal-Garcia et al. 2008; Brown 1996). Soto-Pinto et al. (2010) examined the capacity of several shaded coffee systems, maize systems, and pastures to store carbon in Chiapas, Mexico. They found that *Inga*-shaded organic coffee maintains carbon in the soil organic matter to an equal extent as nearby forests, and that less intensive shaded plantations (organic and non-organic traditional polycultures) maintained more carbon than other land-use types examined. They suggest that these multi-strata coffee agroforests thus make important contributions for reducing emissions by deforestation and degradation (REDD) (Soto-Pinto et al. 2010).

In Brazil, Palm et al. (2005) found that simple shaded coffee systems (1–3 tree species) sequestered an additional 55 t of carbon per hectare in above ground biomass than in unshaded coffee monocultures. In Togo, shaded coffee plantations sequestered 53 additional tons of carbon in above ground biomass compared with an unshaded plantation (Dossa et al. 2008). However, it is important to take into account that intensively managed plantations, which use heavy applications of synthetic fertilizer, release N_2O, another greenhouse gas. This release would decrease the total contribution to climate mitigation from these systems. A recent study comparing N_2O emissions from heavily fertilized unshaded and shaded monocultures in Costa Rica found that shaded plantations released higher levels from having overall higher N from litter and N-fixation by *Inga* shade trees (Hergoualc'h et al. 2008). Thus, the management of fertilization and selection of species in shade coffee plantations will influence the level of climate mitigation provided by these systems.

At a global scale, the climate regulating services provided by shade coffee may become increasingly important as the planet faces more extreme weather events in the face of global climate change. Currently, climatological models predict general drying in parts of the Caribbean and Central America, coupled with stronger and later-season hurricanes (Neelin et al. 2006; Webster et al. 2005). Heavy rain and driving winds can wreak havoc during flowering and fruit bearing periods, the timing of which is coincident with hurricane season. Furthermore, much of the world's coffee-growing regions set fruit in April or May and fruit ripens anywhere from late August through November. If late season extreme-climate events, such as hurricanes, increase in frequency with global climate change, the existence of shade cover will be of even greater importance to buffer these events and thus sustain livelihoods and preserve ecosystem services in the face of global change.

The most representative Global Circulation Models (GCM) of the Fourth Assessment Report (AR4) for the Special Reports on Emission Scenarios (SRES)

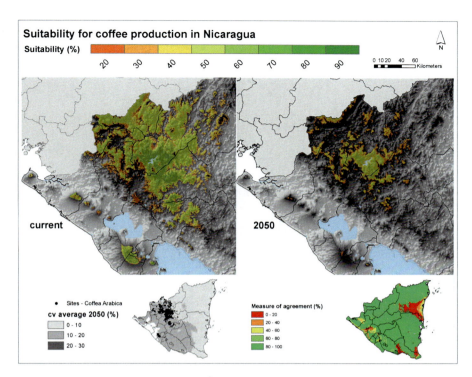

Fig. 7 Predicted (according to MAXENT) suitability for coffee production in the Nicaragua coffee-producing areas today and in 2050 (*large maps*) and the coefficient of variation (*CV*) and Measurement of Agreement for the study area with the points representing the sampled *Coffea arabica* farms (*small map*) (Modified from Läderach et al. (2010))

A2a (business as usual) emission scenario draws a trend of decreasing precipitation and increasing temperature for coffee-producing regions in Nicaragua (Läderach et al. 2010). The results of MAXENT (Phillips et al. 2006), a crop prediction model, indicates an important decrease in the suitability of coffee-producing areas in Nicaragua by 2050 (Fig. 7). There is a general pattern of decrease in the area suitable for coffee and a decrease in suitability within these areas. Suitability for coffee will move upwards on the altitudinal gradient with climate change, with lower-altitude areas having low to no suitability for coffee growing. The areas in 2050 that will still be moderately (40–60%) suitable for coffee production are mainly areas that currently show particularly high (>70%) suitability.

The optimum coffee-producing zone in Nicaragua is currently at an altitude between 800 and 1,400 m above sea level (masl); by 2050 the optimum elevation will increase to between 1,200 and 1,600 masl. Between today and 2050, areas at altitudes between 500 and 1,500 masl will suffer the greatest decrease in suitability and the areas above 1,500 masl the greatest increase in suitability. As the suitable altitude increases, less and less land area will be available at mid-elevation for coffee growing regions, like those in Nicaragua (Fig. 8, green line labeled Area).

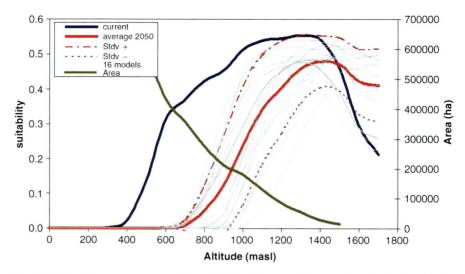

Fig. 8 Relation between current and future (2050) coffee suitability and altitude of coffee (*Coffea arabica*). 'Current' refers to current suitability, 'Average 2050' is the average suitability predicted in 2050, 'Stdv +/–' is the Standard Deviation, '16 models' is the average for 16 GCM, and 'Area' is the area available at each altitude (Modified from Läderach et al. 2010)

The first step in adaptation is to reduce the vulnerability of coffee farmers to climate change. In this regard, use of technical "no regret" measures that strengthen the resilience of the system (e.g. sound agronomy, sustainable management of natural resources) will be beneficial to growers and their livelihoods and may as well minimize the effects of climate change. In areas that will become unsuitable for growing coffee, farmers will need to identify alternative crops. In areas that will remain suitable for coffee, but with some reductions in suitability, agronomic management might be adapted to buffer the impacts of climate change. Drought resistant varieties, irrigation, and shade cover are all useful practices that can be implemented; shade cover can decrease average temperatures by up to 4°C (Vaast et al. 2006).

Areas where coffee is not grown today, but which in the future will become suitable for coffee, need strategic investments to develop coffee production. Account needs to be taken of environmental viability, since higher altitudes are often forest reserves that provide environmental services to the lowland population and to agriculture. The shift in altitude will definitely increase the pressure on land at higher altitudes. In regions that may be forced to abandon coffee, existing supply-chain actors need to think carefully about what their role in this transition may be. There are substantial investments in coffee processing and drying facilities, but it might be possible to use some of these facilities for other, non-coffee crops that are better adapted to projected future climates. In addition to physical infrastructure, many coffee-growing regions boast a highly qualified and specialized group of business services focused on coffee. If they continue to specialize on coffee, they will need to adapt and move to other regions, or if they choose not to move, they will need to begin to work on other crops. This combination of physical and human capacity is a current strength of coffee-growing areas and may well be leveraged to help identify and promote a planned transition to other income sources.

5   Interacting Ecosystem Services and the Socio-Economic Costs and Benefits of Shade Coffee

Farmers cite increases in coffee yields as the main reason for removing shade trees and native vegetation (Staver et al. 2001), but the ecological evidence on the relationship between shade and yield is far from clear. Some studies have demonstrated declines in yield with higher shade cover (Lagemann and Heuveldop 1983; Nolasco 1985), while some have documented increases (Ramírez 1993; ICAFE 1989). Other empirical studies have reported the highest coffee yield at intermediate (approximately 35–50%) canopy cover (Perfecto et al. 2005; Soto-Pinto et al. 2000; Muschler 1997). Because so many factors affect coffee yields, including soil conditions, elevation, precipitation, inputs, coffee variety, and shade, it has been very difficult to make clear statements about the relationship between shade, per se, and yield or even to compare across studies with more quantitative methods (e.g. meta-analysis) (Perfecto et al. 2005). Nonetheless, reviews have demonstrated that increases in shade tree diversity do not directly affect coffee yields (Peeters et al. 2003; Romero-Alvarado et al. 2002), and thus biodiversity and its associated provisioning ecosystem services may be easily promoted by increasing shade tree diversity, if not density, within coffee agroecosystems (Jha and Vandermeer 2010).

As discussed, a number of empirical studies show that plant and animal diversity within shade coffee systems provide pest control, pollination, and erosion control services; however, the potential economic benefit of these services often remains obscure to farmers because producers are infrequently directly rewarded for these services (Giovannucci 2003). In response to this lack of information, a few recent studies have quantified the economic value of ecosystem services in coffee agroecosystems. For example, a study on coffee pollination by native bees in Costa Rica calculated that coffee plants located within 1 km of a Costa Rican forest fragment had increased yields (>20% higher), an amount that totaled $62,000 of added income for the farm studied (Ricketts et al. 2004). This represents substantial benefits to farmers and highlights the importance of maintaining forest fragments in agricultural landscapes, even if small. An additional study conducted in the Blue Mountains of Jamaica documented that pest control services provided by birds to combat the coffee berry borer improved yields between 1% and 14% (Kellermann et al. 2008). In economic terms, this amounted to >$4,000 for farmers of the four small farms investigated, or between 2% and 69% of the per capita gross national income for Jamaica for each farm (Kellermann et al. 2008).

But coffee yields are not the only provisioning service provided by shade coffee systems. One often overlooked factor in assessing relationships between coffee yields and farm revenue is that shaded coffee farms with a diverse assemblage of trees often provide other sources of income to farmers. Shaded coffee farms include additional revenue from timber and non-timber products from the shade trees (Somarriba et al. 2004). In Peru, shade tree products may account for ~30% of revenues for each farm, especially fruits and firewood rather than timber (R. Rice, unpublished data 2002). Escalante et al. (1987) found that fruits from the shade canopy accounted for 55–60% of income, and timber for 3%. In Costa Rica, fruit sales accounted for 5–11% of income from coffee growing areas (Lagemann and Heuveldop 1983). Available products from the shade tree canopy reduce vulnerability to market fluctuations and household dependence on outside products while increasing local commerce. Thus, product diversification can reduce the need to exploit nearby forests. Perhaps most importantly, shade tree canopy products can buffer farmers in tough financial periods, especially when coffee prices are very low (Escalante et al. 1987).

Despite a basic understanding of the independent ecosystem services acquired with shade coffee farms, very little research has examined how ecosystem services may interact. One review, conducted across a number of modern agricultural systems, revealed that most often, the only ecosystem service considered is the production of the marketed commodity, with little thought to regulating services such as water and air filtration, disease suppression, and wildlife habitat (Robertson and Swinton 2005). Recent work has also pointed to the need to consider the multiple ecosystem services present in a particular area in order to promote synergistic services and avoid tradeoffs that may enhance one service at the expense of another (Bennett et al. 2009; Robertson and Swinton 2005). For example, within the shade coffee system, practices used to enhance one regulating service, such as planting fast growing tree species for carbon sequestration, may impact other services, such as the provisioning services provided by the coffee crop, or the regulating service of pollination derived from supporting bees dependent on diverse shade tree canopy. What is needed is an orientation towards understanding the full agro-ecological system and the many ecosystem services provided within it, which will provide a better understanding of how these services are coupled and what potential trade-offs may exist (Robertson and Swinton 2005; Robertson et al. 2004).

Recent research (Raudsepp-Hearne et al. 2010) proposes visualizing ecosystem service 'bundles' that allow for an examination of the different types of ecosystem services, and how each service within the bundle is enhanced or reduced as a result of management interventions. These interactions can be expected to differ depending on the type of shade coffee systems, geographical location and socio-economic context. Méndez et al. (2009) found that a higher density and diversity of shade trees resulted in small-scale, individual farms having a higher potential for provisioning services (e.g. timber, fruit and firewood) than larger, collectively managed cooperatives. However, additional shade tree products came at the expense of lower coffee yields, showing a negative interaction between two different types of provisioning services (Mendez et al. 2009). In addition, these differences in provisioning services did not significantly affect regulating services in the form of above ground C stocks from the shade tree canopy. Henry et al. (2009) conducted an in-depth examination of the interactions between plant biodiversity and regulating (C sequestration) and provisioning (food production) ecosystem services in smallholder farms of Kenya. Although the study only included a few coffee plots, the results showed that plant biodiversity had no effects on C stocks, but that increasing C sequestration by adding more trees would have a negative effect on food production. Similar interactions might be seen in smallholder coffee households that manage different types of agricultural crops in addition to coffee. Future research needs to focus on the trade-offs involved with interacting ecosystem services and the optimal strategies for long-term ecosystem service provision and conservation across multiple shade coffee landscapes.

6   Farmer Livelihoods, Vulnerability and Change

6.1   Sustainable Livelihoods

A livelihoods-based approach seeks an integrated assessment of the way that individuals and households access and use a diversity of assets to "make a living and make it meaningful" (Bebbington 2000). We selected this focus because it links the economic elements of "making a living" – including food security, monetary incomes, and barter – with the cultural dimensions of making it meaningful. Scoones (1998) elaborated a working definition, stating that "A livelihood comprises the capabilities, assets (including both material and social resources) and activities required for a means of living." Planners conducting livelihoods assessments consider the social assets (i.e., participation in a cooperative or other local association, networks of friends and family etc.), natural assets (i.e., the land, water and micro-climates that a household could potentially use), financial assets (i.e., loans and savings), physical assets (i.e., houses and equipment) and potentially many other assets including those related to cultural memory, shared experiences, and local knowledge, as well as human capabilities that are embedded in the relationships that households use to articulate their livelihood projects (Scoones 2009, 1998; Bebbington 1999) (Fig. 9).

Although several scholars initially limited the discussion of livelihoods to a categorization of these different assets, sometimes referred to as the five capitals (social, natural, physical, human and built capital), many community-based researchers and those interested in deeper theoretical work related to development and sustainability were keen to also address contextual variables (Bebbington 1999, 2000). This includes the multi-scale political, economic, and ecological structures and processes that influence the construction or depletion of assets and that intercede in a household or individual's ability to access (Ribot and Peluso 2003) the benefit flows at a particular moment in time (Scoones 2009). The vulnerability context also influences several broader trends, including seasonality (i.e., dry vs. rainy season, and/or particularly cold, hot or wet year), the presence of sudden shocks to a livelihood

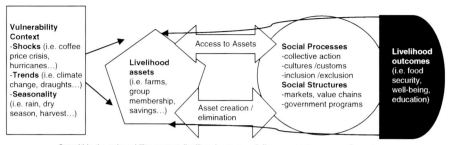

Set within the vulnerability context, livelihood outcomes influence asset access and asset building/diminishing practices and social processes in iterative cycles of continuity and change.

Fig. 9 Livelihoods change framework for coffee smallholders (Modified from Amekawaa et al. 2010)

system (e.g. a flood, hurricane, drought, market or political collapse) and other, ongoing stressors of daily life.

To discuss vulnerability in the context of this review, we draw from a political ecology approach that uncovers the social processes (e.g. economic poverty, exclusion, poor land use planning) and environmental degradation that can transform an external shock or low level stressor (seasonal drought, decreasing real income, or climate change) into a disaster with measurable social and ecological impact (Wisner et al. 2004). An important focus also concerns the different strategies that households use to cope with shocks. Some strategies, such as certain types of farm and livelihood diversification can increase intermediate-term sustainability, while others, such as pulling children out of school to work, can contribute to persistent poverty (Devaux et al. 2009). Households with more sustainable livelihoods are able to cope with and recover from shocks, maintain or enhance their capabilities and assets, and provide more sustainable livelihood opportunities for the next generation (Chambers 1992, 1991). Less vulnerable livelihoods have lower exposure to shocks and stressors and are more capable of mitigating the consequences of the hazards that do affect them.

6.2 Types of Livelihoods

The diversity of coffee-dependent livelihoods ranges from the part time barista making espressos in northern cities to the seasonally employed coffee pickers balancing upon the steep slopes of southern mountain landscapes. Coffee livelihoods also include traders, export managers and farmers, among others. On both ends of the value chain these part-time coffee workers often face structural difficulties. In the United States, most baristas do not have access to basic health care and can often make salaries that are close to the minimum wage. In the coffee growing regions, coffee pickers are often migratory laborers that exist as marginalized members of society (frequently with indigenous origins) living in some of the world's most economically poor countries (Oxfam 2002). Paid by the pound of coffee cherries harvested, these pickers may earn as little as $2–$10/day (Oxfam 2002).

Coffee pickers and rural coffee laborers (who may also do the pruning, fertilizing, and weeding) were among the most vulnerable to the post 1999 coffee crisis (Bacon et al. 2008b; CEPAL 2002). Many did not have access to key livelihoods assets, most of which come from owning land, and were also cut off from access to collectively managed forests or range land. The direct economic impact as larger farms in Central America (over 50 ha) stopped maintaining and often harvesting their farms included the loss of more than 40 million days of work (CEPAL 2002). Although personal observation suggests the importance of coffee shade fruit trees in provisioning workers with food and a cooler environment during the coffee harvest, we are not aware of any systematic studies evaluating the effects of shade vs. sun coffee upon coffee farm worker livelihoods.

A rapidly growing literature concerns the multiple dimensions of coffee farmer livelihoods. Three studies involving farmers in Mexico, Nicaragua, El Salvador and Guatemala found that coffee remains the most important income source and a core component of their livelihood strategies (Méndez et al. 2010a; Bacon et al. 2008a; Jaffee 2007). Previous and ongoing livelihood-oriented studies that assessed the relationships between indigenous identities and organic coffee production have yielded contradictory results. This research shows a positive initial fit (Nigh 1997) and tensions between community life and the increasingly high expectations and reporting requirements associated with the annual organic inspections (Mutersbaugh 2004). Others have documented the gender relations and issues of inequality, exclusion and empowerment among coffee farmers and within smallholder cooperatives (Lyon et al. 2010; Hanson and Terstappen 2009). The research findings to date reveal persistent inequalities as women are continually marginalized and certifications, such as Fair Trade, have not yet delivered on their gender-related goals. Finally, several studies include a comparative analysis addressing several livelihood outcomes, such as food insecurity, education, incomes, and vulnerability among farmers connected to certified networks vs. those selling only to conventional coffee value chains (Méndez et al. 2010a; Arnould et al. 2009).

An overwhelming result of these studies reveals persistent livelihood difficulties among most small-scale coffee farmers in Mesoamerica (Méndez et al. 2010b; Bacon et al. 2008b; Jaffee 2007). Jaffee's insightful book discussed the seasonal, 'hungry' or 'thin months', communicating what many Mexican rural development planners and researchers have long known (Jaffee 2007). Another study involving 469 households in Mesoamerica found that 63% of those interviewed reported that they struggled to meet their basic food needs (Méndez et al. 2010a). A comparative study involving 177 households in northern Nicaragua also documented similar trends and revealed that the average households, including those connected to Fair Trade and organic markets, generated less than a dollar a day per person from their coffee production (Bacon et al. 2008b).

6.3 Do Sustainable Coffee Certifications Improve Farmer Livelihoods?

The post-1999 coffee crisis provided a dramatic natural experiment or common treatment to study the effects on farmers of participation in cooperatives and different certified coffee networks. Two large quantitative studies in Latin America showed that Fair Trade and organic certifications are able to provide some benefits to smallholder farmers, but that these remain small in terms of a broader livelihood perspective (Méndez et al. 2010a; Arnould et al. 2009). Certifications were able to provide higher prices, but since the volumes sold were relatively low, this did not amount to significant increases in income. No effects were observed in terms of improving access to food through purchasing or production, which is one of the persistent challenges of smallholder and cooperative coffee farmers. However, some of the benefits reported included improvements in access to health, credit, and savings. In addition, farmers reported links to international development networks as an important benefit that has the potential to support farmers when combined with other development or environmental support (Méndez et al. 2010a). Although there are nuances, local exceptions and occasionally differences in methodology and interpretation, a summary of the available evidence related to the studies and observations included in this review suggest the following findings concerning the relationships of coffee smallholder livelihoods, certifications, and global markets:

1. The livelihood conditions among smallholders are generally difficult and suffered severely during the post 1999 coffee crisis (Méndez et al. 2010b; Arnould et al. 2009; Jaffee 2007; Bacon et al. 2005). Few studies have been published with data emerging after 2006 when green coffee commodity prices started to increase. The available studies, personal observations from travel to coffee growing regions, interviews, and conferences, as well as preliminary findings from works in progress involving this review's authors suggest that while the more pinching dimensions of the post 1999 coffee price crash such as the humanitarian crisis, broad-based job losses and abandoned coffee farmers have decreased, seasonal hunger, marginalization and vulnerabilities persist (Peyser 2010; Renard 2010).
2. Participation in cooperatives connected to Fair Trade, often partially mitigates exposure and thus livelihood vulnerability to falling coffee commodity prices and – for those that can access the market – it could potentially offer support through international development networks to diminish the negative consequences of other changes to the vulnerability context, such as food shortages, hurricanes, and earthquakes (Jaffee 2007; Raynolds et al. 2007; Bacon 2005a).
3. Farmers affiliated with these cooperatives often have more access to credit and in selected cases are more likely to practice sustainable land management practices (i.e., soil and water conservation practices on the farm, avoidance of pesticides) than their conventional counterparts (Méndez et al. 2010a).
4. Empirical realities, including the persistence of hunger and ongoing gender inequalities and uneven development within coffee growing regions, contradict

the enthusiastic publicity associated with many certifications, including Fair Trade, Rainforest Alliance and Utz Certified (Lyon et al. 2010; Bacon et al. 2008a; Lyon 2008; Jaffee 2007; Fridell 2006; Mutersbaugh 2004).
5. Coffee cooperatives can be effective local organizations for coordinating collective action and have enabled hundreds of thousands of smallholders to retain the title and use of their lands. Those that have developed administrative capacity and accountability to their membership, as well as external partners can also provide valuable technical assistance, leverage international development funding to improve coffee yields and quality, and support a wide array of social development and diversification projects (Raynolds et al. 2007). Examples can be found among several of the pioneer cooperatives in Nicaragua, such as SOPPEXCCA and PRODECOOP, as well as the stronger cooperatives in Mexico and Peru (i.e., CIPECAFE in Peru and CESMACH in Chiapas, Mexico).

6.4   Diversification Within Coffee Production Systems

Coffee production systems and the landscapes into which they are embedded often include other cropping systems. The majority of smallholder coffee farmers are peasant producers farming for subsistence. With few resources other than their own labor and a small plot of land, their "coffee farms" are much more than that. For example, farmers in Mexico and Central America often also cultivate corn-bean-squash systems and manage pastures (Méndez et al. 2010a; Philpott et al. 2007); coffee smallholders in Brazil farm sweet potato, sugar cane, black pepper and various fruit crops (Steward 2007); and farmers in Indonesia normally also cultivate rice and perform aquaculture (Waltert et al. 2005). Within the coffee plots themselves, farmers worldwide often incorporate a high number of different plants, including fruits (e.g. orange, banana, mango, avocado, durian), nuts (e.g. candlenut), wood products for timber or firewood, and additional export crops (black pepper, cinnamon, cloves) (Philpott et al. 2007, 2008b; Rice 2008; Méndez et al. 2007; Michon et al. 1986). In Peru, growers in the Apurimac/Ene river valley make use of up to 13 different species of bananas (*Musa* spp.) alone.

A comparative study conducted in El Salvador and Nicaragua found that households growing shade coffee managed at least four distinct types of plant functional types, including shade trees, agricultural crops, medicinal plants and epiphytes (Table 8) (Méndez et al. 2010b). Plant agrobiodiversity was found in four locations, including shade coffee plantations, homegardens, agricultural plots, and living fences. Shade trees, medicinal plants, and epiphytes were found in several locations, while crops were only found in agricultural plots. Trees were the most species-rich group, with a total of 123 and 106 species in El Salvador and Nicaragua, respectively. Diversity of agricultural crops was similar in both countries, but differences were observed in the types of crops grown and the number of varieties. Nicaraguan households managed thirteen varieties of corn and nine varieties of beans, a higher figure than what was found in El Salvador.

Table 8 Additional crops grown by households cultivating shade coffee in Nicaragua and El Salvador (Modified from Méndez et al. (2010b))

Agrobiodiversity type	Growth habit	No. of species	Uses reported	Value reported by farmers
El Salvador				
Trees	Woody perennial	123	S, FW, Fr, M, T	Firewood obtained from shade trees saved households an average of $71.50 per year in 2002.
Agricultural crops	Herbaceous	7	F, M	62% of the sample (n = 18) reported producing at least 40% of the food used by the family in 1 year
Medicinal plants	Woody perennial, shrubs, herbaceous	119	F, M, FW	Medicinal plants are valued because farmers cannot afford modern medicines or health care.
Nicaragua				
Trees	Woody perennial	106	S, FW, Fr, M, T	Farmers reported an average of $167 per year from firewood sales, in addition to covering their own firewood needs.
Agricultural crops	Herbaceous	7	F, M	Average of 50% of food is produced in these fields
Orchids	Primarily epiphytes	96	O	Aesthetic and ornamental

Uses reported: F = Food; Fr = Fruit; FW = Firewood; M = Medicinal; S = Shade; T = Timber; O = Ornamental

The Nicaraguan coffee farmers also had a higher number of coffee varieties (eight), compared to their El Salvador counterparts (two). Medicinal plants, which were only found in El Salvador, contained a high diversity of species and growth habits (119 species of trees, shrubs, and herbs).

6.5   Farm Size Cooperatives, Livelihoods and Shade

The different livelihood activities of coffee farmers can have implications for the design and management practices of their shade coffee. These practices can in turn influence the associated biodiversity and ecosystem services of a particular farm or landscape. Guadarrama-Zugasti (2008) compared management practices related to agrichemical use between small-scale and large farms. He found that small-scale producers were using lower levels of synthetic pesticides and fertilizers per farm, which resulted in fewer soil and water contamination problems than those observed in larger farms. He then used several indicators to develop a farmer typology, identifying at least eight different production strategies, including 'coffee/corn farmer', 'agricultural worker/coffee farmer', and 'hobby coffee farmer'. These different types of farmers were then associated with varying intensities of agrochemical use, soil erosion, and incomes. The results showed that small-scale farmers that were using practices with low environmental impact were slowly transforming to more intensified management, such as used by larger growers. Methodologies such as this one could prove useful to re-think common shade coffee system classifications and to provide an interdisciplinary synthesis that identifies the most effective interventions. The typologies developed by Moguel and Toledo (1999), which describe shade tree canopies and management regimes, are useful to characterize the biophysical structure of agroecosystems, but may not adequately describe livelihood strategies. The heterogeneity of farmer livelihood strategies is often overlooked (Shulman and Garret 1990) and few studies have connected farmer types with issues of sustainability and technological change (Guadarrama-Zugasti 2008).

Farmer cooperatives have been instrumental for smallholders to negotiate coffee farming and commercialization at different scales. In Nicaragua, strong cooperative unions have been able to become national leaders in coffee production and commercialization, while also embracing environmentally friendly production (Bacon 2005b, 2010). Research in El Salvador and Nicaragua also has found that origin, type, and governance of coffee farmer cooperatives can have a direct effect on management practices and the resulting levels of biodiversity and ecosystem services provided by plantations (Méndez et al. 2009, 2010b). These studies showed that individually managed farms that belonged to farmer associations contained higher levels of shade tree species diversity and uses (i.e., fruit, firewood, timber) than plantations where cooperatives used centralized collective management arrangements.

7   Discussion: Synthesis and Policy Directions

7.1   Need for an Integrated Multi-scaled Interdisciplinary Framework

While local and regional coffee landscapes have broad impacts on both ecosystem services and farmer/worker livelihoods, it can be challenging to simultaneously analyze both impacts within a single coffee value chain. This is partially due to the fact that investigation of each of these areas requires a distinct methodological approach, and that the high number of transactions between farmer/worker and consumer make it difficult to relate coffee revenue to worker livelihoods (for details see Sect. 5). Additionally, although many ecosystem services within agricultural landscapes are appreciated by people, they currently have limited market opportunities, and remain largely unrewarded (Swinton et al. 2007; Robertson and Swinton 2005). Because provisioning services, such as farm products, have market values, these services take precedence over recreational, supporting, and regulating services. Many studies suggest that in order for policy to consider ecosystem services without market value, an alternative valuation technique must be employed (Swinton et al. 2007; Robertson and Swinton 2005).

Within the shade coffee system, there are only a few examples where ecosystem service value has been estimated. One can be seen in the work of Philpott et al. (2008c) who examined both erosion control services and economic yields from a single coffee growing region in Chiapas, Mexico. As described in Sect. 3.4, by examining both economic and ecological data, the study revealed that road conditions, not coffee yields, were most negatively affected by hurricane damage. Thus, the erosion control provided by dense and diverse shade trees had a direct positive economic impact on coffee transport infrastructure, rather than on coffee yield. A number of other studies have also taken the initial step of quantifying the socio-economic gains procured by ecosystem services within coffee farms (described in Sect. 3.6). For example, researchers have revealed that local forest patches increase the pollinating activity in coffee farms, a service calculated to be worth $128.6 USD/ha per year in a Costa Rican farm and $1,860 USD/ha in a Brazilian farm (De Marco and Coehlo 2004). Kellerman et al. (2008) similarly quantified the value of bird-mediated pest control to be worth $44–$105/ha USD.

In many of these examples, however, the research does not specifically discuss the actual economic impacts of ecosystem services on farmer/worker livelihoods. As the science currently stands, we know little about the direct and changing impacts that ecosystem services have on worker/farmer livelihoods (i.e., how people make a living and how they make it meaningful). We suggest that future research utilize a multi-scalar approach to examine both livelihoods and interacting ecosystem services within shade coffee landscapes. For example, the value of biodiverse shade coffee farms is visible not only in coffee yields but also in the contribution to regional and global water conservation and carbon sequestration; however, the costs/benefits of the entire ecological and socio-economic system are not often simultaneously understood. In order to improve farmer livelihoods and promote long-term sustainability in shade coffee landscapes, we need to define goals for these regions (McAffee and Shapiro 2010) and work towards a long-term vision where both livelihoods and ecological sustainability are taken into consideration.

7.2 Biologically Rich Lands, Economically Impoverished People

The evidence from many coffee landscapes confirms the persistent paradox of 'rich lands and poor people' (Peluso 1994). In other words, coffee growing regions often have very nutrient-rich soils, high biodiversity, and ideal climates for crop cultivation, yet the incomes generated from these landscapes are minute. Farm families that sustain some of the most vital ecosystem services in fragile mountain landscapes, such as regional water storage and carbon sequestration, also are among the most socially marginalized and economically impoverished (Bacon et al. 2008b). This paradox reveals the way that the dominant state-backed development models and current configuration of coffee value chains have undervalued and poorly compensated farmers, farm workers and shade coffee landscapes (Trujillo 2008). The empirical evidence shows that millions of coffee farmers continue to struggle for survival despite the major contributions made to producing high quality coffees and generating ecosystem services (Bacon et al. 2008b; Jaffee 2007).

Although resistance and alternative approaches prevail in many places, including many smallholder coffee landscapes, such as the Zapatistas in Chiapas (Watts 2007; Fox 1994), a raft of Neoliberal policies remains the dominant trend in many coffee growing communities (Topik et al. 2010). As defined in Sect. 3.7 of this chapter, the Neoliberal influence is visible through the passage of free trade agreements, the collapse of the International Coffee Agreement (that governed the markets from 1962 to 1989), the rollback of state invested agriculture and rural development, and the fraying of social safety nets (Talbot 2004). Although the dissolution of bureaucratic, frequently ineffective and occasionally corrupted state-backed marketing boards and coffee extension agencies opened the spaces for rapid growth of market-based sustainability certifications, both organic and fair trade systems originated in social movements outside and prior to this Neoliberal shift in coffee market governance, and – at least initially – represented alternative approaches to agricultural production, trade, and consumption (Bacon 2010).

The empirical evidence shows direct and indirect benefits associated with the rise of sustainable coffee certifications (Méndez et al. 2010a; Jaffee 2007; Philpott et al. 2007; Bacon 2005a), but persistent hunger and livelihood insecurities remain the dominant trend. Thus far, much of the biodiversity and many of the valuable ecosystem services have persisted at both farm and regional scales (e.g. Perfecto et al. 2007). However, without a change in the predominant coffee value chain there is reason to believe that the vulnerability of the livelihoods of shade coffee farmers will continue to increase. Accelerating climate change, volatile markets, and inequalities in the coffee value chain (among others) could potentially overwhelm the local resiliency that these systems have demonstrated the past half century. The challenge before us is to fundamentally re-think the current approach. This includes a re-orientation of the strategic and technological approaches with the participation of a wide diversity of stakeholders prior to the selection of the most promising policy directions. This process could generate investments and actions that are commensurate with effective action at multiple scales.

7.3   Political Possibilities and Policy Options

Business as usual will likely lead to the continued decline of many diverse shade coffee production systems in the Americas, resulting in social and physical landscape transformations that the regions cannot afford. While non-governmental and private concerns have made inroads and some headway in promoting the shade coffee concept, it could be argued that governments need to act upon the opportunity before them. It is worth noting, for instance, that the United Nations' Millennium Development includes one assessment addressing poverty and another focusing on environmental sustainability (Goals 1 and 7, respectively). Meeting these goals with the support of governments and the international community in coffee producing countries could potentially turn the tide to favor the conservation and enhancement of shade coffee livelihoods and landscapes. For many producing countries facing rural poverty and biodiversity threats, the increasing support of shade coffee consumption is an opportunity to boost rural income and support native biodiversity, especially when compared to the unsustainable alternatives of deforestation and pasture establishment. Our review of politics and policies that influence the future of shade coffee is guided by three normative goals: (1) improved rural livelihoods, (2) cultural survival, and (3) the conservation and recovery of vital ecosystem services. A critical starting point are the current practices, knowledge systems and organizations (often cooperatives and ejidos, but sometimes NGOs, coffee roasters, state agencies and others) that have helped sustain important ecosystem services and culturally diverse coffee farmers. While the current configurations of coffee value chains and government regulation also are a necessary starting point, this need not constrain the horizons of political possibility. In fact, the coffee industry is full of innovative examples and partnerships geared towards the improvement of livelihoods, the strengthening of local organizations and the conservation of ecosystem services (Linton 2005). Two recent examples include farmer exchanges that were undertaken with the support of coffee industries. First, farmers from Nicaragua visited Peru to learn best practices for organic compost making, and in the second case, farmers and cooperative leaders from Rwanda were able to visit Nicaragua to learn about the strategies used to build strong smallholder cooperatives and improve coffee quality (Bacon et al. 2008b). However, these specific examples could be part of a broader proposal to involve a wider range of farms and farm workers. The following sections touch upon several of the strategic themes to consider for all stakeholders interested in maintaining or promoting sustainable coffee production.

7.3.1 Ensure That 'Sustainability' Covers Basic Human Needs

Among international industry associations, many of which lobby for decreased regulation and dodge critical issues of social and environmental sustainability, the specialty coffee industry stands out for its efforts to promote sustainability through certifications, pledges to global development, and direct farm and community level investments (Dicum and Luttinger 1999). It has served as an effective platform for launching many mainstream initiatives from certified organic and Fair Trade products to social responsibility partnerships for education and environmental conservation in coffee growing communities.

Despite these industry, non-profit and governmental commitments to sustainability in the specialty coffee sector, the research shows that many smallholder farmers continue to negotiate hunger on an annual basis (Méndez et al. 2010b). Although the paradox of hungry farmers and farm workers is hardly monopolized by small-scale coffee producers, witness for example, the high levels of hunger and rural poverty that plague California's Central Valley, one of the world's most agriculturally productive landscapes (Harrison et al. 2002). Thus, evidence from both 'developed' and 'developing' countries suggests the need to re-focus efforts upon the basic needs and environmental justice / social equity dimensions of sustainability (Shiva 2008; Agyeman et al. 2003).

The Brundtland Report, which established one of the few commonly accepted definitions of sustainability, recognized the fundamental importance of food, water, housing, education and health especially among the most economically poor and marginalized (WCED 1987). "Sustainable development is development that meets the needs of the present without compromising the ability of future generations to meet their own needs." In other words, truly sustainable systems require an evaluation of farmer livelihoods, ecosystem services, and well-being. Existing sophisticated marketing schemes for socially just and ecologically beneficial coffee will soon begin to ring hollow if the basic needs of farmers and workers are not addressed.

Longer-term solutions to these challenges will address global scale processes such as the structure and governance of coffee value chains (Bacon 2010; Topik et al. 2010; Daviron and Ponte 2005), state investments (Talbot 2004), and climate changes (Lin et al. 2008) that are re-shaping the vulnerability context in which small-scale coffee growing communities must operate. The same global challenges continue to undermine the sustainability of global food systems across a wide diversity of foods and commodities (Perfecto et al. 2009; Watson and Herren 2009). The responses to these challenges will shape the future of food and agriculture with their profound corollary affect upon ecosystem services, rural livelihoods, and food security. With these global trends in mind, the following sections consider several strategies that could improve livelihoods and sustain ecosystems in coffee growing communities. Two effective strategies to address both hunger and rural livelihood vulnerability are through diversification and sustainable intensification.

7.3.2   Farm and Livelihood Diversification

For both economic and ecological security, the diversification of crops and livelihoods is essential for coffee producers (Rice 2008). As discussed in Sect. 5, maintaining a diverse array of crops provides farmers with (1) alternative income sources in case of crop losses, (2) income across the growing season, (3) reductions in pest pressure, and (4) food for home consumption. Diversification of shade trees can also help farmers garner ecosystem services (Méndez et al. 2009). As reviewed in Sect. 4, the planting and diversification of natural vegetation and shade trees provides fertilization, erosion control, and habitat for pollinators and pest-predators. Livelihood diversification could include on-farm and off-farm income sources, such as honeybee management, timber harvesting, construction, and the sale of crafts. This could be further enhanced with training, small-scale capital investment, and mentoring to innovate rural enterprises focused on processing agricultural products for storage and sale, accelerating communication, and in some cases, community-based rural tourism. Examples of incipient agro-ecotourism projects in shade coffee communities can be found in Central America. Starting in 2003, cooperative unions in Matagalpa, Nicaragua, launched an agro-ecotourism project with the support of researchers and the NGO Lutheran World Relief (Bacon 2005b). The project has received more than 1,200 visits from Fair Trade networks, foreign universities, and solidarity organizations. However, despite these accomplishments, the farmers face persistent challenges, including an insufficient number of visitors to cover the costs of the program, a factor that could be addressed with increased advertising (Méndez et al. 2010b).

Intensification efforts have focused on strategies to increase yields and decrease food loss from storage and crop loss from drought. One of the most effective, albeit costly strategies for increasing yields is through irrigation, especially in areas dependent on rain-fed agriculture. Second and third strategies are soil fertility improvement and selection and sharing of heirloom and local seed varieties (especially corn, beans, rice and other subsistence crops) that are locally desired and resistant to extreme weather and changing precipitation patterns (Méndez et al. 2010b). Multi-cropping with local seeds can be encouraged by funding of local seed banks and extension services aimed at subsistence crop cultivation. Many communities have long histories of local subsistence crop cultivation, but little has been recorded about the implementation of these practices. Civil society and local group involvement is necessary to resurrect these practices for a diversified farming system as evidenced through activities such as the non-profit support for the Mesoamerican farmer to farmer movement (Holt-Giménez 2006). Although civil society investment remains important, especially to develop innovative and pilot community-level initiatives that support diversification, intensification, afforestation and food security (Pretty 2002), the structural drivers affecting persistent hunger, the fraying rural safety net for health and educational opportunities, and broader scale investments may require a new type of state-led regulation and investment (Watson and Herren 2009; Bacon et al. 2008b).

7.3.3 Revive Strategic State Action

National, state and local governments together with the citizens and residents of coffee growing communities are fundamental stakeholders in sustainable community development in coffee growing regions. The state also remains a central participant in creating, coordinating and enforcing the political, economic and agri-environmental standards that structure important components of the coffee value chain (Bacon 2010; Talbot 2004). Since 1989, the role of national governments in directly influencing global coffee markets and prices paid to the producers (through the International Coffee Agreement) and organizing international marketing and production practices has decreased as most governments adopted Neoliberal approaches (Topik et al. 2010; Talbot 2004). In many cases, rural poverty rates have increased together with accelerating rates of environmental destruction (Heynen 2007). This is not to say that top down state control and a closed communist government offers an effective solution to sustainability challenges in these regions. Recent evidence reveals high levels of environmental contamination and social marginalization accumulated in many of the post-socialist states, such as Vietnam and Hungry, now in "transition" (O'Rourke 2004). However, the transition to a form of capitalism nearly devoid of state regulation and dominated by transnational firms and national elites taking advantage of the cheap privatization of state agencies and power vacuums to dramatically accelerate the rerates of natural resource exploitation maybe the worst of both worlds (O'Rourke 2004). When coupled with the evidence of persistent livelihood vulnerabilities in coffee growing regions, the conclusion is that states must regain their active roles in providing basic services to their populace and protecting ecosystem services if the negative results of a hands-off policy are to be curtailed and reversed.

Investment in the rural sector in ways that support sustainable coffee production is one way states could begin addressing multiple problems within their borders. Incentives directed toward farmers who maintain diverse shade within coffee plots would better assure the longevity of such management practices and allow producers to make a living while being good stewards of the land. Creation of regulations aimed at preserving biodiversity via agroforestry promotion could be a strong pillar in these efforts. Likewise, establishing and streamlining an infrastructure that supports farmers' efforts and brings in foreign exchange at the same time, would work to the benefit for both the state and its people.

Most national governments claim that jurisdiction and elaborate plans play a key role in contributing to longer term food security, rural education, and health care in coffee growing communities and elsewhere. For example, according to the FAO Special Rapporteur, "the right to food" is now alive in 24 constitutions as well as in different national policies, food security institutions, and courts (http://www.srfood. org/index.php/en/component/content/article/684-revising-the-cfa-five-proposals-for-the-dublin-consultation). This right has been embedded into the constitutions of many of the world's top 20 coffee exporting countries. More targeted investments include current programs such as Mexico's rural subsidy programs to assure school attendance and reduce vulnerability in rural areas, which now generate up to 20% of the income among coffee growing households (C. Bacon, personal communication with T. Barham). The current challenge may be one of financing these programs but also a deeper coordination connecting rural civil society, international development agencies and firms within the coffee value chain to create a more innovative and fairer partnership with coffee growing communities (Bacon et al. 2008b).

7.3.4 Improve Certification Systems

Given the existing coffee infrastructure, the most commonly employed method to 'integrate' ecosystem service acquisition and farmer/worker livelihoods is via farm-scale coffee certification. The ecological and socioeconomic benefits of certification (e.g. fair trade, organic, bird friendly, Sect. 4, Table 4) vary substantially between certification types, primarily because certification systems set different ecological standards, offer different economic incentives to different agents (directly to growers vs. to certification agencies), and differ in the price premium provided (Bacon et al. 2008a; Raynolds et al. 2007; Calo and Wise 2005). While organic and fair trade certification may raise coffee export prices (Bacon et al. 2008a), certification alone cannot provide incentives for optimal biodiversity conservation within coffee farms (Bacon et al. 2008a; Jaffee 2007; Philpott et al. 2007). Furthermore, organic certification alone often fails to cover the additional costs associated with certification and maintenance (Calo and Wise 2005). Fair trade premiums have yielded mixed effects, with some studies citing high returns (Calo and Wise 2005), while others demonstrate that fair trade premiums do not provide workers with higher wages or greater security than those working in uncertified coffee farms (Valkila and Nygren 2010). Finally, recent research has documented that when discounted for inflation, the real price premiums and minimum prices delivered to farmers by the leading sustainability certifications have declined during the past decades and with it the prospects for providing a strong incentive for more sustainable management (Bacon 2010).

Among the most important benefits of Fair Trade is the establishment of smallholder cooperatives that have gained a competitive foothold in export markets. In most cases, the colonial history of coffee has excluded these organizations from direct access. If these cooperatives are accountable to their members, business partners, and development agencies, they can emerge as a vital defense assuring smallholder access to land and advancing local development. Thus far, many smallholder cooperatives have been successful at establishing slightly better pay for their members and more secure markets for their coffee (Bacon et al. 2008a; Jaffee 2007). However, creating and maintaining these cooperatives so that they are accountable to their membership is a collective action challenge. Meeting this challenge involves a combination of community organization, support from state agencies (Fox 1996), and, in the case of certifications, non-profit investment in order to meet standards and improve the capacity of the certification industry (Bacon et al. 2008a). Thus, state, universities, certification agencies and socially responsible coffee firms could be involved with farmers to contribute to building alternative cooperative models and participatory certification initiatives that more effectively deliver benefits to both coffee drinkers and coffee-growing regions (Jaffe and Bacon 2008).

Another challenge to the existing certification system is that only a few certifications are currently available, thus farms that provide substantial ecosystem services, but do not qualify for the specifics of existing certifications, are left out. Without drastically changing the certification system, a number of changes could be made to make the process more effective. First, costs to farmers of inspection and certification are too high, especially within the Fair Trade system (e.g. Philpott et al. 2007). This is partially due to the monopoly held by FLO-CERT, which is the only Fair Trade certification agency in the world. Unfortunately, this semi-independent agency is plagued by poor management. Fair Trade retail sales have now topped 3 billion dollars, easily covering its operating costs; yet it continues to increase fees charged to producers.

If other Fair Trade certifying agencies were allowed into the market, all certifiers would have to compete in order to provide certification for producers, forcing them to charge producers more reasonable fees. Additionally, coffee producers could seek a combined certification approach (i.e., both fair trade and organic) which might help balance out the costs and returns of both certification systems (Philpott et al. 2007; Calo and Wise 2005). However, given the current cost of certification, this may be prohibitive, especially for small land-owners that do not produce large quantities of coffee (Calo and Wise 2005). Third, the initial costs of certification and transition could be subsidized by government agencies, or could be paid by the farmers only after the first years of profit are secured. University extension could also play a critical role in aiding in this initial transition stage, be providing government subsidized support and services.

Finally, the certification system could also be revised so that it does not discount the involvement of small land-holders. For example, the price of certification could be proportionate to the amount of land in cultivation. This may be tricky as certification costs need to minimally cover the expense of employing experts who must visit the farms periodically. However, it is possible that multiple individual farms could coordinate certification visits and thus reduce costs. In order to evaluate the feasibility of these alternatives, it is important that we fully understand the time, manpower, and cost involved with each step in the process of certification. Future work could explicitly explore the financial, institutional, and community support needed to transition a single farm from non-certified to Fair Trade, organic, or biodiversity friendly coffees.

7.3.5   Compensation for Ecosystem Services

Another, more direct method to secure both ecosystem services and farmer livelihoods is via Payments or Compensation for Ecosystem Services (PES) which provides payments from the beneficiaries directly to the land holders (reviewed in Engel et al. 2008). These payments could reward landowners who preserve water filtration, erosion control, pest-control, and pollination services within shade coffee landscapes, without forcing them to pay certification fees. While PES are not designed to single-handedly regulate land management, they may be used for providing incentives, especially in conjunction with extension services that provide land-holders with management information (Engel et al. 2008), such as Mexico's national certification initiative, Certimex (Calo and Wise 2005). Two examples of existing 'hybrid' PES programs are Mexico's 'Payment for Ecological Services-Hydrological (PSA-H), and the 'Program for the Development of Markets for the Ecosystem Services of Carbon Sequestration, the Derivatives of Biodiversity, and to Promote the Introduction and Improvement of Agroforestry Systems' (PSA-CABSA) (McAfee and Shapiro 2010), both of which are administered by the National Forestry Commission (CONAFOR). The PES for PSA-H, is paid by communities living downstream of the forest fragments and is based on the local opportunity costs of cutting forest. According to recent estimates, the annual payment of

$18.2 USD per ha is enough for more than 40% of forest owners to prefer conserving forests to cutting them (Jaramillo 2002).

However, the PES system has been shown to face many challenges, especially for smallholders and poor rural communities (Rosa et al. 2004). Studies reviewing the efficacy of PES practices have provided a number of critiques, most stemming from the fact that PES relies on a neoliberal framework, where nature is converted into a tradable commodity (McCauley 2006). This is because the practices of marketing and measuring commodity values do not often fit with the unpredictability and unquantifiable quality of nature, the social and cultural practices of potential stakeholders, or the policies of local or state governments (McAfee and Shapiro 2010). Many worry that PES policies will only further exploit the poor (Lovera 2004) or will disrupt their relationships with the landscape (Barreda 2004). The PSA-H and the PSA-CABSA of Mexico received substantial opposition from the farming communities, who viewed the ecosystem services as valuable contributions not only to regional and global markets, but also to local peasant livelihoods. They insisted that these ecosystem services are produced not only by nature, but also by the *campesino* communities who manage the landscapes. While the PSA-H was highly federally controlled, it only benefitted a small portion of land owners, and involved little input from community members. The more successful PSA-CABSA additionally involved a coalition of farmers and cooperative leaders, provided incentive for a wider range of land-managers, and supported environmental restoration that specifically also provided farmers with livelihood security (McAfee and Shapiro 2010). Thus, in the development of a sustainable PES system, it is essential that local stakeholders are involved and that restoration practices are also linked to farmer livelihoods (Rosa et al. 2004).

Deciding the value of a particular ecosystem service, such as erosion control, within agricultural landscapes may be challenging. Current employed practices for ecosystem valuation in agriculture are relatively rudimentary (Robertson and Swinton 2005; Gutman 2003; Daily 1997). According to a recent review, for ecosystem services that are currently unvalued, but have a measurable consumer-driven demand, there are three basic models that can be used to determine value: (1) the 'travel cost' method, where value is determined from the amount consumers would spend to gain access, (2) the 'hedonic price analysis' method, where the value is determined by the estimated contribution to the overall real-estate value of the land, and (3) the 'averting expenditures' method, where value is estimated based on the price consumers are willing to pay to avoid exposure to harmful outputs from the ecosystem (Robertson and Swinton 2005).

Assessing value for services that completely lack any connection to existing markets is even more challenging. One potential method is that of 'stated preference', which relies on surveys asking consumers how much they would be willing to pay for a service (e.g. Freeman 1993). While this method is increasingly utilized, one drawback is that it requires consumers to be educated about the environmental service being evaluated (Robertson and Swinton 2005). A second method takes the opposite approach, which relies on surveys asking producers what they would be willing to accept to provide the service. One example of this method is the erosion control value that the World Agroforestry Centre (ICRAF) began implementing in the early 2000s (Jack et al. 2009). In this program, a number of workshops on erosion control techniques were conducted in a coffee farming village in Sumatra. Farmers were then polled by a private auctioning system in order to find out the payment needed for them to conduct the erosion control techniques on their land (Jack et al. 2009). By keeping the actual bids private, farmers were not 'out-bidding' one another. Instead, the bids were then used to come up with an appropriate uniform price for the management practices across villages.

Based on these examples, it is clear that PES may involve long periods of negotiation and research, and most importantly, it is necessary to incorporate many stakeholders in the discussion. This means that federal and international policies for PES systems need to be locally-developed for each coffee-growing region. Adequate information needs to be provided to all stakeholders, and substantial discussion of the action plan must take place before policy decisions are made. Though this may sound daunting, this process conducted on a small-scale could save large amounts of time, money, and strife in the long run.

8   Conclusions

Studies in agroforestry systems have been highly useful in making conceptual and theoretical strides in the field of ecology (Greenberg et al. 2008). Agroforestry studies have provided ideal locations in which to determine and distinguish the impacts of local vegetation and landscape factors on biodiversity (e.g. Jha and Vandermeer 2010; Tscharntke et al. 2008), to examine relationships between biodiversity and ecosystem services (e.g. Klein et al. 2008), and to study tropical spatial ecology, difficult in more heterogeneous tropical forests (Perfecto and Vandermeer 2008b). Studies in agroforests are some of the first to examine relationships between biodiversity of vertebrates and ecosystem function, specifically demonstrating the importance of a diversity of vertebrate predators for providing ecosystem services (Philpott et al. 2009; Van Bael et al. 2008). Thus, the insights from coffee studies have also enriched the ecological literature.

Insights from research with coffee producers and their organizations have enriched the social sciences. First, the willingness of many smallholder cooperatives to permit researchers to enter deep into their histories and thus facilitate detailed research with their members. Coffee has emerged as an important test case for assessing the effects of different certification programs and more broadly assessing several potentially alternative forms of globalization. This research on the coffee value chain links global tendencies – including countertendencies – with local outcomes and continues to pioneer many approaches that are later used to assess other value chains (such as cacao and cotton) (Ponte 2008; Talbot 2004). Finally, the shade coffee systems of Mesoamerica offer an empirical research arena for participatory and interdisciplinary research collaborations (Méndez et al. 2010b; Bacon et al. 2008b; Rice and Ward 1996). However, there is much work ahead to integrate innovative ecological research in shade coffee (Lin 2010; Perfecto and Vandermeer 2008b; Philpott et al. 2004) with analysis on livelihoods, community and value chains (Bacon et al. 2008b; Mutersbaugh 2004). This points to the need to develop more interdependent, interdisciplinary, and generative research approaches to better understand these dynamic systems.

In order to support ecological and livelihood supporting coffee systems, it is essential that we consider all stakeholders in the coffee production chain (Fig. 4). The first set of players, involved at the local spatial scale, include coffee farms, farmers, laborers, and the communities they comprise. Some of these individual communities form cooperatives to secure the community's role in the local coffee market and to streamline the transition between growers and buyers. The management of the coffee farm depends primarily on the practices of the farmers and laborers, and these practices can have major impacts on the ecosystem services garnered (e.g. pollination, pest-control, fertilization). At a larger, regional scale, government agencies, NGO's, and universities play a critical role in regulating and modifying coffee management practices, distribution processes, and coffee prices. However, the influence of these players (should) also reach back to the practices and people involved at the local scale of coffee cultivation. Regional landscapes are comprised of a mosaic of communities, many of which support different crops, cattle, or urban centers. Regions benefit greatly from the enhanced water storage, erosion control, and resilience offered by biodiverse shaded coffee landscapes. A wide range of organisms with extended migration patterns (e.g. migratory birds) benefit from large clusters of biodiverse shade coffee communities. Thus, sound stewardship of land at a regional scale is of critical importance.

Finally, at the largest spatial scale, a number of key players have colossal impact on coffee land management and livelihoods. These include world government and private trade organizations, government and private certifiers, roasters, distributors, and consumers. Organizations and consumers, in their willingness to consider the broader impacts of their consumption, determine the profit margins for global distributors. One of the most challenging realizations of this review is that individuals and landscapes that generate important ecosystem services at the local farm scale do not necessarily harvest the benefits in terms of income, incentives, and opportunities. Only a small portion of the benefits reaped by retailers in specialty (e.g. organic) markets actually reaches individuals who participate in coffee production in the early stages. Further, key regulating ecosystem services provided by shade coffee, such as water storage, water filtration, erosion control, and carbon sequestration are not rewarded in current markets. The lack of direct compensation to farmers threatens current and future coffee ecosystems and farmer livelihoods. In summary, in order to build sustainable and livelihood-serving shade coffee landscapes, it is essential that we (1) incorporate worker livelihoods and well-being into global concepts of sustainability, (2) encourage farmers to diversify their coffee systems for greater resilience to risk and global change, and (3) improve certification and potential payment systems in order to compensate shade coffee farmers for the innumerable services that their shade coffee landscapes provide.

Creating stronger linkages between farmers, community members, certifiers, global agencies, researchers and consumers will allow for greater transparency and response to the ecological processes and well-being of all stakeholders in the global coffee production system.

Acknowledgements We would like to thank the editors and reviewers for their helpful comments and suggestions. We are especially grateful for Louise Jackson for several rounds of engaged, critical and constructive edits. We would also like to express our gratitude to the coffee farmers of Mexico, Nicaragua, El Salvador, Guatemala, Peru, Indonesia, and Costa Rica, for their support and permission to conduct research on their land. Christopher M. Bacon is grateful for support from the S.V. Ciriacy-Wantrup postdoctoral fellowship during the early phases of preparing this manuscript.

References

Agyeman J, Bullard RD, Evans B (2003) Just sustainabilities: development in an unequal world. MIT Press, Cambridge

Akiyama T, Baffes J, Larson D, Varangis P (2003) Commodity market reform in Africa: some recent experience. Econ Syst 27:83–115

Alegre C (1959) Climates et cafeiers d'Arabie. Agron Trop 14:25–48

AMECAFE (2010) Data prepared and provided by Jose Luis Benvidez of the Asociacion Mexicana de la Cadena Productiva del Café, A.C, 7 July 2010

Amekawaa Y, Sseguyaa H, Onzerea S, Carranzaa I (2010) Delineating the multifunctional role of agroecological practices: toward sustainable livelihoods for smallholder farmers in developing countries. J Sustain Agr 34:202–228

Anacafe (2008) Green book: Guatemalan coffees. Guatemalan National Coffee Association, Guatemala City

Armbrecht I, Gallego MC (2007) Testing ant predation on the coffee berry borer in shaded and sun coffee plantations in Colombia. Entomol Exp Appl 124:261–267

Arnould EJ, Plastina A, Ball D (2009) Does fair trade deliver on its core value proposition? Effects on income, educational attainment, and health in three countries. J Public Pol Market 28:186–201

Bacon C (2005a) Confronting the coffee crisis: can fair trade, organic and specialty coffees reduce small-scale farmer vulnerability in northern Nicaragua. World Dev 33:497–511

Bacon CM (2005b) Confronting the coffee crisis: Nicaraguan farmers use of cooperative, fair trade and agroecological networks to negotiate livelihoods and sustainability. PhD dissertation, University of California-Santa Cruz, Santa Cruz

Bacon CM (2010) Who decides what is fair in fair trade? The agri-environmental governance of standards, access, and price. J Peasant Stud 37:111–147

Bacon C, Méndez VE, Brown M (2005) Participatory action-research and support for community development and conservation: examples from shade coffee landscapes of El Salvador and Nicaragua. Center for Agroecology and Sustainable Food Systems (CASFS), University of California-Santa Cruz, Santa Cruz

Bacon CM, Méndez VE, Flores Gomez MA, Stuart D, Díaz Flores SR (2008a) Are sustainable coffee certifications enough to secure farmer livelihoods? The millenium development goals and Nicaragua's fair trade cooperatives. Globalizations 5:259–274

Bacon CM, Mendez VE, Gliessman S, Goodman D, Fox J (2008b) Confronting the coffee crisis: fair trade, sustainable livelihoods and ecosystems in Mexico and Central America. MIT Press, Cambridge

Bair J (2009) Frontiers of commodity chain research. Stanford University Press, Stanford

Bakermans MH, Vitz AC, Rodewald AD, Rengifo CG (2009) Migratory songbird use of shade coffee in the Venezuelan Andes with implications for conservation of Cerulean warbler. Biol Conserv 142:2476–2483

Barreda A (2004) Invasiones invisibles, subsidios perversos, guerra continua. Hojarasca, La Jornada 20–23

Bebbington AJ (1999) Capitals and capabilities: a framework for analyzing peasant viability, rural livelihoods and poverty. World Dev 27:2021–2044

Bebbington A (2000) Reencountering development: livelihood transitions and place transformations in the Andes. Ann Assoc Am Geogr 90:495–520

Beer J, Muschler R, Kass D, Somarriba E (1998) Shade management in coffee and cacao plantations. Agroforest Syst 38:139–164

Bennett E, Peterson G, Gordon L (2009) Understanding relationships among multiple ecosystem services. Ecol Lett 12:1–11

Biderman J (1982) Class structure, the state and capitalist development in Nicaraguan agriculture. PhD dissertation, University of California-Berkeley, Berkeley

Borkhataria R, Collazo J, Groom M (2006) Additive effects of vertebrate predators on insects in a Puerto Rican coffee plantation. Ecol Appl 16:696–703

Brenner N (2001) The limits to scale? Methodological reflections on scalar structuration. Prog Hum Geog 25:591–614

Brown S (1996) Present and potential roles of forest in the global climate change debate. Unasylva 47:3–10

Calo M, Wise TA (2005) Revaluing peasant coffee production: organic and fair trade markets in Mexico. Global Development and Environment Institute, Medford

Canadell JG, Raupach MR (2008) Managing forests for climate change mitigation. Science 320:1456–1457

Cannell M (1976) Crop physiological aspects of coffee bean yield: a review. Kenya Coffee 41:245–253

Cannell M (1983) Coffee. Biologist 30:257–263

Carr M (2001) The water relations and irrigation requirements of coffee. Exp Agric 37:1–36

CEPAL (2002) Globalización y desarrollo. CEPAL, Santiago

Chambers R (1991) In search of professionalism, bureaucracy and sustainable livelihoods for the 21st century. Ids Bull-I Dev Stud 22:5–11

Chambers R (1992) The self-deceiving state. Ids Bull-I Dev Stud 23:31–42

Charrier A, Lashermes P, Eskes A (2009) Botany, genetics and genomics of coffee. In: Wintgens JN (ed.) Coffee: growing, processing, sustainable production. Wiley-VCH, Weinheim, pp 25–60

Clarence-Smith W (2003) The coffee crisis in Asia, Africa, and the Pacific, 1870–1914. In: Clarence-Smith W, Topik S (eds.) The global coffee economy in Africa, Asia and Latin America, 1500–1989. Cambridge University Press, Cambridge, MA, pp 100–119

Cook O (1901) Shade in coffee culture. United States Department of Agriculture, Division of Botany, Washington, DC

Costanza R, D'Arge R, de Groot R, Farber S, Grasso M, Hannon B, Limburg K, Naeem S, O'Neill RV, Paruelo J, Raskin RG, Sutton P, van den Belt M (1997) The value of the world's ecosystem services and natural capital. Nature 387:253–260

Craswell ET, Sajjapongse A, Howlett DJB, Dowling AJ (1997) Agroforestry in the management of sloping lands in Asia and the Pacific. Agroforest Syst 38:121–137

Cruz-Angon A, Sillett TS, Greenberg R (2008) An experimental study of habitat selection by birds in a coffee plantation. Ecology 89:921–927

D'Haeze D, Deckers J, Raes D, Phong T, Loi H (2005) Environmental and socioeconomic impacts of institutional reforms on the agricultural sector of Vietnam land suitability assessment for Robusta coffee in the Dak Gan region. Agric Ecosyst Environ 105:59–76

Daily GC (1997) Nature's services: societal dependence on natural ecosystems. Island Press, Washington, DC

DaMatta F (2004) Ecophysiological constraints on the production of shaded and unshaded coffee: a review. Field Crop Res 84:99–114

DaMatta FM, Chaves ARM, Pinheiro HA, Ducatti C, Loureiro ME (2003) Drought tolerance of two field-grown clones of *Coffea canephora*. Plant Sci 164:111–117

Daviron B, Ponte S (2005) The coffee paradox: global markets, commodity trade, and the elusive promise of development. Zed Books, London

Davis AP, Govaerts R, Bridson DM, Stoffelen P (2006) An annotated checklist of the genus *Coffea* L. (Rubiaceae). Bot J Linn Soc 152:465–512

De la Mora A, Livingston G, Philpott SM (2008) Arboreal ant abundance and leaf miner damage in coffee agroecosystems in Mexico. Biotropica 40:742–746

De Marco P, Coelho FM (2004) Services performed by the ecosystem: forest remnants influence agricultural cultures' pollination and production. Biodivers Conserv 13:1245–1255

Decazy F, Avelino J, Guyot B, Perriot JJ, Pineda C, Cilas C (2003) Quality of different Honduran coffees in relation to several environments. J Food Sci 68:23–56

Devaux A, Horton D, Velasco C, Thiele G, López G, Bernet T, Reinoso I, Ordinola M (2009) Collective action for market chain innovation in the Andes. Food Policy 34:31–38

Dick CW, Jones FA, Hardy OJ, Petit RJ (2008) Spatial scales of seed and pollen-mediated gene flow in tropical forest trees. Trop Plant Biol 1:20–33

Dicum G, Luttinger N (1999) The coffee book: anatomy of an industry from crop to the last drop. The New Press, New York

Donald PF (2004) Biodiversity impacts of some agricultural commodity production systems. Conserv Biol 18:17–37

Dossa EL, Fernandes ECM, Reid WS, Ezui K (2008) Above- and belowground biomass, nutrient and carbon stocks contrasting an open-grown and a shaded coffee plantation. Agroforest Syst 72:103–117

Eakin H, Tucker C, Castellanos E (2006) Responding to the coffee crisis: a pilot study of farmers' adaptations in Mexico, Guatemala and Honduras. Geogr J 172:156–171

Elevitch CR, Wilkinson KM (2000) Agroforestry guide for the Pacific islands. Permanent Agriculture Resources, Holualoa, 240 pp

Engel S, Pagiola S, Wunder S (2008) Designing payments for environmental services in theory and practice: an overview of the issues. Ecol Econ 65:663–674

Escalante F (1995) Coffee and agroforestry in Venezuela. Agroforest Today 7:5–7

Escalante F, Aguilar R, Lugo P (1987) Identifacion, evaluacion y distribucion espacial de especies utilizadas como sombra en sistemas tradicionales de cafe (*Coffea arabica*) en zonas del estado Trujillo, Venezuela. Venezuala Forestal 3:50–62

FAO (2008) Food and Agriculture Organization. Resource document. http://faostat.fao.org/site/567/DesktopDefault.aspx?PageID=567#ancor. Accessed Oct 2010

FAO (2009) FAOSTAT: crops. Resource document. http://faostat.fao.org/site/567/default.aspx#ancor. Accessed Oct 2010

FAO (2010) FAOSTAT: crops. Resource document. http://faostat.fao.org/site/570/default.aspx#ancor. Accessed Oct 2010

Fox J (1994) The roots of Chiapas. Econ Polit Weekly 29:119–1122

Fox J (1996) How does civil society thicken? the political construction of social capital in rural Mexico. World Dev 24:1089–1103

Freeman A III (1993) The measurement of environmental and resource values: theory and methods. Resources for the Future, Washington, DC

Fridell G (2006) Fair trade and neoliberalism – assessing emerging perspectives. Lat Am Perspect 33:8–28

Gallina S, Mandujano S, Gonzalez-Romero A (1996) Conservation of mammalian biodiversity in coffee plantations of Central Veracruz. Agroforest Syst 33:13–27

Gillison AN, Liswanti N, Budidarsono S, van Noordwijk M, Tomich TP (2004) Impact of cropping methods on biodiversity in coffee agroecosystems in Sumatra, Indonesia. Ecol Soc 9:16–23

Giovannucci D (2003) Emerging issues in the marketing and trade of organic products. In: Monograph in the proceedings of the OECD workshop on organic agriculture, OECD, Paris, Sept 2002

Giovannucci D, Liu P, Byers A (2008) Adding value: certified coffee trade in North America. In: Liu P (ed.) Value-adding standards in the North American food market – trade opportunities in certified products for developing countries. FAO, Rome, pp 33–49

Goodman D (2008) The international coffee crisis: a review of the issues. In: Bacon CM, Mendez VE, Gliessman S, Goodman D, Fox J (eds.) Confronting the coffee crisis: fair trade, sustainable livelihoods and ecosystems in Mexico and Central America. MIT Press, Cambridge

Greenberg R, Bichier P, Angon AC, Reitsma R (1997a) Bird populations in shade and sun coffee plantations in Central Guatemala. Conserv Biol 11:448–459

Greenberg R, Bichier P, Sterling J (1997b) Bird populations in rustic and planted shade coffee plantations of Eastern Chiapas, Mexico. Biotropica 29:501–514

Greenberg R, Bichier P, Angon AC, MacVean C, Perez R, Cano E (2000) The impact of avian insectivory on arthropods and leaf damage in some Guatemalan coffee plantations. Ecology 81:1750–1755

Greenberg R, Perfecto I, Philpott SM (2008) Agroforests as model systems for tropical ecology. Ecology 89:913–914

Guadarrama-Zugasti C (2008) A grower typology approach to assessing the environmental impact of coffee farming in Veracruz, Mexico. In: Bacon CM, Méndez VE, Gliessman SR, Goodman D, Fox JA (eds.) Confronting the coffee crisis: fair trade, sustainable livelihoods and ecosystems in Mexico and Central America. MIT Press, Cambridge, pp 127–154

Guhl A (2004) Coffee and landcover changes in the Colombian coffee region landscape 1970–1997. Ensayos, Bogota

Gutman P (2003) Macroeconomics for sustainable development. Program Office, World Wildlife Fund, Washington, DC

Guyot BD, Gueule JC, Maniez JJ, Perriot J, VL Giron (1996) Influence de l'altitude et de l'ombrage sur la qualité des cafés Arabica. Plantations, Recherche, Développement 3:272–283

Hanson L, Terstappen V (2009) Collaboration on contentious issues: research partnerships for gender equity in Nicaragua's fair trade coffee cooperatives. J Agromedicine 14:105–111

Harrison G, DiSogra C, Manalo-LeClair G, Aguayo J, Yen W (2002) Over 2.2 million low-income California adults are food insecure; 658,000 suffer hunger. UCLA Healthy Policy Research Brief

Harvey D (2005) A brief history of neoliberalism. Oxford University Press, Oxford

Henry M, Tittonell P, Manlay RJ, Bernoux B, Albrecht A, Vanlauwe B (2009) Biodiversity, carbon stocks and sequestration potential in aboveground biomass in smallholder farming systems of western Kenya. Agric Ecosyst Environ 129:238–252

Hergoualc'h K, Skiba U, Harmand JM, Henault C (2008) Fluxes of greenhouse gases from Andosols under coffee in monoculture or shaded by *Inga densiflora* in Costa Rica. Biogeochemistry 89:329–345

Heynen N (2007) Neoliberal environments: false promises and unnatural consequences. Routledge, London

Holt-Giménez E (2006) Campesino a campesino: voices from Latin America's farmer to farmer movement for sustainable agriculture. Food First Books, Oakland

ICAFE (1989) Instituto del Café de Costa Rica. Manual de recomendaciones para el cultivo del café. Ministerio de Agricultura y Ganadería, San Jose

Illy E (2002) The complexity of coffee. Sci Am 6:86–91

International Coffee Organization (ICO) (2000) Monthly statistics. http://www.ico.org/prices/m1.htm. Accessed July 2010

IPCC (2007) Impacts, adaptation and vulnerability. Contribution of Working Group II to the Fourth Assessment Report of the IPCC. Intergovernmental Panel on Climate Change, Wembley

Jack BK, Leimona B, Ferraro PJ (2009) A revealed preference approach to estimating supply curves for ecosystem services: use of auctions to set payments for soil erosion control in Indonesia. Conserv Biol 23:359–367

Jaffe R, Bacon C (2008) From differentiated coffee markets toward alternative trade and knowledge networks. In: Bacon CM, Mendez VE, Gliessman S, Goodman D, Fox J (eds.) Confronting the coffee crisis: fair trade, sustainable livelihoods and ecosystems in Mexico and Central America. MIT Press, Cambridge, pp 311–336

Jaffee D (2007) Brewing justice: fair trade coffee, sustainability, and survival. University of California Press, Berkeley

Jaramillo L (2002) Estimación del costo de oportunidad del uso de suelo forestal en ejidos a nivel nacional. DGIPEA working paper. Instituto Nacional de Ecología, México

Jha S, Dick CW (2008) Shade coffee farms promote genetic diversity of native trees. Curr Biol 18:1126–1128

Jha S, Dick CW (2010) Native bees facilitate gene flow across shade coffee landscapes. Proc Natl Acad Sci 107:13760–13764

Jha S, Vandermeer J (2010) Impacts of coffee agroforestry management on tropical bee communities. Biol Conserv 143:1423–1431

Johnson MD (2000) Effects of shade-tree species and crop structure on the winter arthropod and bird communities in a Jamaican shade coffee plantation. Biotropica 32:133–145

Johnson MD, Levy NJ, Kellermann JL, Robinson DE (2009) Effects of shade and bird exclusion on arthropods and leaf damage on coffee farms in Jamaica's Blue Mountains. Agroforest Syst 76:139–148

Kay M, Franks TR, Smith L (1997) Water: economics, management and demand. E & FN Spon, London

Kellermann J, Johnson M, Stercho A, Hackett S (2008) Ecological and economic services provided by birds on Jamaican Blue Mountain coffee farms. Conserv Biol 22:1177–1185

Klein AM, Steffan-Dewenter I, Tscharntke T (2003a) Bee pollination and fruit set of *Coffea arabica* and *C. canephora* (Rubiaceae). Am J Bot 90:153–157

Klein AM, Steffan-Dewenter I, Tscharntke T (2003b) Fruit set of highland coffee increases with the diversity of pollinating bees. Proc R Soc Lond B Biol Sci 270:955–961

Klein AM, Steffan-Dewenter I, Tscharntke T (2003c) Pollination of *Coffea canephora* in relation to local and regional agroforestry management. J Appl Ecol 40:837–845

Klein AM, Cunningham SA, Bos M, Steffan-Dewenter I (2008) Advances in pollination ecology from tropical plantation crops. Ecology 89:935–943

Kotecha S, Hieu P, Kuit M, Von Enden J, Swinkels R (2003) Vietnamese coffee competitiveness and policy options to improve it. Conference summary found at. www.xttmnew.agroviet.gov.vn/en/stories/Coffee_Policy_brief.pdf

Kremen C, Williams NM, Aizen MA, Gemmill-Herren B, LeBuhn G, Minckley R, Packer L, Potts SG, Roulston T, Steffan-Dewenter I, Vazquez DP, Winfree R, Adams L, Crone EE, Greenleaf SS, Keitt TH, Klein AM, Regetz J, Ricketts TH (2007) Pollination and other ecosystem services produced by mobile organisms: a conceptual framework for the effects of land-use change. Ecol Lett 10:299–314

Läderach P, Oberthür T, Niederhauser N, Usma H, Collet L, Pohlan J (2006) Café especial: factores, dimensiones e interacciones. In: Pohlan J, Barrera J (eds.) El cafetal del futuro: realidades y visiones. Shaker, Aachen, pp 141–160

Läderach P, Oberthür T, Pohlan J (2009) Management of intrinsic quality characteristics for specialty coffee – a framework developed and tested in coffee growing regions across Latin America. Dr. Müller Verlag, Saarbrück, Germany, 157 pp

Läderach P, Lundy M, Jarvis A, Ramírez J, Pérez P, Schepp E, Eitzinger KA (2010) Predicted impact of climate change on coffee supply chains. In: Leal Filho W (ed.) The economic, social and political elements of climate change. Springer, Berlin, pp 703–724

Lagemann J, Heuveldop J (1983) Characterization and evaluation of agroforestry systems. CATIE, Tumalba

Lara-Estrada LD (2005) Efectos de la altitud, sombra, producción y fertilización sobre la calidad del café (*Coffea arabica* L. var. *caturra*) producido en sistemas agroforestales de la zona cafetalera nor-central de Nicaragua. Agroforestry. Centro Agronómico Tropical de Investigación y Enseñanza, Turrialba

Larsen A, Philpott SM (2010) Twig-nesting ants: the hidden predators of the coffee berry borer in Chiapas, Mexico. Biotropica 42:342–347

Lewin B, Giovannucci D, Varangis P (2004) Coffee markets: new paradigms in global supply and demand. World Bank, Washington, DC

Lin BB (2007) Agroforestry management as an adaptive strategy against potential microclimate extremes in coffee agriculture. Agric For Meteorol 144:85–94

Lin BB (2010) The role of agroforestry in reducing water loss through soil evaporation and crop transpiration in coffee agroecosystems. Agric For Meteorol 150:510–518

Lin BB, Perfecto I, Vandermeer J (2008) Synergies between agricultural intensification and climate change could create surprising vulnerabilities for crops. Bioscience 58:847–854

Linton A (2005) Partnering for sustainability: business–NGO alliances in the coffee industry. Develop Prac 15:600–614

Lomeli-Flores J (2007) Natural enemies and mortality factors of the coffee leaf miner *Leucoptera coffeella* (Guerin-Meneville) (Lepidoptera: Lyonetiidae) in Chiapas, Mexico. PhD dissertation, Texas A&M University, College Station

Lovera S (2004) Environmental markets impoverish the poor. Resource document. Ecosystem Marketplace. http://ecosystemmarketplace.com/pages/article.opinion.php?componentid=2268&componentversionid. Accessed Oct 2010

Lyon S (2008) We want to be equal to them: Fair-trade coffee certification and gender equity within organizations. Hum Organ 67:258–268

Lyon S, Bezaury JA, Mutersbaugh T (2010) Gender equity in fair trade-organic coffee producer organizations: cases from Mesoamerica. Geoforum 41:93–103

Magalhaes A, Angelocci L (1976) Sudden alterations in water balance associated with flower bud opening in coffee plants. J Hortic Sci 51:419–421

Manrique AJ, Thimann RE (2002) Coffee (*Coffea arabica*) pollination with africanized honeybees in Venezuela. Interciencia 27:414–416

Marsden T, Munton R, Ward N, Wahtmore S (1996) Agricultural geography and the political economy approach: a review. Econ Geogr 72:361–375

Marston S (2000) The social construction of scale. Prog Hum Geogr 24:219–242

McAfee K, Shapiro EN (2010) Payments for ecosystem services in Mexico: nature, neoliberalism, social movements, and the State. Ann Assoc Am Geogr 100:579–599

McCarthy J (2004) Privatizing conditions of production: trade agreements as neoliberal environmental governance. Geoforum 35:327–341

McCauley DJ (2006) Selling out on nature. Nature 443:27–28

McCreery D (1995) Wage labor, free labor and vagrancy laws: the transition to capitalism in Guatemala. In: Roseberry W, Gudmundson L, Kutschbach MS (eds.) Coffee, society, and power in Latin America. Johns Hopkins University, Maryland, pp 1920–1945

Méndez VE, Gliessman SR, Gilbert GS (2007) Tree biodiversity in farmer cooperatives of a shade coffee landscape in western El Salvador. Agric Ecosyst Environ 119:145–159

Méndez VE, Shapiro EN, Gilbert GS (2009) Cooperative management and its effects on shade tree diversity, soil properties and ecosystem services of coffee plantations in western El Salvador. Agroforest Syst 76:111–126

Méndez VE, Bacon CM, Olson M, Morris KS, Shattuck AK (2010a) Agrobiodiversity and shade coffee smallholder livelihoods: a review and synthesis of ten years of research in Central America. Prof Geogr 62:357–376

Méndez VE, Bacon C, Olson M, Petchers S, Herrador D, Carranza C, Trujillo L, Guadarrama-Zugasti C, Cordón A, Mendoza A (2010b) Effects of fair trade and organic certifications on small-scale coffee farmer households in Central America and Mexico. Renewable Agric Food Syst 25:236–251

Michon G, Mary F, Bompard JM (1986) Multistoried agroforestry garden system in West Sumatra, Indonesia. Agroforest Syst 4:315–338

Millenium Ecosystem Assessment (MEA) (2005) Ecosystems and human well-being: synthesis. Island Press, Washington, DC

Moguel P, Toledo VM (1999) Biodiversity conservation in traditional coffee systems of Mexico. Conserv Biol 13:11–21

Muriel SB, Kattan GH (2009) Effects of patch size and type of coffee matrix on ithomiine butterfly diversity and dispersal in cloud-forest fragments. Conserv Biol 23:948–956

Muschler RG (1997) Shade or sun for ecologically sustainable coffee production: a summary of environmental key factors. In: Proceedings of the 3rd scientific week, CATIE, Turrialba, 3–7 Feb 1997, pp 109–112

Muschler RG (2001) Shade improves coffee quality in a sub-optimal coffee zone of Costa Rica. Agroforest Syst 51:131–139

Mutersbaugh T (2004) Serve and certify: paradoxes of service work in organic-coffee certification. Environ Plann D 22:533–552

Mutersbaugh T (2005) Fighting standards with standards: harmonization, rents, and social accountability in certified agrofood networks. Environ Plann A 37:2033–2051

Neelin JD, Munnich M, Su H, Meyerson JE, Holloway CE (2006) Tropical drying trends in global warming models and observations. Proc Natl Acad Sci USA 103:6110–6115

Nigh R (1997) Organic agriculture and globalization: a Maya associative corporation in Chiapas, Mexico. Hum Organ 56(4):427–435

Nolasco M (1985) Cafe y sociedad en Mexico. Centro de Ecodesarrollo, Mexico City, 454 pp

Nunes M, Bierhuizen J, Ploegman C (1968) Studies on the productivity of coffee, I: effect of light, temperature, and CO_2 concentration on photosynthesis of *Coffea arabica*. Acta Bot Neerl 17:93–102

Nutman F (1937) Studies on the physiology of *Coffea arabica*, I: photosynthesis of coffee leaves under natural conditions. Ann Bot 1:353–367

O'Brien TG, Kinnaird MF (2003) Caffeine and conservation. Science 5619:587

O'Rourke D (2004) Transition environments: ecological and social challenges to post-socialist industrial development. In: Peet R, Watts M (eds.) Liberation ecologies: environment, development, social movements. Routledge, London

Oxfam (2001) Bitter coffee: how the poor are paying for the slump in coffee prices. www. globalexchange.org/campaigns/fairtrade/coffee/OxfamPriceReport.pdf

Oxfam (2002) Mugged: poverty in your coffee cup. www.oxfam.org.uk/resources/papers/ mugged.html

Palm C, Van Noordwijk V, Woomer P, Alegre J, Arevalo L, Castilla C, Cordeiro D, Hariah K, Kotto-Same J, Moukam A, Parton W, Ricse A, Rodriques V, Sitompul S (2005) Carbon losses and sequestration alter land use change in the humid tropics. In: Palm CA, Vosti SA, Sanchez PA, Ericksen PJ (eds.) Slash-and-burn agriculture: the search for alternatives. Columbia University Press, New York, pp 41–63

Peet R, Watts M (2004) Liberation ecologies: environment, development, social movements. Routledge, London

Peeters LYK, Soto-Pinto L, Perales H, Montoya G, Ishiki M (2003) Coffee production, timber, and firewood in traditional and *Inga*-shaded plantations in Southern Mexico. Agric Ecosyst Environ 95:481–493

Peluso NL (1994) Rich forests, poor people: resource control and resistance in Java. University of California Press, Berkeley, 336 pp

Perfecto I, Vandermeer J (2006) The effect of an ant-hemipteran mutualism on the coffee berry borer (*Hypothenemus hampei*) in southern Mexico. Agric Ecosyst Environ 117:218–221

Perfecto I, Vandermeer J (2008a) Biodiversity conservation in tropical agroecosystems – a new conservation paradigm. Ann NY Acad Sci 1134:173–200

Perfecto I, Vandermeer J (2008b) Spatial pattern and ecological process in the coffee agroforestry system. Ecology 89:915–920

Perfecto I, Rice RA, Greenberg R, VanderVoort ME (1996) Shade coffee: a disappearing refuge for biodiversity. Bioscience 46:598–608

Perfecto I, Vandermeer JH, Bautista GL, Nunez GI, Greenberg R, Bichier P, Langridge S (2004) Greater predation in shaded coffee farms: the role of resident neotropical birds. Ecology 85:2677–2681

Perfecto I, Vandermeer J, Mas A, Pinto LS (2005) Biodiversity, yield, and shade coffee certification. Ecol Econ 54:435–446

Perfecto I, Armbrecht I, Philpott SM, Soto-Pinto L, Dietsch TV (2007) Shaded coffee and the stability of rainforest margins in Latin America. In: Tscharntke T, Leuschner C, Zeller M, Guhadja E, Bidin A (eds.) The stability of tropical rainforest margins, linking ecological, economic and social constraints of land use and conservation. Springer Environmental Science Series, Heidelberg

Perfecto I, Vandermeer J, Wright A (2009) Nature's matrix linking agriculture, conservation and food sovereignty. Earthscan, London

Peyser R (2010) Hunger in the coffee lands. Specialty Coffee Association of America, Anaheim

Phillips S, Anderson R, Sapohire R (2006) Maximum entropy modeling of species geographic distributions. Ecol Model 190:231–259

Philpott SM, Maldonado J, Vandermeer J, Perfecto I (2004) Taking trophic cascades up a level: behaviorally-modified effects of phorid flies on ants and ant prey in coffee agroecosystems. Oikos 105:141–147

Philpott SM, Bichier P, Rice R, Greenberg R (2007) Field-testing ecological and economic benefits of coffee certification programs. Conserv Biol 21:975–985

Philpott SM, Bichier P, Rice RA, Greenberg R (2008a) Biodiversity conservation, yield, and alternative products in coffee agroecosystems in Sumatra, Indonesia. Biodivers Conserv 17:1805–1820

Philpott SM, Lin BB, Jha S, Brines SJ (2008b) A multi-scale assessment of hurricane impacts on agricultural landscapes based on land use and topographic features. Agric Ecosyst Environ 128:12–20

Philpott SM, Arendt WJ, Armbrecht I, Bichier P, Diestch TV, Gordon C, Greenberg R, Perfecto I, Reynoso-Santos R, Soto-Pinto L, Tejeda-Cruz C, Williams-Linera G, Valenzuela J, Zolotoff JM (2008c) Biodiversity loss in Latin American coffee landscapes: review of the evidence on ants, birds, and trees. Conserv Biol 22:1093–1105

Philpott S, Soong O, Lowenstein J, Pulido A, Tobar-Lopez D, DeClerck F (2009) Functional traits and ecosystem services: bird predation on arthropods in tropical agroecosystems. Ecol Appl 19:1858–1867

Ponte S (2002) The 'Latte Revolution'? Regulation, markets and consumption in the global coffee chain. World Dev 30:1099–1122

Ponte S (2008) Greener than thou: the political economy of fish ecolabeling and its local manifestations in South Africa. World Dev 36:159–175

Prendergast M (1999) Uncommon grounds: the history of coffee and how it transformed our world. Basic Books, New York

Pretty JN (2002) Agri-Culture: reconnecting people, land, and nature. Earthscan Publications, London

Ramírez L (1993) Producción de café (*Coffea arabica*) bajo diferentes niveles de fertilización con y sin sombra de *Erythrina poeppigiana* (Walpers). In: Cook OF (ed.) *Erythrina* in the new and old worlds. Nitrogen Fixing Tree Association, Paia, pp 121–124

Raudsepp-Hearne C, Peterson GD, Bennett EM (2010) Ecosystem service bundles for analyzing tradeoffs in diverse landscapes. Proc Natl Acad Sci USA 107:5242–5247

Raw A, Free JB (1977) Pollination of coffee (*Coffea arabica*) by honeybees. Trop Agric 54:365–370

Raynolds LT, Murray D, Heller A (2007) Regulating sustainability in the coffee sector: a comparative analysis of third-party environmental and social certification initiatives. Agric Hum Values 24:147–163

Renard M (2010) The Mexican coffee crisis. Lat Am Perspect 37:21–33

Ribot JC, Peluso NL (2003) A theory of access. Rural Sociol 68:153–181

Rice R (2000) Noble goals and challenging terrain: organic and fair trade coffee movements in the global marketplace. J Agric Environ Ethics 14:39–66

Rice R (2003) Coffee production in a time of crisis: social and environmental connections. SAIS Rev 23:221–245

Rice RA (2008) Agricultural intensification within agroforestry: the case of coffee and wood products. Agric Ecosyst Environ 128:212–218

Rice PD, McLean J (1999) Sustainable coffee at the crossroads. Consumer's Choice Council, Washington, DC

Rice R, Ward J (1996) Coffee, conservation, and commerce in the Western Hemisphere. The Smithsonian Migratory Bird Center and the Natural Resources Defense Council, Washington, DC

Ricketts TH, Daily GC, Ehrlich PR, Michener CD (2004) Economic value of tropical forest to coffee production. Proc Natl Acad Sci USA 101:12579–12582

Robertson GP, Swinton SM (2005) Reconciling agricultural productivity and environmental integrity: a grand challenge for agriculture. Front Ecol Environ 3:38–46

Robertson GP, Broome JC, Chornesky EA, Frankenberger JR, Johnson P, Lipson M, Miranowski JA, Owens ED, Pimentel D, Thrupp LA (2004) Rethinking the vision for environmental research in US agriculture. Bioscience 54:61–65

Romero-Alvarado Y, Soto-Pinto L, Garcia-Barrios L, Barrera-Gaytan JF (2002) Coffee yields and soil nutrients under the shades of *Inga* sp. vs. multiple species in Chiapas, Mexico. Agroforest Syst 54:215–224

Roncal-Garcia S, Soto-Pinto L, Castellanos-Albores J, Ramirez-Marcial N, de Jong B (2008) Agroforestry systems and carbon stocks in indigenous communities from Chiapas, Mexico. Interciencia 33:200–206

Rosa H, Kandel S, Dimas L (2004) Compensation for environmental services and rural communities: lessons from the Americas. Int Forest Rev 6:187–194

Roubik DW (2002) Tropical agriculture – the value of bees to the coffee harvest. Nature 417:708–708

Salinas-Zavala CA, Douglas AV, Diaz HF (2002) Interannual variability of NDVI in northwest Mexico. Associated climatic mechanisms and ecological implications. Remote Sens Environ 82:417–430

Sayre N (2005) Ecological and geographical scale: parallels and potential for integration. Prog Hum Geogr 29:276–290

Schroth G, Krauss U, Gasparotto L, Aguilar JAD, Vohland K (2000) Pests and diseases in agroforestry systems of the humid tropics. Agroforest Syst 50:199–241

Scoones I (1998) Sustainable rural livelihoods: a framework for analysis. Institute for Development Studies (IDS), Brighton

Scoones I (2009) Livelihoods perspectives and rural development. J Peasant Stud 36:171–196

Shiva V (2008) Soil not oil: environmental justice in a time of climate crisis. South End Press, Cambridge

Shulman M, Garret P (1990) Cluster analysis and typology construction: the case of small-scale tobacco farmers. Sociol Spectrum 10:413–428

Somarriba E, Harvey C, Samper M, Anthony F, Gonzalez J, Staver C, Rice R (2004) Biodiversity in coffee plantations. In: Schroth G, Fonseca G, Harvey C, Gascon C, Vasconcelos H, Izac AMN (eds.) Biodiversity in coffee plantations. Island Press, Washington, DC, pp 198–226

Soto-Pinto L, Perfecto I, Castillo-Hernandez J, Caballero-Nieto J (2000) Shade effect on coffee production at the northern Tzeltal zone of the state of Chiapas, Mexico. Agric Ecosyst Environ 80:61–69

Soto-Pinto L, Perfecto I, Caballero-Nieto J (2002) Shade over coffee: its effects on berry borer, leaf rust and spontaneous herbs in Chiapas, Mexico. Agroforest Syst 55:37–45

Soto-Pinto L, Anzueto M, Mendoza J, Ferrer GJ, de Jong B (2010) Carbon sequestration through agroforestry in indigenous communities of Chiapas, Mexico. Agroforest Syst 78:39–51

Staver C, Guharay F, Monterroso D, Muschler RG (2001) Designing pest-suppressive multistrata perennial crop systems: shade-grown coffee in Central America. Agroforest Syst 53:151–170

Steward C (2007) From colonization to "environmental soy": a case study of environmental and socio-economic valuation in the Amazon soy frontier. Agric Hum Values 24:107–122

Sturgeon T (2000) Turnkey production networks: a new American model of industrial organization? Industrial Performance Center, Working paper, MIT, Cambridge

Swinton SM, Lupi F, Robertson GP, Hamilton SK (2007) Ecosystem services and agriculture: cultivating agricultural ecosystems for diverse benefits. Ecol Econ 64:245–252

Talbot JM (2004) Grounds for agreement: the political economy of the coffee commodity chain. Rowman and Littlefield Publishers, Lanham

Topik S, Talbot J, Samper M (2010) Introduction globalization, neoliberalism, and the Latin American coffee societies. Lat Am Perspect 171:5–20

Trujillo L (2008) Coffee production strategies in a changing rural landscape: a case study in Central Veracruz, Mexico. In: Bacon CM, Mendez VE, Gliessman SR, Goodman D, Fox JA (eds.) Confronting the coffee crisis: fair trade, sustainable livelihoods and ecosystems in Mexico and Central America. MIT Press, Cambridge, pp 69–98

Tscharntke T, Sekercioglu C, Dietsch T, Sodhi N, Hoehn P, Tylianakis J (2008) Landscape constraints on functional diversity of birds and insects in tropical agroecosystems. Ecology 89:944–951

Ukers W (1922) All about coffee. The Tea and Coffee Trade Journal Company, New York

USDL (2010) US Department of Labor, Bureau of Labor Statistics. Resource document. http://data.bls.gov/cgi-bin/dsrv. Accessed Oct 2010

Vaast P, Bertrand B, Perriot J-J, Guyot B, Genard M (2006) Fruit thinning and shade improve bean characteristics and beverage quality of coffee (*Coffea arabica* L.) under optimal conditions. J Sci Food Agric 86:197

Valkila J, Nygren A (2010) Impacts of fair trade certification on coffee farmers, cooperatives, and laborers in Nicaragua. Agric Hum Values 27:321–333

Van Bael SA, Philpott SM, Greenberg R, Bichier P, Barber NA, Mooney KA, Gruner DS (2008) Birds as predators in tropical agroforestry systems. Ecology 89:928–934

Van der Vossen HAM (2005) A critical analysis of the agronomic and economic sustainability of organic coffee production. Exp Agric 41:449–473

Varangis P, Siegel P, Giovannucci D, Lewin B (2003) Dealing with the coffee crisis in Central America: impacts and strategies. In: World Bank Policy Research #2993. World Bank, Washington, DC

Vélez M, Bustillo A, Posada F (2001) Hormigas de la zona central cafetera y perspectivas de su uso en el control de *Hypothenemus hampei* (Ferrari) (Coleoptera: Scolytidae). Resúmenes XXVIII Congreso (ed. by Sociedad Colombiana de Entomología), Pereira

Waltert M, Mardiastuti A, Muhlenberg M (2005) Effects of land use on bird species richness in Sulawesi, Indonesia. Conserv Biol 18:1339–1346

Watson R, Herren R (eds) (2009) International assessment of agricultural knowledge, science and technology for development (IAASTD): synthesis report. Island Press, Washington

Watts MJ (2007) What might resistance to neoliberalism consist of? In: Heynen N, McCarthy J, Prudham S (eds.) Neoliberal environments: false promises and unnatural consequences. Routledge, London, pp 273–279

WDPA Consortium (2004) National IUCN protected areas, global. Downloadable Data. UNEP, IUCN, Oct 2004

Webster PJ, Holland GJ, Curry JA, Chang HR (2005) Changes in tropical cyclone number, duration, and intensity in a warming environment. Science 309:1844–1846

Williams R (1994) States and social evolution: coffee and the rise of national governments in Central America. University of North Carolina, Chapel Hill

Wilson K (1999) Coffee, cocoa and tea. CABI, Wallingford

Wisner B, Blaikie P, Cannon T, Davis I (2004) At risk – natural hazards, people's vulnerability and disasters. Routledge, London

World Commission on Environment and Development (WCED) (1987) Our common future. Oxford University Press, Oxford

You LWS (2005) Assessing the spatial distribution of crop production using a cross-entropy. Int J Appl Earth Obs 7:310–323

Ecosystem Services from Smallholder Forestry and Agroforestry in the Tropics

Travis Idol, Jeremy Haggar, and Linda Cox

Abstract Smallholder forestry and agroforestry systems in the tropics provide essential products and services for millions of producers, their surrounding communities, national and international consumers, and global society. The diversity of products provided by these systems meet the needs of smallholder producers for fuelwood, food, animal fodder, and other household and farm needs; they provide additional income to supplement major commodity crops; and they offer flexibility in production and income to buffer against falling commodity prices, crop failure, or other sources of financial or economic difficulty. The sustainability of these systems is increasingly dependent upon sources of income beyond the sale of conventional products, such as price premiums from sustainability certification and agro- and ecotourism. Payments for ecosystem services (PES) programs have been developed at multiple levels to provide incentives for smallholders to conserve and enhance tree cover and management practices to provide ecosystem services such as watershed protection and carbon sequestration. Ecotourism provides an enterprise-based strategy to engage producers in conservation and enhancement of these services. This review evaluates the ability of smallholder systems to support the provision of ecosystem services (ES) and the capacity of smallholders to participate in support programs and take advantage of other emerging opportunities to support smallholder enterprises.

T. Idol (✉) • L. Cox
Department of Natural Resources and Environmental Management,
University of Hawaii-Manoa, 1910 East West Rd, Honolulu, HI 96822, USA
e-mail: idol@hawaii.edu; lcox@hawaii.edu

J. Haggar
Natural Resources Institute, University of Greenwich, Central Avenue,
Chatham Maritime, Kent ME4 4TB, UK
e-mail: jeremyhaggar@gmail.com

W.B. Campbell and S. López Ortíz (eds.), *Integrating Agriculture, Conservation and Ecotourism: Examples from the Field*, Issues in Agroecology – Present Status and Future Prospectus 1, DOI 10.1007/978-94-007-1309-3_5,
© Springer Science+Business Media B.V. 2011

Findings suggest that there are a variety of smallholder forestry and agroforestry systems that support the provision of ES as well as the sustainability of smallholder enterprises, but only a few common systems have been well-studied. By themselves, smallholder systems can support ES at the local scale, such as protection against runoff and erosion and retention and cycling of nutrients. The mosaic of smallholder systems within a watershed or a landscape can support plant and animal biodiversity, water quality, and carbon sequestration – both avoiding losses and providing for net gains. However, the limited size and great diversity of smallholder systems impose significant constraints on provisioning of ES at larger scales and thus the ability of individual producers to obtain sustainability certification or qualify for PES programs. Ecotourism offers an alternative, enterprise-based approach for compensating smallholders for ecosystem services. As with PES programs, individual smallholders have limited capacity to participate in and influence the structure and contract terms of ecotourism enterprises. Promoting community-based resource management as a part of project or enterprise development activities is recommended to coordinate the provision of ecosystem services across multiple landowners, share resources and expertise, distribute costs and returns equitably, and strengthen the social capital necessary to carry out successful enterprises. This requires participatory approaches to train producers in adaptive management and help them articulate their own needs and the types of systems that can meet household needs and support ES. Such an approach is key to slowing the tide of rural out-migration and farm consolidation that are the real threats to the sustainability of smallholder systems and the communities they support.

1 Introduction

Pedro Sanchez published a seminal paper in the journal *Agroforestry Systems* entitled "Science in Agroforestry" (Sanchez 1995). This paper highlighted the ecological, socioeconomic, and management complexities of agroforestry systems and thus the need for rigorous and multi-disciplinary science to understand their potential to promote sustainability in agriculture. It also emphasized the need for profitability as an essential component of sustainability, mirroring the earlier Brundtland Report on sustainable development (World Commission on Environment and Development 1987). In the 15 years since this paper was published, there has been a concerted effort to study smallholder forestry and agroforestry systems in the tropics with regard to their sustainability, their contribution to the well-being of the world's resource-poor farmers, and their adaptability in the face of changing socioeconomic pressures and environmental conditions.

Around the same time that Sanchez's paper was published, several working groups convened after the 1992 Earth Summit in Rio de Janeiro to develop sustainability certifications for temperate and tropical forest management operations. Certification was based on an evaluation of the ability of management plans and practices to meet specific criteria and indicators of environmental, economic, and social aspects of sustainability. Although multiple forestry certification systems exist, the most widely used system in the world, including the tropics, is from the Forest Stewardship Council (FSC), with over 41 million hectares certified worldwide in 2004. Yet, less than 20% of this is in the tropics. Sustainability certification requires a level of professionalism and documentation that is often difficult for smallholders to achieve, but FSC and other major certifiers do have 'family forest' programs that try to accommodate non-professional forestland owners and forest managers.

After the publication of the first Millennium Ecosystem Assessment (MEA) report in 2005 (Hassan et al. 2005), the focus has been on understanding and more importantly quantifying how smallholder systems affect the conservation and sustainable use of the world's ecosystems. The linkages between ecosystems and human well-being are now known as 'ecosystem services' (ES), and these include all of the productive, supporting, and regulating aspects of agroforestry and other human-managed ecosystems that Sanchez highlighted in his paper. In line with the social aspects of sustainability, the MEA added cultural ES to this collection of mainly biophysical properties and processes. For many tribal and traditional smallholders, connections to the land and ways of life are a strong part of their personal and cultural identity and traditions.

The importance of specific ecosystem services to the well-being of the greater community was recognized long before the MEA was convened. Governments and non-governmental organizations (NGOs) have developed ways to compensate landowners and producers for the conservation or enhancement of these services at provincial, national, and even international levels. The most famous examples in tropical areas are payments for carbon sequestration and watershed protection. As well, both governments and NGOs have developed programs to purchase land or conservation easements (i.e., development rights) to conserve areas deemed important for various ES, such as support for biodiversity or even continued agricultural production. The ability of smallholders to participate in these payments for ecosystem service (PES) programs varies widely, based on the ES of interest and the program specifications. Developing and adapting programs to facilitate and encourage smallholder participation is a logistical and financial challenge for large programs but can provide important financial incentives to sustain smallholder production and maintain or adopt more environmentally sustainable management practices, including greater tree cover and diversity.

As Sanchez recommended, the research on agroforestry and related smallholder systems over the past 15 years has been multi-disciplinary. The full range of natural and social science disciplines have been involved in studying smallholder systems with respect to quantifying various ES; understanding their linkages to larger biophysical, social, and political environments; identifying the challenges and opportunities for sustainability; and recommending management and policy changes that can promote the sustainability of these systems, including compensation to producers for their contributions to ES. Truly *inter*-disciplinary studies have been fewer, but there is a growing set of published information that can be used to evaluate the ES and sustainability of smallholder systems from an interdisciplinary perspective. There are a few comprehensive interdisciplinary efforts, such as the MEA, that attempt to summarize and document as much of this information as possible within individual and synthesis reports (URL: www.millenniumassessment.org). However, these reports cover the entire spectrum of the world's ecosystems and human linkages, of which smallholder systems are only a part.

The purpose of this review is to assess the current science and understanding of smallholder forestry and agroforestry in the tropics, with a focus on the ecosystem services generated by these systems, using the framework of the MEA. We chose forestry and agroforestry systems specifically because incorporation of trees within smallholder production systems holds great potential for improving certain ES. The inherent biophysical structure and function of trees can improve ES such as habitat for wildlife, soil stabilization, nutrient recycling, and carbon sequestration. As well, management practices often change to accommodate trees, including reduced soil disturbance and increased diversity of understory plants, that improve ES. Finally, trees can be used strategically within a smallholder system to conserve and enhance ES such as creating or retaining riparian buffers; providing habitat for pest predators or crop pollinators; reducing pressures on forest reserves for timber and fuelwood; creating windbreaks or terraces to slow wind and water erosion; and supporting the development of certified organic, eco-friendly or sustainable crop production systems. Although more and more of the world's food and fiber production comes from large-scale and intensified agriculture and forestry operations, smallholder systems and family farms still provide the majority of the world's production for most of the diversified agricultural products and continue to contribute significantly to staple food crops like rice, sorghum, and pulses, especially in the tropics. Much of the tropical and especially the developing world rely heavily on wood for heating and cooking; the collection and sale of fuelwood and charcoal in these regions is overwhelmingly a small-scale enterprise. Although the logging and manufacturing of wood products in the tropics is largely the domain of private businesses, smallholder plantations do contribute to local industries. In addition, many smallholder communities utilize and manage forest areas for wood and non-wood forest products, so they affect and are affected by what happens to these forests.

Finally, one of the major motivations of the MEA is to understand the world's ecosystems and their contribution to human well-being within the context of changing climate, socioeconomic development, human populations, and disturbance patterns. Smallholder systems are paradoxically both adaptable and vulnerable to these changes. Individual smallholder producers tend to be resource-poor and thus vulnerable to changing markets, weather patterns, socioeconomic conditions, natural disasters, and government policies. With little capital or cash on hand and labor limited mainly to household members and temporary day laborers, small changes can make or break individual producers. On the other hand, smallholder systems as a whole are highly adaptable *because* they require relatively little land, labor, equipment or capital investment. In other words, the barriers to initiating new or changing existing smallholder systems are relatively small. In addition, smallholders can draw upon traditional knowledge, practices, and resources that are generally robust to stress and limiting climatic or soil conditions. Indeed, traditional knowledge and resources are often developed specifically to ensure survival of the system rather than to maximize productivity of one or a few crops. This clearly can be a challenge for intensified modern agriculture, but it can be an advantage when attempting to design more sustainable production systems or optimize various ES.

This review is broken down into two major sections. The first describes the variety of ES promoted in smallholder forestry and agroforestry systems. These include those for which benefits are mainly on-farm, such as income generation, and those that are mainly off-farm, such as wildlife habitat. Similarly, some ES are generated largely by the actions of a single farm, while others depend greatly on the aggregate structure and function of a community of smallholder systems. The second section describes several major mechanisms by which smallholders are or can be compensated financially for the provision of ecosystem services. As with the provision of ecosystem services, certain mechanisms are relevant and achievable mainly at the single-producer level, while others require or are best achieved within a community-based or cooperative approach. The goal of both provision of and compensation for ecosystem services is to promote the sustainability of smallholder production systems, the livelihood of these communities, and the larger society of which they are a part, including the global community. While our focus is on individual smallholders and the surrounding community, the influence of larger societies and external forces are critical to their sustainability. By recognizing the interdependency of these often rural and sometimes disconnected communities with the larger society, it is hoped that their sustainability can be promoted within a context that promotes ecosystem services rather than continuing to focus on socioeconomic development that separates production systems and community livelihoods from environmental conservation.

2   Smallholder Forestry and Agroforestry in the Tropics

Smallholder forestry and agroforestry in the tropics (and temperate zones) are quite common, especially where a large percentage of the population still relies upon agriculture or forestry to meet some or all of their subsistence needs. Smallholder agriculture generally is defined as landholdings <5 ha. Smallholder forestry has been delimited to areas <100 ha. For this review, we do not follow any strict convention on size but rather use a comparative framework to distinguish smallholders from larger landholdings or production systems. We also include community-managed forests that may exceed 100 ha in total, since they often support or are managed by many households that individually are considered smallholder producers.

Similarly, we do not set any strict requirements for what constitutes a forest or agroforest land cover type but rather rely upon a comparative approach to evaluate a continuum of production systems. Forestry is any land management system that attempts to generate products and services from forests. Agroforestry is a land management system that attempts to integrate agriculture and forestry within the same land area. These can be simultaneous or sequential systems. Simultaneous systems attempt to find space and share resources required for plant growth among the crops and trees. Tree and crop species selection, tree spacing, and management of tree crowns are common practices used to maintain acceptable growth among the tree and crop components. Sequential systems generally shift from agriculture to forestry, since trees over time will dominate the capture of light and other resources. In shifting cultivation, this cycle repeats, utilizing the ability of forests to accumulate and recycle nutrients to improve soil fertility and quality for future cropping cycles.

For both, trees provide important products, as well as services. Smallholder forestry and agroforestry systems in particular exist along a continuum of ecological conditions, desired products and services, and management practices. At one end, there are intact native forests in which useful trees and understory plants are selectively retained and nurtured through competition control, overstory thinning and outplanting to maintain their growth and reproduction within the forest (Trauernicht and Ticktin 2005). As well, crops can be integrated within single-species tree plantations to provide alternative products and income, especially during the early years of tree growth (Witcomb and Dorward 2009). At the other end are relatively open agricultural systems in which select trees are planted in specific arrangements and managed primarily to benefit the growth of the crop, such as N-fixing hedgerows; vegetated terraces to conserve soil and water; windbreaks; living fences; and shade for crop species that are sensitive to excessive solar radiation, high temperature, or low relative humidity. In between are a range of systems in which the degree of tree cover, crop production, and management of trees, crops, and their interactions vary.

One of the most common crops used in tropical agroforestry systems is coffee (*Coffea* L.). It is both a crop of worldwide importance and cultivation as well as one that can survive and produce yields under a range of shade levels and management intensities. Davidson (2005) described a range of coffee agro-ecosystems in Mexico from those planted under otherwise native forest and traditional polyculture shade systems to coffee under monoculture tree plantations or full-sun. This range of systems is not unique to Mexico or to coffee agroecosystems, but the descriptions in Davidson (2005) provide a useful set of categories that correspond to both ecological conditions and management systems.

Livestock also are important components of many smallholder systems, and their integration into agriculture, forestry and agroforestry is similarly varied. Direct grazing of managed forests is common in India (Government of India 2005). Dry-season grazing of crop residues and forest and shrubland is common in monsoonal or semi-arid places like India and sub-Saharan Africa (Sherperd 1992). Consideration of livestock management is beyond the scope of this review, but it is a key component of smallholder-managed landscapes and is a controversial issue when attempting to conserve forests or native grasslands (Government of India 2005; Brockington 2002) or integrate trees into agricultural systems.

Modern smallholder forestry and agroforestry development have centered on meeting the needs of smallholders within an increasingly limited land base, both growing rural populations and out-migration of rural producers to cities, and decline in on- and off-farm resources such as soil and water quality and local forest cover and health. Household or community woodlots have been promoted to provide local sources of fuel and construction wood. Living fences, windbreaks, and border plantings of fuelwood trees are alternatives where land is limiting. To conserve soil and water on cultivated hillslopes, vegetated terraces have been encouraged, especially with N-fixing tree or shrub species. These multi-purpose hedgerows check soil erosion, allow for build-up of terraces, and provide a source of green manure for crops. Alley-cropping with similar tree or shrub species to provide green manure for food staple crops like maize has had mixed success; intercropping with N-fixing ground cover species tends to provide similar benefits without the labor required to manage tree or shrub species. In shifting cultivation systems, fallows improved with N-fixing or fast-growing trees help restore soil fertility and quality under pressures of reduced fallow periods. Commercial tree plantations have been promoted within sequential agroforestry systems, such as taungya, and in simultaneous systems with shade-tolerant crops, like coffee or cacao (*Theobroma cacao* L.). Diversification of tree species in smallholder forestry and agroforestry systems, especially of native species, has been promoted primarily to conserve biodiversity (Spellerberg and Sawyer 1996), but other research has explored the effects on overall productivity and nutrient recycling as well (Bigelow et al. 2004).

3 Ecosystem Services from Smallholder Forestry and Agroforestry

Smallholder forestry and agroforestry systems are increasingly being recognized for their potential and actual contributions to ecosystem services (ES). The Millennium Ecosystem Assessment (Garrity 2004) groups ecosystem services into four categories: provisioning (e.g. production of food or fiber), regulating (control of climate or pests and diseases), supporting (e.g. nutrient cycling and plant pollination), and cultural (e.g. spiritual, recreational, or aesthetic). Most agroforestry systems have been evaluated for the ES that benefit the producer directly, and these cover all four of the major categories. Conversely, forestry systems have been studied extensively for the ES that benefit larger stakeholder groups. Part of this difference is due to the fact that while forestry and agroforestry exist along a continuum they are conceptually defined near the extremes. Agroforestry is considered as an extension of agricultural production: relatively simple, highly managed and frequently disturbed, and focused on product and income generation. Ecosystem services, similarly, are focused on supporting the production system and thus the producer. Indeed, much of modern agroforestry development has been motivated by the goal of supporting the sustainability of agricultural production by generating ES such as soil and water conservation, pest and weed control, and diversified products. Outside of monoculture plantations, forests are considered to be essentially natural ecosystems that inherently provide ecosystem services that benefit non-producers on-site and contribute to ES at larger spatial (and temporal) scales. Given that forestry and agroforestry systems exist along a continuum, especially at the smallholder scale, the ability of agroforestry to provide ES that benefit larger groups of stakeholders and the direct benefits of forests to smallholders have increasingly been topics of study and evaluation.

The ability of any one system to provide ES depends upon the underlying climatic, geographic and topographic setting; the plant species composition and their spatial and temporal arrangement; the management of the system as a whole; the location of the system within a larger landscape; and the surrounding matrix of land cover and land uses. Much smallholder research and documentation has focused on ES at the farm scale, such as production of crops and trees, conservation of soil and water, on-farm biodiversity, and carbon sequestration. Comparisons of ES in small-holder systems with surrounding land uses, such as intact forest or open agricultural land, has provided some perspective from which to evaluate smallholder systems within a larger spatial context. However, few studies have actually evaluated small-holder systems within an explicit landscape context. This is unfortunate, as many ES like biodiversity and watershed function operate at multiple spatial and temporal scales that cross individual land units and thus are affected by the aggregate conditions and processes occurring within and among smallholder systems and the surrounding land cover and land use types. This information is key to the development of land management policy and planning strategies at larger spatial scales. For example, certain types of forestry and agroforestry activities are allowed within 'buffer zones' surrounding core protected forest areas of Biosphere Reserves (Cedamon et al. 2005). The assumption is that these activities provide livelihood alternatives to exploitation of resources within the forest reserve while contributing more to regulating and supporting ES than open-area agriculture or grazing.

Below we compare the ability of various smallholder forestry and agroforestry systems to generate ES within most of the categories outlined by the Millennium Ecosystem Assessment. Examples will draw heavily from coffee agroforestry systems and smallholder woodlots. Coffee is a worldwide commodity that is produced primarily in smallholder systems. It is tolerant of a wide range of shade levels and thus is produced in many of the smallholder categories listed above, including tree plantations and secondary forests. In addition, there are multiple sustainability certification systems well-established for coffee production that include criteria and indicators for achieving a range of ES across the four major categories. At the forestry end of the spectrum, we focus mainly on individual woodlots which have been successfully developed and promoted in many areas of the tropics, especially the semi-arid tropics, where local wood supplies are often limiting. They are grown for fuelwood, timber, and non-timber forest products (NTFPs) and provide for both household needs and income. As with coffee, there are several sustainability certification systems focused on timber production.

4 Comparison of Ecosystem Services from the Two Systems

4.1 Productivity of Crops, Trees, and Non-Wood Forest Products

Productivity within smallholder agriculture is generally considered to be less than in large-scale systems because smallholders often use lower inputs, may rely on lower-yielding traditional crop varieties, and increasingly are relegated to lower-quality sites. However, most of the world's diversified agricultural products are grown primarily in smallholder systems and so represent most of the world's supply. Some of these crops also are grown primarily in agroforestry systems. Coffee production, for example, covers 0.9 million hectares of Central America (CEPAL 2002). The majority comprises shaded coffee agroforestry systems (Galloway and Beer 1997), with 94% of the coffee having shade cover in Nicaragua (MAGFOR 2002), and 91% of coffee area having trees integrated into the plantations in Costa Rica (CATIE 2002). Intensive management of modern and even traditional varieties and land races of coffee can generate maximum yields, but the required supplemental irrigation and fertilization are beyond the capacity of most smallholders. Shade thus offers a compromise between maximizing coffee yields and minimizing management input requirements to sustain production.

In addition, trees can provide direct products and services to the producer. Coffee agroforestry systems across Central America typically provide fruits, fuelwood and timber to the farmers. Fuelwood is typically provided by legume shade trees, either *Inga* P. Mill. or *Gliricidia* Kunth; only in Costa Rica where *Erythrina* L. are the dominant shade trees is fuelwood not an important product. Most fuelwood production is for household or farm needs, and in some cases portions are sold (Westphal 2008; Schibli 2001). The most important fruits are dessert bananas, cooking bananas and plantains (*Musa* L.). Over 70% of smallholders integrate *Musa* spp. with coffee production. As a part of the staple diet, much of the production is for home consumption, but they also provide income throughout the year. This is especially important during periods when coffee prices are low. The importance of other fruits varies greatly, but usually they are for home consumption, although a few farmers (about 10%) produce sufficient numbers for sale and generate significant income.

Timber species also are common in agroforestry systems. Many studies have shown that trees generally benefit from the wide spacing and management inputs and practices typically associated with agroforestry systems (Thulasidas and Bhat 2009; Ceccon 2005; Haggar et al. 2003; Norgrove and Hauser 2002; Nissen et al. 2001; Somarriba et al. 2001), although at least one study has shown trees do better under plantation conditions (Swamy and Puri 2005). In one study in the Yucatan region of Mexico, both timber trees (*Swietenia macrophylla* King and *Cedrela odorata* L.) and fruit trees generally grew better in intercropped systems than in pure plantations (Haggar et al. 2004). Smallholders growing trees for timber production often choose fast-growing species such as laurel (*Cordia alliodora* [Ruiz & Pav.] Oken) in Costa Rica, *Gmelina arborea* Roxb. in the Philippines (Cedamon et al. 2005; Bertomeu 2004), *Falcataria moluccana* (Miquel) Barneby and Grimes in Southeast Asia and the Pacific region (Roshetko et al. 2004) and more broadly *Acacia mangium* Willd. and select *Eucalyptus* L'Hér. species. Annual growth rates of these trees during the first 5–10 years often equal or exceed 3 m in height and 3 cm in stem diameter at breast height (dbh). In an assessment of laurel planted for timber production with relatively wide spacings in agroforestry systems in Costa Rica, Somarriba and Beer (1987) calculated an optimal rotation age of 34 years, with a total volume of 300–700 m^3 ha^{-1}. This translates to a mean annual increment of 10–20 m^3 ha^{-1}, which is as high as or higher than the maximum rates reported by the FAO (2001) or the US Forest Service (Burns and Honkala 1990) for laurel trees in plantations.

Table 1 Individual tree and stand characteristics at 5 years for laurel in smallholder systems of Costa Rica (Modified from Somarriba et al. (2001))

Measure	Production system					
	CLP	TA	NC	LP	OC	PP
Stem density (stems ha^{-1})	58	466	234	192	151	933
DBH (cm)	28.5	24.8	22.2	18.1	16.8	15.6
Height (m)	19.1	17.9	17.9	14.8	14.5	13.4
Stem vol. (m^3 stem^{-1})	0.31	0.25	0.19	0.08	0.08	0.05
Rank	*1*	*2*	*3*	*4*	*5*	*6*
Stand vol. (m^3 ha^{-1})	18.0	116	44.5	15.4	12.1	46.6
Rank	*5*	*1*	*3*	*4*	*6*	*2*

CLP cacao-laurel-plantain, *TA* taungya system, *NC* new cacao field, *LP* line planting of trees, *OC* old cacao field, *PP* pure plantation

In some agroforestry studies, productivity measurements have focused on individual trees rather than the entire stand (Bertomeu 2004; Norgrove and Hauser 2002; Somarriba et al. 2001), perhaps due to the limited size of smallholder fields and the sometimes non-uniform planting arrangement of trees. Individual tree growth is not the key metric for timber production but rather the total productivity of all trees within the system. As an example, Somarriba et al. (2001) reported individual tree growth, including volume, for laurel trees growing in a variety of smallholder systems but did not include stand or system-level production. Using reported density and survival data of the trees, the ranking of sites with respect to productivity per area is different than when ranked by individual tree growth (Table 1). The pure plantation site had the lowest stem diameter and tree height at 5 years, but due to dense planting and high survival, the stand produced the second-highest total volume. Sites with lower production potential, therefore, may be better suited to pure tree plantations to maximize timber production, but better-quality sites have the potential to support both crop and timber production with relatively wide tree spacings.

In this set of systems, the taungya design produced the optimal balance of individual tree and stand growth at half the density of the pure plantations. As a temporal agroforestry system, the goal of a taungya system is for the trees to eventually form a closed canopy (fully stocked) tree plantation. Crops are grown for several years after tree planting until canopy closure. An interesting variation on the taungya system from southern Cameroon was reported by Norgrove and Hauser (2002). *Terminalia ivorensis* Chev. trees planted as relatively dense plantations were thinned at either 6 or 17 years. This allowed for underplanting of plantain (*Musa*) and/or tannia (*Xanthosoma sagittifolium* [L.] Schott) until the tree canopies refilled the gaps created by thinning. In their study, the greater thinning intensity (final density of 64 trees ha^{-1}) resulted in more rapid growth of remaining trees but at the expense of total stand basal area and biomass compared to the unthinned plantation. However, the lighter thinning treatment (final density of 192 trees ha^{-1}) resulted in similar basal area and total biomass as the plantation but with significantly larger average tree size. As might be expected, productivity of plantain and tannia were better in the greater thinning treatment.

Planting of trees in woodlots or other arrangements to provide fuelwood has been promoted widely for smallholders to provide more accessible local sources and reduce pressures on off-site forest lands. Local wood supplies are increasingly scarce in the semi-arid tropics where lack of rainfall limits natural productivity and regeneration. The result has been poaching of wood from protected forests, use of lower-quality alternative fuel sources such as crop residue and animal dung that would be better used as soil cover and organic matter inputs, traveling longer distances to find available wood supplies, or purchasing wood from local vendors. Woodlot species are generally selected for fast early growth. Productivity within smallholder systems is generally tied to appropriate species selection and good nursery management to ensure vigorous seedlings are planted. Rotations are generally short (10 years or less), and silvicultural interventions between planting and harvest are minimal. Even within the restrictions of smallholder systems and semi-arid conditions, most woodlots have proven to be highly beneficial in terms of meeting landowner wood needs. In central Ghana, fuelwood plantings accounted for one-third of all the land area under agroforestry in 1994 (Anane and Twumasi-Ankrah 1998). Fuelwood plantings span a range of systems, from managed secondary forest, to fallow plantings, to dedicated woodlots, to intercropped agroforestry systems, to a few scattered trees (Bensel 1995). Trees used for fuelwood are quite diverse, especially from secondary forests (Klock 1995) and often are useful for other purposes, such as fruit or nut production, timber and N-fixation (Kalinganire 1996; Bensel 1995). Intentional fuelwood plantings are often carried out in areas where off-site supplies are limited and/or must be purchased from local vendors, such as sub-Saharan Africa (Ngetich et al. 2009; Akinnifesi et al. 2008; Ramadhani et al. 2002; Tefera et al. 2001; Kalinganire 1996) and heavily deforested or intensively managed forest areas of the Philippines (Bensel 2008; Arnold and Contreras 1979). Otherwise, fuelwood planting or use of trees for fuel from smallholder farms is done according to individual landowner needs and the types of agricultural or forestry systems being managed (e.g. Klock 1995).

Reforestation of deforested or degraded lands through initiation of smallholder forestry and agroforestry systems is increasingly common in tropical areas. Bensel (2008) has argued that such "reclamation" of degraded land has reversed the trend of forest degradation and deforestation in Cebu Province in the central Philippines. This reversal appears to be driven in large part by commercial demand for fuelwood and other wood products rather than subsistence needs. Some of these are pure forest plantations (Garrity and Mercado 1993), but others are intercropped at least partially (Haggar et al. 2003) or as fully integrated agroforestry systems (Bensel 1995), depending upon the potential of the site to support agricultural crops (Haggar et al. 2003) and the marketability of timber versus other products (Hoch et al. 2009). As already shown, intercropping can improve individual tree growth and provide incentives for smallholders to reforest. In Southeast Asia, particular attention has been paid to restoration of wastelands dominated by the grass *Imperata cylindrica* (L.) Beauv. (Foresta and Michon 1996; Macdicken et al. 1996; Menz and Grist 1996). Experimental work and models of restoration pathways demonstrating the feasibility of these systems have been generally confirmed in operational examples with smallholders (Roshetko et al. 2002; Foresta and Michon 1996). Thus, smallholder forestry and agroforestry represent viable options for restoring wastelands that may not recover without active management.

Incorporation of native species within reforestation programs is promoted for biodiversity conservation and enhancement, but often there is little known about the potential growth rates or proper silvicultural management of these species. Screening and seed selection can help to identify trees and genetic sources that have sufficient growth rates and are compatible with various smallholder or larger-scale production systems. In Sarapiqui, Costa Rica, a native species reforestation program run by the Organization for Tropical Studies evaluated 67 native and 17 exotic tree species for reforestation of degraded pastures over a 10-year period (Butterfield 1995). Among the native species, ten had growth rates similar to the most productive exotic species (*Acacia mangium* and *Gmelina arborea*). Other exotic taxa (*Pinus* L., *Eucalyptus* and *Tectona grandis* L. f.) had poor growth and survival, and even *A. mangium* and *G. arborea* were prone to stand die-back due to pests and diseases (Haggar et al. 1998). One of the major limitations of the native species was poor form when grown on open sites, but this could be improved with selection of seed sources and early formative pruning. Among the native species evaluated it was possible to identify different species groups that would be appropriate for different kinds of production systems (Haggar et al. 1998) including: reforestation of degraded pastures (e.g. *Vochysia guatemalensis* J.D. Sm.), mixed timber plantations (e.g. *Jacaranda copaia* [Aubl.] D. Don with *Calophyllum brasiliense* Cambes), agroforestry systems (*Dipteryx panamensis* [Pittier] Record & Mell), or biomass plantations (*Goethalsia meiantha* [J.D. Sm] Burret), thus demonstrating the potential of native species to meet different productive needs. Prior to this work, only one native tree species and various exotic species were approved for reforestation in the Atlantic zone of Costa Rica. Afterward, eight to ten native tree species were approved for reforestation in the region.

Communally managed forests are common in parts of Mexico, South and Southeast Asia, and sub-Saharan Africa. They span a range of sizes, structures, conditions, management objectives, allowable resource uses, and local ownership and control (Balooni and Inoue 2007; FAO 1986; Arnold and Contreras 1979). Common forest land is almost always multi-purpose, with a mixture of timber and non-timber forest products (NTFPs), including dry season forage for livestock (Sherperd 1992). Collectively, the NTFPs are generally more important for household use and sale than is the timber. Commercial timber production from commonly held forest lands is practiced in many areas and is generally done via concession, (i.e., through contracting with a logging company). As with forests generally, the variability in community-managed forests precludes any generalizations about tree or stand productivity. Mexico's community-managed (*ejido*) forests are often held up as a successful example of deliberate and sustained timber production with strong community control over use and accrual of benefits from harvesting (Bray et al. 2003; Kiernan 2000). In other areas, such as the Solomon Islands, timber companies negotiate directly with local government councils or community leaders to harvest from communal forests, often resulting in exploitation of the resource and little active silviculture (Kabutaulaka 2005; Bennett 2000) due to poor information on actual timber values and lack of expertise in forest management. Although the volume of timber extracted is generally recorded as a part of regulated business transactions (Montgomery 1995), individual site productivity is rarely measured or recorded, and silvicultural practices are usually neglected or are relatively simple.

Smallholder tree plantations and agroforestry plantings of timber trees, however, are important sources of wood for local mills in some areas, such as the Philippines (Bertomeu 2006, 2004; Magcale-Macandog et al. 2006; Anyonge and Roshetko 2003; Garrity and Mercado 1993), the Amazon basin (Sears et al. 2007), and Kenya (Holding et al. 2006). As mentioned previously, individual tree growth and even stand volume growth can be as good as or better than in plantations. Wood quality, however, may not be as good as in well-managed plantations. Wide and irregular spacing or lack of early pruning or selective thinning can result in trees with poorer stem form and more visual defects (Thulasidas and Bhat 2009).

4.2   Income Generation

Although productivity is a relatively straight-forward metric for quantification and comparison, for smallholders it is the contribution of these products to household livelihoods that is of importance. Maximizing production of a single crop or product is rarely the goal in smallholder systems. Multipurpose trees and diversity on the farm generally provide more stability and flexibility in production choices and help farmers manage risk associated with changing market conditions, weather patterns, or crop performance. In southern Africa, fodder trees, woodlots, and fruit trees have proven to be profitable for smallholders (Akinnifesi et al. 2008). For community-managed forests, the NTFPs and uses of the forest provide the primary benefits to smallholders, including sale of marketable products, such as açaí palm fruits (*Euterpe oleracea* Mart.) in the Amazon basin of Brazil (Muñiz-Miret et al. 1996).

One of the most widely grown crops in agroforestry systems is coffee; mainly as varieties and hybrids of *Coffea arabica* L. (Arabica) and *Coffea canephora* Pierre ex Froehner (Robusta) (formerly *C. robusta* L.). Coffee agroforestry systems provide livelihoods for hundreds of thousands of households. In most regions, such as Central America, the majority of producers are smallholders (CEPAL 2002). Little information is available on why farmers maintain high tree diversity in coffee agro-forests and what the economic benefits are of doing so. An analysis of different case studies indicate average income from trees and other shade plants in the coffee agroforestry system rarely surpasses 20% of gross income (Table 2). Although this average hides considerable variation, in each group of farmers there is always a minority that obtains substantial income from fruit, banana or plantain production associated with the coffee, or in the case of Pacific Guatemala, from timber sales. In this region, timber sales combined with fuelwood production provide the equivalent of 38% of the value of coffee production. Probably of greater significance to coffee agroforestry smallholders is the range of products provided by coffee agroforestry systems for household use. In separate studies in Nicaragua, Schibli (2001) and

Table 2 Range of income (US $) from products associated with coffee in coffee agroforestry systems (Modified from Haggar (2008), (2006), Guharay et al. (2005), Martinez (2005), Schibli (2001), and Westphal (2000))

Location	Fuelwood	Bananas	Fruit	Timber	% Income
Guatemala					
San Marcos	120/ha	70/ha	0	nd	18%
Huehuetenango	200/ha	15/ha	0	nd	12%
Ocosito	30/ha	10/ha	0	77/ha	35%
Honduras					
Chiquimula	160/ha	223/ha	21/ha	nd	17%
El Paraiso	4/ha	48/ha	1.5/ha	0	11%
Ocotepeque	24/ha	9/ha	1/ha	nd	1.5%
El Salvador					
Metapan	21/ha	nd	3/ha	nd	6%
Nicaragua					
North	0	135/farm	38/farm	nd	14%
South	9/farm	32/farm	125/farm	nd	15%
Costa Rica	0	120/ha	nd	0	11%

nd not determined

Westphal (2000) found 80–90% of household needs for cooking bananas and fuelwood, basic products for food security and energy, were met from within the coffee agroforestry system. In these cases, the economic benefits are reduced expenditures for household goods and greater self-sufficiency.

Although sales of tree products may not be the major source of income for most coffee smallholders, trees do represent important sources of income for other smallholders. Along the TransAmazon highway in Brazil, timber sales by smallholders are used to provide income during times of financial stress (Amacher et al. 2009) rather than as a regular stream of income or a windfall at the end of a defined rotation. In these situations, trees require practically no management and represent mainly an opportunity cost to use the land for other purposes. In contrast, Mexico has between 300 and 500 community forest enterprises that manage forests specifically for commercial timber production (Bray et al. 2003). In Southeast Asia, smallholder timber plantation development in the 1980s and 1990s was driven in large part by commercial demand for fast-growing trees (Garrity and Mercado 1993). Similar opportunities increased tree planting in agroforestry systems in the Brazilian Amazon (Smith et al. 1996).

Fuelwood plantings are attractive for smallholders as much for the income potential as for the ability to provide a household supply of wood. In Ghana, smallholders employ a shamba (temporal) agroforestry design on a 10-year rotation in which trees are intercropped with maize and cassava for 3–4 years and then used for charcoal and fuelwood for the remainder (Anane and Twumasi-Ankrah 1998). The cost-benefit ratio has been estimated as approximately 2.5. In Tanzania, smallholders employ a similar system on a 5-year rotation, using mostly legume trees (Ramadhani et al. 2002). In this situation, land is not limiting, and trees can be managed outside the cropping season when labor demands are reduced. Farmers use the wood grown in these systems for their own needs, so the financial benefit is the reduction or elimination of the need to purchase fuelwood. The potential returns to land and labor for growing fuelwood trees were higher than in a maize-fallow system. The ability to use degraded lands or land not suitable for crop production to manage tree plantations is attractive to many smallholders (Jagger and Pender 2003). Where crop production is a necessity due to land limitations, boundary plantings of trees can still provide significant wood for household needs and income generation (Witcomb and Dorward 2009).

One of the inherent barriers to smallholder forestry is the investment period required to establish and manage these systems before products and economic benefits are generated (e.g. Scherr 1995). Agroforestry systems have the potential to buffer this opportunity cost by providing intermediate products while the trees are maturing. There also are generally net economic benefits for smallholders by combining understory crops and NTFP harvesting in timber plantations, even on degraded lands (Menz and Grist 1996). In a comparison of pine plantations and related agroforestry designs in West Java, Indonesia, combining food crops with trees yielded the best financial returns over a 15-year period, and resin-harvesting from the trees benefited producers over a 30-year rotation (Sopandi and Rule 2000). Over shorter time scales, returns from intercropping may not cover the increased management costs as compared to tree plantations (Haggar et al. 2003). However, for smallholders, growing staple crops provides for subsistence needs on land that could otherwise be used for more intensive food production.

The spatial and temporal mixture of trees, crops, and NTFPs can be driven as much by market prices and availability of labor and land as by the technical and agricultural feasibility of managing specific systems. Smallholders will experiment with a wide variety of species mixtures and system designs in response to market demands and household needs. Smith et al. (1996) noted 108 agroforestry configurations with 72 crop species among 136 smallholder fields in the Brazilian Amazon. Typically, trees become a more attractive option relative to cash crops as labor becomes more limiting (Nissen et al. 2001). Conversely, where land availability is limited, forests that are traditionally maintained for NTFPs or fuelwood gathering may be cleared for food production (Ngetich et al. 2009).

The diversity of products from smallholder systems is mirrored in the diversity of markets for these products. Rural smallholders with limited infrastructure or transportation options generally sell the majority of their products in local markets. Some of these products are targeted at local markets, such as traditional varieties of rice and other food staples and most NTFPs. In these cases, smallholders, generally women, take the products to market and sell them directly (Balooni and Inoue 2007). For those products targeted at larger markets, smallholders can be at a disadvantage. Isolation, lack of transportation, lack of storage capacity for perishable products like fruits or vegetables, and lack of knowledge of market prices combine to make smallholders dependent upon local buyers and thus vulnerable to exploitation

(Balooni and Inoue 2007). Even when buyers are paying fair market prices, the inability to store seasonal products means that most smallholders are harvesting and selling individual products all at the same time, leading to a glut in supply and thus a depression in prices. For those products that can be stored and shipped to global markets, such as coffee, prices are, of course, still subject to global fluctuations in supply and demand. Specialized, low-volume products like gum arabic (*Acacia senegal* [L.] Willd.) may be particularly vulnerable to these fluctuations, even in the presence of rising demand (Elmqvist et al. 2005).

Smallholder production and income can be promoted through direct support programs. In the Philippines, the World Bank financed the development of small-holder tree farms to supply pulpwood for local paper mills. The technical support, reliability of a local market, and reasonable prices for the product all supported the financial viability of the enterprise. However, a focus merely on production and away from sustaining yield threatened to undermine the sustainability of the enterprise. As well, farmers were locked into a single species-product-market system, making them vulnerable to external problems, such as typhoon damage to the trees and government price controls on product value (Hyman 1983). Thus, while such enterprises can be initially beneficial for smallholders, they simultaneously make them dependent upon a specialized production and marketing system in which sustainability is largely outside their control.

Lack of an integrated production-marketing plan can easily work against small-holders. In Orissa State, India, several private and government-sponsored tree-growing initiatives have largely failed to benefit smallholders. A private tree plantation initiative for smallholders required farmers to leverage the future yield of the trees to pay for the seedlings and inputs required for planting. When the trees failed to achieve the volume growers were told to expect after 10 years, the value of the trees was not sufficient to pay off the loans to the mill that initiated the project (V. Das, personal communication). Even government-sponsored projects, such as fruit and nut tree planting, that provide technical support to ensure proper growth and yield, may not benefit smallholders if market access is through a single buyer and products are sold all at the height of seasonal production and harvesting (Balooni and Inoue 2007).

In Costa Rica, Haggar et al. (2003) reported that smallholders engaged in refor-estation often incorporate crops into parts of the plots. Although trees benefited from weed control and crop production provided income to farmers, this did not cover the increased management costs. Farmers still considered intercropping a benefit because it allowed them to reforest land that otherwise they would reserve for crop production. In the Yucatan Peninsula in Mexico big-leaf mahogany (*Swietenia macrophylla* King) and Spanish cedar (*Cedrela odorata* L.) grew better in enrichment plantings or as plantations with understory crops than in pure planta-tions (Haggar et al. 2004). Enrichment plantings had lower investment and mainte-nance costs, but the understory plantings provided additional benefits from crop production. These systems also incorporated a range of ten different fruit trees. Growth of some of the native fruit trees such as *Manilkara zapota* (L.) van Royen and *Pimienta doica* L. was better in enrichment planting conditions, but the majority (e.g. *Persea americana* Mill. and *Byrsonima crassifolia* [L.] Kunth) grew better under the plantations. These cases demonstrate that the option of incorporating crops into reforestation systems can significantly improve tree growth and provide incentives and benefits for smallholder farmers to reforest. But in both cases the majority of agroforestry options did not generate a net economic benefit within the first 3 years.

4.3   Biodiversity Conservation: Ecological and Functional (Agricultural) Importance

4.3.1   Agricultural and Native Plant Biodiversity

In humid and sub-humid climates, forests are the natural vegetative cover. The structural and functional diversity of plants within a forest generally supports the greatest biological diversity for plants, animals, and microorganisms. Thus, maintaining or promoting tree and forest cover within agricultural landscapes is generally seen as a way to conserve or enhance biodiversity. Within the context of smallholder systems, this can be achieved in a number of ways. Although intensification of agricultural production has been blamed for accelerating conversion of forests to cropland on a global scale, defenders point out that increasing yield per unit area and encouraging 'settled agriculture' for smallholders may conserve forest cover by reducing land requirements for agricultural production and moving producers away from shifting cultivation. Incorporation of trees in smallholder production systems can lead to net increases in forest cover. Examples include establishment of smallholder tree plantations in degraded areas of Southeast Asia (Foresta and Michon 1996), planting of trees in woodlots for fuelwood production in Africa (Bensel 2008), and integration of useful trees with crops in agroforestry systems of Central America (Montagnini et al. 2005).

Often, smallholder forestry and agroforestry systems occur near remaining forest patches or large forested areas. For example, coffee agroforestry in the mid-upper watersheds of Central America often borders forest remnants and protected areas. These are generally within the Mesoamerican biological corridor, highlighting the environmental importance of these systems for the protection of water sources and conservation of biodiversity (DeClerk et al. 2007). The diversity and environmental importance of trees in coffee agroforestry systems in Central America and southern Mexico has been highlighted by several authors (e.g. Méndez et al. 2007; Philpott and Dietsch 2003; Moguel and Toledo 1999; Greenberg et al. 1997; Perfecto et al. 1996).

Individual smallholder forestry and agroforestry systems support a variable number of species, depending upon the type of system and the overall landscape context. Collectively, however, total species richness within smallholder systems can be quite high. Across 6 regions in southwestern Bangladesh, Kabir and Webb (2008)

Table 3 Summary of tree species composition found in coffee agroforestry systems in Central America (Modified from Haggar (2006, 2008), Méndez et al. (2007), Guharay et al. (2005), and Virginio Filho (2005))

Site	Farms	Spp./plot[a]	Shade	Fruit	Timber	Other	Total
Guatemala							
San Marcos	16	14	3	14	16	11	44
Ocosito	36	9	6	10	13	14	43
Huehuetenango	25	11	4	11	8	15	38
El Salvador							
Tacuba	52	12–22	12	15	16	80	123
Nicaragua							
San Ramon	16	26	9	22	10	73	114
Las Sabanas	7	16	5	9	11	36	61
Costa Rica	120	*nd*	19	17	24	54	114

nd not determined

[a] Farmer plots evaluated varied from 0.25 to 1.00 ha

recorded 419 plant species (59% native) within 402 homegardens. Klock (1995) recorded 180 tree species within 108 woodlots and shifting cultivation plots in two villages in the Philippines. Greenberg et al. (1997) documented 180 bird species in coffee plantations in Chiapas, Mexico, exceeded only by intact forest.

Table 3 summarizes results from tree species richness inventories in coffee agroforestry systems in Central America. Within a given region, between 40 and 100 tree species are typically found in coffee systems, with 8–26 species per farm based on inventories of 0.25–1.0 ha. From the repeated inventories in Nicaragua there appears to be a considerable turnover of low-frequency species. A compilation of different studies of tree diversity in coffee across Mesoamerica by DeClerk et al. (2007) found a total of 627 species from 98 families, of which 92% were native. However, 37 of the families were represented by just one species, and most species were only infrequently found. Legumes (family Fabaceae) represented 125 of the species. The average number of species per farm was only 4.6, although the surveyed plots were typically only 1,000 m² (0.1 ha).

As in the studies presented above, smallholder coffee plantations are typically more diverse than larger land-holdings, which rarely have more than five species (De Clerk et al. 2007). Larger land-holdings tend to be dominated by a few multi-purpose legume trees such as *Inga*, *Erythrina* or *Gliricidia sepium*, though usually with other species scattered at lower densities. Even in the cases presented above where larger land-holdings have shade systems derived from natural forest, as in Pacific Guatemala, they tend to be less diverse than those of smallholders. This can be seen in the contrast between the adjacent Ocosito catchment and San Marcos in Guatemala. These locations share the same ecological conditions, but smallholders dominate in San Marcos, compared to the much larger land-holdings in Ocosito. Although the total number of species is similar, the smallholder farms in San Marcos average 14 species per plot compared to 9 per plot in Ocosito.

Small-scale plantations in a degraded forest landscape can provide sites for establishment of understory tree and shrub species and potentially facilitate the regeneration of forest cover (Parrotta et al. 1997). At the La Selva Biological Station in the Atlantic region of Costa Rica, Powers et al. (1997) studied natural regeneration in abandoned pastures compared to small-scale plantations reforested with different species. Although there were differences between reforestation species, there was generally greater species richness and/or regeneration density in the reforestation plantations than in abandoned pasture. Overall, 550 species of vascular plants were identified in an area of 9 ha, 126 of which were woody species. This represented one-third of the known flora in the La Selva Biological Station. Results from the experimental plantations were similar to plantations managed by farmers, although regeneration was affected by farm management such as cattle grazing (Haggar et al. 1997). In both studies, woody regeneration was dominated by secondary forest shrub and tree species, with a low frequency of seedlings of primary forest species.

4.3.2   Wildlife Biodiversity

Conversion of relatively unmanaged and intact forest to agricultural production systems generally represents a loss of structural and overall plant diversity, as well as a potentially large change in species composition (e.g. Martinez et al. 2009). As might be expected, faunal diversity also tends to decline. In some cases, the species richness of agroforests can be as high, if not higher than, surrounding forests (Cassano et al. 2009; Maas et al. 2009; Beukema et al. 2007; Harvey and González Villalobos 2007). The similarity of species composition ranges widely, with averages from 25% to 65%, depending upon the organismal group (e.g. bats, birds, insects, herbaceous plants, or trees) (Bhagwat et al. 2008; Beukema et al. 2007). Fragmentation of forests, greater openness, and more intensive or frequent disturbance associated with management activities tend to favor common or invasive species and provide poorer habitat for forest interior and especially rare plant and animal species. Conversely, larger trees, greater tree cover, and lower management intensity result in more similar levels of biodiveristy (Bisseleua et al. 2009; Bhagwat et al. 2008; Beukema et al. 2007). Active management of diverse agroforests can conserve important tree species that could otherwise be lost through uncontrolled exploitation (Cassano et al. 2009). Despite the potential loss of biodiversity compared to unmanaged forests, agroforestry and smallholder plantations are promoted in buffer areas around protected forest, specifically within biosphere reserves (Cedamon et al. 2005), as alternatives to completely open pastures or agricultural cropping systems.

Within coffee and cacao plantations there is a relatively consistent pattern of greater plant and animal diversity in diverse, shade-grown versus sun-grown systems (Clough et al. 2009b; Moguel and Toledo 1999; Perfecto et al. 1996). Complex agroforestry systems also can support greater animal diversity than mono-specific shade (Clough et al. 2009b; Greenberg et al. 1997). Calvo (2004) studied biodiversity in the coffee agroforestry plantations in Guatemala and found more bird species, greater species diversity and a more even species distribution in diversified or forest-shaded coffee than in *Inga*-shaded coffee. Although most bird species were typically of forest edge or secondary forest, some primary forest species were found in the forest-shaded coffee. There also was greater diversity and cover of epiphytes in the trees of the forest-shaded coffee, which was related to a greater diversity of bird species. Birds in the forest shaded coffee were typically nectar-feeders, while those in *Inga*-shaded coffee were insect-feeders.

On-farm tree planting also can increase forest cover and species diversity within smallholder landscapes. Cedamon et al. (2005) documented 88 different species planted by farmers in four rural communities in Leyte Province of the Philippines, with a maximum of 100 trees per farm. As with coffee farms, most of the trees planted (83%) came from just 10 species. In traditional shifting cultivation systems in the Philippines, Klock (1995) recorded 180 different tree species in managed secondary forest plots. As with management intensity, there is generally a tradeoff between biodiversity of the agroforestry system and the productivity of the 'crop' species. Lawrence (1996), for example, found an inverse correlation between plant species diversity and rubber productivity in agroforestry gardens in Indonesia. Reduced abundance of the crop species and increasing competition with other plants is inevitable with more diverse production systems. Finally, on-farm tree-planting indirectly benefits biodiversity by reducing pressure on intact and/or protected forest areas for forest products (Ramadhani et al. 2002). Bhagwat (2008) cites several studies that showed increased health and availability of forest resources in protected areas where communities practiced agroforestry versus open agriculture.

Even if individual smallholder systems are too small or lack sufficient diversity to support high levels of plant or animal diversity, they can be important components within a landscape that supports biodiversity at this larger scale. Smallholder forestry and agroforestry are considered to be important for maintaining biodiversity in highly deforested landscapes (Boffa et al. 2008; Beukema et al. 2007). In landscapes with <20% tree cover, tree cover and diversity in smallholder systems can serve directly as refugia for forest plant and animal species (Cassano et al. 2009). However, the real value of agroforestry systems for biodiversity is when they are integrated into a landscape mosaic of forested and agricultural land uses. Smallholder forestry plantations and agroforests are important for providing connectivity among remnant forests (Cassano et al. 2009; Uezu et al. 2008; Chacon Leon and Harvey 2006). Structuring smallholder systems to serve as biological corridors within the landscape has been recommended to maximize this function (Weerd and Snelder 2008). Encouraging heterogeneity in smallholder systems can also support greater landscape-level biodiversity (Hoehn et al. 2010; Tscharntke et al. 2008).

Finally, plant and animal diversity in smallholder systems may decline as a function of distance from remnant forests (Uezu et al. 2008), but variation within a landscape can have a stronger influence on biodiversity than distance from intact forest *per se* (Boffa et al. 2008). Perhaps more importantly, tree plantations and agroforests in buffer zones and forest margins can extend the range of suitable habitat for forest species (Bhagwat et al. 2008). Given this, smallholder forestry and agroforestry are allowable and even encouraged land use practices in the buffer zones around United Nations-designated Biosphere Reserves (Mehta and Leuschner 1997).

An increasing number of studies also have documented effects of land management on soil biodiversity. The diversity of arthropod groups like termites and ants within managed landscapes have been the subject of conservation-oriented research (Gillison et al. 2003; Jones et al. 2003; Eggleton et al. 2002). The functional diversity of soil fauna and microorganisms is considered to be highly important for regulating and supporting ecosystem services such as nutrient cycling, detoxification, and carbon sequestration. Research has shown that within the soil decomposer community, despite high functional redundancy, certain species are highly important to observed functional capacity (Wolters 2001). There is a recognized link in both forested and agricultural systems between the structural and species diversity of plants and the diversity of soil organisms and soil function (Gillison et al. 2003; Altieri 1999). George (2006) compared soil fauna diversity under different land uses and found significantly greater diversity of species in shaded than unshaded coffee and greater diversity in organic vs. conventional management. Furthermore, diversity in the organic shaded system was equal to or greater than that in nearby forest. Many of the influences on soil diversity are related to agronomic practices, including soil cover, individual species presence, plant litter deposition and organic matter inputs (mulch, manure, compost, etc.), pesticide use, and tillage or soil disturbance (Barros et al. 2003; Eggleton et al. 2002; Altieri 1999). Benefits from healthy and diverse soil biological communities will impact a variety of ecosystem services, such as nutrient cycling, water infiltration, and carbon sequestration.

The major caveat to the potential of smallholder systems to support biodiversity is that on-farm species composition, tree cover, and management practices are dynamic; they change over time. Pardini et al. (2009) undertook a comprehensive survey of the biodiversity of multiple organismal groups in the Atlantic forests of Bahia State, Brazil, where cacao agroforestry is mixed with mature forest. While the existing agroforests and secondary forests had high species diversity, changes in tree species composition in smaller remnant forests and especially the forest understory suggested that future changes in forest composition would compromise the sustainability of present diversity. In Indonesia, cacao production often follows a cycle of establishment in partially cleared or secondary forest followed by intensification and increasing overstory clearing to increase yields and incomes (Steffan-Dewenter et al. 2007). Clough et al. (2009a) documented this process in detail, showing that such intensification in the short-term encourages immigration to an area and expanded cultivation. Over a 20–30 year time frame, increased pest and disease pressures as well as aging cacao trees greatly reduce yield. For the next generation of farmers, cutting down and replanting the cacao trees is the only option to overcome poor yields. Instead, they often choose to abandon existing plantations and emigrate to new areas of forest where the cycle is repeated. The potential for forest recovery in these abandoned and degraded plantations was not studied, but the concern is that new forest clearing and plantation establishment are generally outpacing natural recovery of abandoned sites. Such dynamic changes in land use and diversity, of course, can also work in reverse. Earlier examples were cited of smallholder reforestation of degraded landscapes in Central America (Haggar et al. 2003) and Southeast Asia (Bensel 2008; Garrity and Mercado 1993). Thus, while smallholder systems and practices that support biodiversity and associated ecosystem services should be encouraged, a comprehensive approach to sustainability must include the flexibility to change based on internal and external factors. At the landscape scale, it is important to plan or evaluate land use over time and space, taking into consideration the typical developmental or cyclical changes likely to occur in particular forestry or agroforestry systems.

4.4   Soil and Water Conservation

Water runoff and soil erosion in agricultural, forest and grassland systems has been extensively studied and modeled. Based on the Universal Soil Loss Equation (USLE) model and subsequent modifications, the major factors influencing these processes include: rainfall intensity and duration; the inherent erodibility of the soil; topographic slope steepness and length; cropping practices such as tillage; and conservation practices such as crop residue cover, contour plowing, crop rotation, and intercropping (Wischmeier and Smith 1965). Soil and water conservation in agriculture has generally focused on modifying cropping and conservation practices, as these are the factors most influenced by management. The concept of 'conservation agriculture' has coalesced around three general practices: continuous organic soil cover, minimal soil disturbance, and appropriate crop rotation (Hobbs 2007). Although this concept is meant to promote the overall sustainability of agriculture, these practices are grounded in soil and water conservation. In undisturbed forests and grasslands, runoff and erosion are generally quite low because both systems maintain continuous organic soil cover, have high surface roughness, and experience minimal soil disturbance, at least when compared to conventional agricultural tillage. The conservation of soil and water in these ecosystems is considered to be important for watershed functions such as groundwater recharge and water quality; clean, reliable stream flows; reduction in flooding risk; healthy aquatic biota; and maintenance of healthy coastal and near-shore marine ecosystems such as coastal wetlands and coral reefs (Fares and El-Kady 2008). These functions are recognized as ecosystem services. In certain environments, such as the high volcanic islands of the Hawaiian archipelago, these are by far the most economically valuable and socially important ecosystem services that forests provide (Kaiser and Roumasset 2002).

Because smallholder agriculture and forestry, especially along the frontier of existing forests and grasslands, is situated on steeper slopes, soil and water conservation practices are vital for watershed-related ecosystem services. Beyond the standard practices included in the concept of conservation agriculture, terrace formation on sloping lands is a common and ancient practice in large- and small-scale agriculture to promote soil and water conservation. By breaking up the slope steepness and length, terraces dramatically lower the impact of these factors. Terraces are often created through building low rock walls ("bunds") along the slope contour and either allowing soil to naturally build up behind the wall or actually digging into the slope behind the rock wall and spreading the soil out between the walls. Alternatively, woody plants are established along the contours in dense line plantings to capture eroding soil and create terraces. Terraces do not eliminate the slope factor, so advanced terrace structures include water diversion channels to concentrate runoff toward low-lying areas where the water can be stored temporarily to promote infiltration (Treacy and Denevan 1994).

In the context of smallholder production systems, both agroforestry and tree-planting generally have been promoted to conserve soil and water, especially on steep slopes. Terraces also are promoted where annual crops are planted or where management may disturb the soil or disrupt soil cover. Planting terrace boundaries with woody multi-purpose trees or shrubs is a common recommendation (Sheng 1989). These vegetated barriers have proven effective, although on steep slopes, erosion may occur on the immediate downslope side of the plants, compromising terrace integrity (Johnson et al. 1982). Tree cover *per se* can be beneficial on hillslopes. The fine roots of most plants promote aggregation of soil particles, increasing soil macropore space and thus surface infiltration. The woody and perennial root systems of trees both anchor the tree and hold the soil in place, which can be important in reducing severe erosion or mass wasting on slopes or inherently unstable soils (Ziemer 1981; Swanson and Dyrness 1975). Given that maintaining trees generally requires minimal tillage or related soil disturbance, the maintenance of plant root networks provides continuous stability.

Communally managed forests typically occupy the steeper slopes and other areas within a watershed that are more vulnerable to runoff and erosion. These areas are less accessible, more difficult to farm, and clearly less sustainable for crop production, so they may be kept in forest cover simply due to their low potential for crop production. Where traditional social and cultural institutions remain strong, such forests may be officially protected against clearing or intensive agricultural use. Stands of trees or forest may be designated as "sacred groves" by local religious leaders. Often these groves surround a religious shrine, place of worship, or dwelling for priests, monks, or gurus (Bhagwat and Rutte 2006). Most smallholders in these communities also recognize the watershed value of these protected forests, so religious sanctions typically reinforce practical regulations accepted and enforced by community members.

Finally, it is worth noting that the protective function of tree cover on runoff and erosion can be severely compromised if adequate soil cover is not maintained. 'Throughfall', the rainfall that penetrates the tree canopy and reaches the soil, can have a similar if not higher erosive potential as rain drops that reach the soil unimpeded from the atmosphere (Brandt 1988; Mosley 1982). Tree leaves can concentrate water on their surface, increasing rain drop size. Tall tree canopies allow for sufficient acceleration of these rain drops to generate a similar or greater force on the soil surface. Thus, soil cover as live vegetation or plant litter (crop residue or tree litterfall) is critical for absorbing the energy of the rain drops and slowing their descent to the mineral soil.

At the watershed scale, water discharge from streams generally increases as vegetative cover, especially leaf area, declines (Hamilton and King 1983). This is due to a reduction in evapotranspiration (ET) with lower leaf area. In forests, harvesting of the trees usually results in increased stream water yield for several years until leaf area is restored through plant regeneration. Conversion of forest to crop land generally results in a more permanent decline in leaf area and thus longer-term increases in water yield. Replacement of forest with agroforestry also increases water yield (Verbist et al. 2005), again likely due to reduced leaf area and thus ET.

The net effect of increasing vegetative and soil cover on groundwater recharge is difficult to estimate. Decreased runoff and increased infiltration must be balanced against increased ET. Recharge also requires percolation of soil water below the effective rooting zone. Narain et al. (1998) estimated that *Eucalyptus* and *Leucaena* trees in forestry and agroforestry systems exploited soil water to a depth of 3.0 m as compared to 1.5 m for crops. Although this was estimated to result in better water use efficiency at the system level, it was not expected to result in greater groundwater recharge. Deep-rooted trees may also reach relatively shallow water tables (<10 m deep), tapping directly into groundwater sources. During the rainy season, soil macropores created by the coarse root channels of trees may allow for greater percolation and groundwater recharge, resulting in increased stream flow and spring discharge during the dry season (Sandström 1998). This would benefit upland users, regardless of the overall effect on downstream water yield.

This may partially explain the conventional wisdom that forest cover increases stream flow. Most evidence at the watershed scale suggests increased forest cover decreases overall water yield (Bruijnzeel 2004). Although total water yield may be lower, increased groundwater recharge and maintenance of stream flows during dry periods are critical for smallholders. They typically rely upon relatively shallow wells and surface water for irrigation and household needs. Structural interventions that increase infiltration and groundwater recharge at the local scale, such as earthen check dams or runoff diversions into temporary reservoirs, can result in significant benefits to smallholders in seasonally dry environments (Sreedevi et al. 2006). Integrating trees within specific runoff and erosion deterrence structures can enhance structural durability without greatly increasing demand for water. Within the larger watershed, this should help to reduce fluctuations in stream flow, improve water quality, and actually preserve remaining forest land by improving the productivity and sustainability of existing agricultural areas.

4.5   Soil Fertility and Nutrient Cycling

Trees in agroforestry systems have been widely recognized for their role in sustaining intra-system nutrient cycling in coffee and cacao agroforestry systems (Beer 1998; Nair et al. 1995). Trees take up large quantities of nutrients that are largely allocated to nutrient-rich but short-lived tissues such as leaves, fine roots, and reproductive tissues (flowers, fruits, and seeds). Turnover of these tissues recycles the nutrients as organic 'litter'. Decomposition and release of the nutrients back into a mineral form completes the nutrient cycle. Trees with deep roots are able to take up nutrients that are not accessible to crops, re-depositing them on or near the soil surface. As well, nitrogen-fixing trees increase the site N capital through atmospheric fixation (Vitousek and Sanford 1986). Nevertheless, trees also adsorb and sequester nutrients in perennial tissues such as branches, stems, and coarse roots, and even short-lived tissues represent competing sinks with crop plants. Furthermore, if tree biomass is harvested for use off-site, it represents a net loss from the system.

Although the principles of nutrient cycling in smallholder forestry and agroforestry are well-recognized, there have not been many studies that have quantified nutrient balances in these managed systems. Because many of these systems use N-fixing trees, the effects of these species on the N cycle are of great importance and research interest. Dommergues (1987) found N fixation by legume trees were generally between 10 and 100 kg N ha^{-1} year^{-1} depending on species and production system, and Roskoski (1982) estimated N fixation of up to 40 kg N ha^{-1} year^{-1} in coffee shaded by *Inga* in Mexico. Improvements in availability of nitrogen measured as nitrogen mineralization in the presence of legume trees have been demonstrated in some cases (e.g. Zuluaga 2004; Babbar and Zak 1994; Haggar et al. 1993). Direct N fixation by legume trees, however, is generally less important than the recycling of N-rich plant litter. Litterfall from *Erythrina poeppeginia* trees used as coffee shade in Costa Rica may return over 200 kg N ha^{-1} year^{-1}, more than enough to meet annual crop demand. Pruning trees to manage shade levels can also return large pulses of N. Youkhana and Idol (2009) estimated N returns of 170 kg ha^{-1} year^{-1} from *Leucaena* trees pollarded twice per year in an experimental shade coffee system. The availability of the N and other nutrients to crop plants during the current cropping cycle may be quite low, around 20% (Palm 1995), but increases in soil organic matter and N can be significant over just a few years (Youkhana and Idol 2009), suggesting that longer-term cycling of nutrients should be enhanced.

Frequent harvesting of trees for use off-site, however, can reduce site nutrient capital. Munguia et al. (2007) found that regular harvesting of firewood from shaded coffee led to a negative nutrient balance. Even where N-fixing trees are used, soil nutrient balances depend upon the proportion of plant (and especially tree) biomass exported as farm products (Shepherd et al. 1995) or transferred to other locations (e.g. as mulch for open-grown crops) (Youkhana and Idol 2009).

In the tropics and subtropics, phosphorus (P) can be as limiting as N to crop and tree production. Palm (1995) reviewed alley-cropping trials with legume trees across various soil types and climates and concluded that regular pruning returned sufficient nutrients to meet crop P demands. Szott and Melendez (2001) found that labile soil P fractions were higher in forest and agroforestry systems as compared to sole cropping, with or without fertilization. Trees may be able to access sparingly soluble soil P better than crop plants through association with mycorrhizal fungi or organic acid exudation (Miyasaka and Habte 2001). However, studies in low-P agroforestry systems suggest competition for available P is high and will negate any benefit to the crop of increased access to soil P (Radersma and Grierson 2004; McGrath et al. 2000).

One frequently mentioned potential benefit of trees in agroforestry systems is their potential to root more deeply than crops, taking up nutrients not available to the crop plants and then recycling them as aboveground litterfall or near-surface root turnover (Jobbágy and Jackson 2004). Despite the enthusiasm, there are few studies that support this "deep mining" hypothesis (Schroth et al. 2001). Mechanistic studies of nutrient flux and root distribution suggest that competition within the shared soil volume is likely much greater than any facilitative effect of deep nutrient capture (Gillespie 1989). Nevertheless, scientists have been developing models and making recommendations based on tree root architecture and other characteristics, assuming deep rooting both reduces competition for surface soil nutrients and maximizes the potential for deep mining (Akinnifesi et al. 1998; Van Noordwijk and Lusiana 1998; Schroth 1995; Van Noordwijk and Purnomosidhi 1995). Across rainfall and soil nutrient gradients in West Africa, Kessler and Breman (1991) showed there was limited potential for increasing nutrient availability via nutrient redistribution (deep rooting and nutrient recycling) and a real risk of increased nutrient (and water) competition under limiting conditions. Schroth et al. (2001) pointed out that the potential benefits of deep mining are limited in more deeply weathered tropical soils such as Ultisols and Oxisols. Avoidance of direct competition through spatial partitioning of root systems in agroforestry systems certainly is not to be discouraged, especially for multipurpose trees, but assumptions of net benefits to crop nutrient status or productivity through recycling of nutrients through deep rooted trees has simply not been observed in many agroforestry systems.

Although deep mining may not be a common direct benefit in agroforestry systems, the tighter nutrient cycles associated with tree cover and forests are well-recognized. In general, forests and tree plantations show low rates of leaching of limiting nutrients and lower leaching overall than grassland or agricultural systems (Imbach et al. 1989). The protective function of trees to reduce soil erosion also conserves nutrients. In shifting cultivation systems, the initial increase in soil pH and base cation status due to burning is utilized for crop production, but these improvements are quickly lost due to leaching, as well as removal of harvested yield. Incorporating trees at this early stage of the cycle can maintain favorable soil properties for longer periods (Alfaia et al. 2004). Finally, the deeper rooting habit of trees can be important for mitigating leaching of fertilizer nutrients. Several agroforestry studies have shown that deeper-rooting trees capture and take up nutrients leached below the crop rooting zone. This has been termed the "safety-net" function of tree roots (Cadisch et al. 1997; Van Noordwijk et al. 1996). It may be especially important under heavy fertilization (Allen et al. 2004; Nair and Graetz 2004) or irrigation (Lehmann et al. 1998).

Nutrient use efficiency (NUE) is a concept used to evaluate the capacity of production systems to function under nutrient-limiting conditions. There are multiple definitions of NUE, depending upon the processes and scales of interest (Ewel and Hiremath 1998). At the leaf level, photosynthetic NUE can be defined as the maximum rate of photosynthesis at a given leaf nutrient content. At the whole plant level, the inverse of nutrient concentration is a simple calculation of NUE because it scales total biomass by internal nutrient content. Because many plants, especially woody perennials, have the capacity to retranslocate nutrients internally, another measure of NUE is defined as biomass production divided by nutrients absorbed from the soil. At the field or ecosystem scale, this can be indexed as the biomass produced divided by the nutrients lost in litterfall or root turnover. A related measure of NUE divides biomass production by an index of soil nutrient availability.

For intensified agriculture, NUE may be defined as growth response per unit fertilizer added. One of the key objectives of crop breeding during the Green Revolution was to develop varieties that could take advantage of high levels of nutrient availability by increasing production of harvestable yield. Plant morphological and physiological adaptations to low nutrient availability generally constrain their maximum productivity at high nutrient availability; thus, 'efficiency' can be defined at both ends of the spectrum of nutrient availability. Smallholder forestry and agroforestry systems generally operate under low nutrient input levels, so NUE is most relevant at the low end of the nutrient availability spectrum. At the field scale, conservation or retention of nutrients already within the system is or should be a key concern. In general, plants that conserve nutrients internally promote higher system-level nutrient retention. This, however, has to be traded off against the need for available nutrients in the soil to supply short-cycle crop plants that will be harvested on an annual or seasonal basis.

Several strategies are available to smallholders to maximize NUE without undermining crop productivity. The use of N-fixing trees is a common method of actually increasing site N capital and substituting for external N inputs. Within experimental tree plantations, the right combination of timber and N-fixing service trees can increase timber production compared to monocultures (Binkley et al. 1992). This may lead to limitations of other nutrients, especially P, so nutrient balances must be considered. Woody fallows, especially when combined with the use of N-fixing trees, are partially a nutrient conservation and rehabilitation strategy. Woody plants build up nutrient capital in plant biomass and litter that can be liberated at the beginning of the next cropping phase. Incorporation of trees as a permanent part of the production system ensures continuous live plant cover and the potential for nutrient uptake, minimizing nutrient losses. Pruning and pollarding of trees to manage shade levels and canopy cover also provides a source of organic nutrients that generally are mineralized faster than from natural litterfall. Even where fertilizers are added, the 'safety net' function of deep-rooted trees can increase site nutrient conservation and thus long-term nutrient use efficiency. In open-grown agricultural systems, relay cropping, cover cropping, green manures, and minimum tillage are analogous practices to minimize nutrient losses and augment site nutrient capital, but they cannot match the longevity, rooting depth, and biomass potential of trees and woody plants.

4.6   Carbon Sequestration: Avoided Loss and Increased Net Capture

Carbon (C) sequestration has become a topic of great interest for agriculture and forestry because of the provisions of the Kyoto Protocol that make plant biomass C an exchangeable commodity. Primary forests contain the largest stocks of biomass C, averaging 300 Mg ha^{-1} in the wet tropics (Palm et al. 1999). Selective logging generally results in about a 50% loss of biomass C, and conversion to other land uses can reduce standing stocks even further (Lasco 2002). Because the C contained in plants is directly proportional to biomass, greater tree cover and larger trees equate to more biomass C. In forests not utilized for production, conservation of tree cover and especially larger trees is key to maintaining standing stocks. In tree plantations and complex agroforestry systems, some of the C lost from the primary or secondary forest is regained as trees grow and mature, but expected maximum levels of biomass C are around 100 Mg ha^{-1} (Palm et al. 1999). Annual sequestration rates vary widely. A review published by the Alternatives to Slash-And-Burn Climate Change Working Group reported sequestration rates of 2–10 Mg ha^{-1} year^{-1} (Palm et al. 1999), with maximum C accumulation occurring in 5–25 years, depending upon system type and rotation length. A more recent review reported sequestration rates from less than 1 to greater than 15 Mg ha^{-1}, depending upon tree density, growth rate, and type of management (Nair et al. 2009). Because tree biomass tends to increase exponentially with stem diameter, it is better from a C sequestration perspective to maintain and promote shade, windbreak protection or other service functions of agroforestry trees with fewer but larger individuals.

Detailed comparisons of C stocks have been made for shade coffee systems in Central America (Fig. 1). For open-grown coffee, aboveground C stocks maximize at approximately 10 Mg ha^{-1}. Managed shade systems, whether with legume shade trees (*Inga* or *Erythrina*), fruit trees or timber species can achieve approximately 30 Mg ha^{-1}. In secondary forests or systems with free-growing forest trees, carbon stocks may be as high as 60–80 Mg ha^{-1}. This still falls somewhat short of intact forest C stocks of 100–200 Mg ha^{-1}. There are examples of coffee grown in the cleared understory of old-growth forests of Hawaii where aboveground tree biomass C is in the 100–200 Mg ha^{-1} range (Elevitch et al. 2009). Because coffee productivity tends to decline at shade levels above 50%, there is an eventual tradeoff between crop production and the size of biomass C stocks. However, maintaining a lower density of mostly large trees can minimize this tradeoff.

As mentioned previously, cacao agroforestry systems, especially in western Africa, are generally established in the shade of relatively intact or secondary forest because of the protection needed by seedlings. Carbon stocks may be quite high as a forest understory planting, but maintaining tree cover over time is of real concern. Because of self-shading that occurs as plants mature, the tendency is to reduce tree cover over time to increase crop yields. Thus, a system that initially appears favorable for balancing crop production and C sequestration may actually be unfavorable over the productive cycle of the crop (~30 years). Once the forest has been mostly cleared and the cacao trees begin to decline in productivity, these sites may be cleared for agriculture, grazing, or simply abandoned as degraded systems.

Biomass carbon in tree plantations may generally exceed that in crop-oriented agroforestry systems due to greater tree density. However, the trees in smallholder plantations are typically grown on a 5–20 year rotation, depending upon the desired product and tree species. Thus, the permanence of the biomass C sequestered depends mainly upon the intended use of the timber. Where the end-product is fuelwood, most of this C will essentially be released back to the atmosphere during combustion. Paper and packaging materials may be recycled but also are not expected to have a long lifespan. Construction wood, furniture, fence posts, and

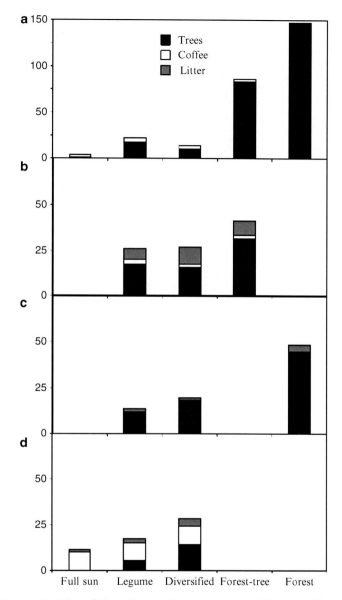

Fig. 1 Carbon stocks (Mg ha^{-1}) in coffee agroforestry systems and forest in different regions of Central America under varying shade types. (**a**) Guatemala (Modified from Medina et al. (2005)); (**b**) Nicaragua (Modified from Suarez (2002); (**c**) Turrialba, Costa Rica (Modified from: Mena-Mosquera 2008); (**d**) Perez-Zeledon, Costa Rica (Modified from: Miguel Magaña et al. 2004). For Turrialba, C stocks in plant biomass include trees + coffee

other solid wood products may last for years to decades, but this ranges widely. In agroforestry systems, retaining large old trees for shade or wildlife habitat or growing fruit trees that may last decades can provide long-term biomass C sequestration at lower tree densities. Plantation establishment on deforested or degraded lands, however, may be considered a net positive for C sequestration, since the processes of tree planting and growth provide short-term benefits that can be extended forward in time as reforestation expands to new areas.

Carbon stocks in root systems are more difficult to quantify, but there are an increasing number of studies that have excavated and estimated root C. Fine root biomass and C have been estimated extensively, but much like leaves, their short lifespan makes them important mainly as a source of inputs to soil organic matter. Coarse root biomass is generally quantified for roots with a minimum diameter of 5 mm. A global dataset of forest biomass suggests that for mature stands, aboveground biomass is greater than root biomass (Cairns et al. 1997). The root:shoot ratio across this dataset ranged from 1:9 to 2:1; forests with higher overall biomass (>400 Mg ha^{-1}) tended to have lower root:shoot ratios (less than 1:3). A regression equation to predict root biomass using aboveground biomass accounted for a large proportion of the variability ($R^2 = 0.83$), but the 95% prediction intervals were quite wide, especially for the upper limits of root biomass. For a forest with aboveground biomass of 200 Mg ha^{-1}, the predicted root biomass ranged from 25 to 75 Mg ha^{-1}. For a stand with aboveground biomass of 300 Mg ha^{-1}, the predicted root biomass was approximately 30–125 Mg ha^{-1}.

For agroforestry systems in which trees may be grown for shorter periods as fallows or pruned and coppiced regularly to manage shade levels or to provide useful products, root biomass may be a greater proportion of total biomass. Kaonga and Bayliss-Smith (2009) estimated total tree biomass in coppiced and non-coppiced fallows in eastern Zambia. For fallows that ranged from 2 to 10 years in age, the range of root:shoot ratios for C was 1:4 to 3:1, suggesting roots constitute a larger proportion of C stocks on average for younger trees and certainly for trees that are periodically coppiced. For other agroforestry systems, root:shoot ratios have ranged from 1:1 to 1:4 at 5–10 years of age (Swamy and Puri 2005; Oelbermann et al. 2004; Youkhana and Idol 2011a).

On a global scale, the C stored in soil organic matter (SOM) greatly exceeds that in terrestrial vegetation (Nair et al. 2009). For mature forests, plant and soil C may be approximately equal, within the range of 100–400 Mg ha^{-1}. Agroforestry systems, because they generally have lower biomass C, are expected to have larger soil than biomass C stocks. In complex agroforestry systems, soil C may be similar to primary forest (Palm et al. 1999). Other agricultural cropping systems generally have much lower soil C stocks than intact forests (Lal 2005; Palm et al. 1999) or grasslands (Conant et al. 2001). Conversion of cropland back to forest or grassland generally increases soil C (Guo and Gifford 2002). However, average long-term rates of soil C sequestration are quite variable; ranging from −50 to +1,500 kg ha^{-1} year^{-1} (Post and Kwon 2000). Because soils contain large stocks of organic C (100–400 Mg ha^{-1}), measuring a change of just 5% is equivalent to a sequestration rate of 5–20 Mg ha^{-1}. This is comparable to the annual net primary production of terrestrial ecosystems. Given the potentially high spatial variability of soil C, it can be difficult to capture even these modest changes in soil C over short time periods. In addition, one of the major mechanisms of soil C stabilization is complexation with silt and clay-sized particles. This influence of soil texture can overwhelm any effects of changes in land cover or management practices (Kaonga and Bayliss-Smith 2009). Given these factors, it is not surprising that some studies have failed to find significant differences in soil C due conversion of cropland to agroforestry (Kaonga and Bayliss-Smith 2009) or tree plantations (Giardina et al. 2004). Where soil C sequestration does increase with a shift to agroforestry or tree planting, the theoretical maximum should be comparable to that from intact forests. Guo and Gifford (2002) reported that long-term (>50 years) re-establishment of secondary forests may be necessary to completely recover soil C and that tree plantations did not sequester or maintain soil C stocks as high as intact or mature secondary forest.

For agroforestry, just conserving existing soil C against further losses associated with cropping system management can be significant (Oelbermann et al. 2006). Tree presence alters soil management, especially tillage, reducing disturbance-related losses. Maintaining soil cover and stability reduces erosion losses. The addition of N fixing trees within a system can actually increase soil C as compared to the use of non-fixing species (Resh et al. 2002). Detectable changes in total soil C may take 10 years or more to become manifest (Young 1997), but certain practices, such as mulching of tree pruning residues, may significantly increase soil C after only 2–3 years (Youkhana and Idol 2009).

For trees to actually increase soil C stocks, it appears that more than just increased organic matter inputs may be needed. Trees must increase the stability of existing or new organic matter (i.e., they must reduce the rate of soil C loss). The major stabilizing mechanisms for soil organic matter include encapsulation of particulate organic matter (POM) within soil aggregates and physical complexation of SOM with silt and clay-sized particles. In addition, there are biochemically recalcitrant compounds, such as lignin and polyphenols, that are released from decaying organic matter or are formed as by-products of microbial decomposition (Six et al. 2002). Degryze et al. (2004) have shown that trees tend to promote better soil aggregation, which provides physical protection to otherwise labile POM. The protection of C within these aggregates is hypothesized to be on the order of years to decades (Six et al. 2002), but aggregates can be destroyed by tillage or compaction associated with crop production. Youkhana and Idol (2011b) have shown that over a 3-year period, addition of N-fixing tree mulch within an agroforestry system significantly increased most soil C fractions in the top 20 cm, including POM protected in fine aggregates (53–250 μm) and silt + clay associated SOM. Under afforestation of former cropland, Degryze et al. (2004) found no significant increase in total C but greater sequestration of POM-C in fine aggregates in the top 7 cm of soil. Old-field succession did increase total soil C, both as POM in fine aggregates and as silt + clay associated SOM. In these cases, it was not clear whether increases in stable C fractions were due to additions of new C or protection of existing C. Where there is mainly replacement of C from one land use type to another (Giardina et al. 2004; Townsend et al. 1995), such questions may be practically unimportant. However, if soil C is to be included in C sequestration accounting, then there may be demands to only include C in more stable fractions that match the general time frame of C offset projects (e.g. 30 years or longer).

The permanence of overall soil C sequestration in agroforestry and tree plantations is at least partially dependent upon the continued inclusion of trees in the system. Where forestry or agroforestry can increase more stable C fractions, this can provide 'legacy' effects on soil C, buffering potential losses due to changing land use or major disturbances. However, for temporal agroforestry systems, such as shifting cultivation or taungya and shamba systems, it is unclear how the relatively short-term changes in vegetation cover and land management, especially tillage or soil disturbance, affect soil C over single or multiple rotations. Previous estimates suggested declines of 15–27% (Detwiler and Hall 1988), similar to estimates for complete conversion of forest to open-grown cropland (Murty et al. 2002). Losses measured in individual studies vary from almost minimal to approximately 20 Mg ha^{-1} (Kotto-Same et al. 1997 and references cited therein). These estimates are much less than loss of biomass C, but a long-term concern for soil C may be that repeated cycles of clearing, burning, and cropping degrade soil properties that support plant productivity and thus organic matter inputs to the soil.

Nitrogen-fixing trees in particular appear to be important for achieving net gains in soil C with conversion of cropland to agroforestry or tree plantations. Several studies comparing plantations of N-fixing and non-fixing trees have shown significant gains of soil C only under the N-fixing trees (Resh et al. 2002; Garcia-Montiel and Binkley 1998). This may be due to a greater C allocation belowground by N-fixing trees (Binkley and Ryan 1998), greater litter inputs, or increased protection of existing soil C via organic N additions (Binkley 2005). Resh et al. (2002) showed that slightly more than half of the increase in soil C under N-fixing trees was a result of greater retention of existing soil C. Regardless of the mechanism, because many agroforestry systems incorporate N-fixing trees for both their product and service functions, this provides opportunities to sequester C in soil organic matter, with a significant proportion as physically or biochemically protected SOM.

5   Compensation for Ecosystem Services

5.1   Flexibility in Production and Income Sources

The major products from smallholder forestry and agroforestry systems are not generally unique or distinct but rather represent a broad range of commodities also grown in larger or more intensified systems. Some, like coffee, may be mostly grown in smallholder systems, but others, like staple food crops, are now grown mostly in large intensified monocultures. Even for staple food crops, smallholders generally rely on their harvest as a significant source of income and not just to meet subsistence needs. Thus, smallholders are subject to the same regional, national, or global trends in commodities markets as larger scale farms.

The global trend over the past 50 years towards agricultural intensification has affected smallholder agroforestry systems, including shaded coffee and cocoa, as well as large-scale agricultural operations. From the 1970s through the 1990s, there was a trend to remove shade from coffee and cocoa plantations (e.g. Samper 1999). The crash in coffee prices between 2000 and 2004 (added to two previous price drops in the 1990s), and reduction in economic and technical support to coffee producers halted the process of intensification and, at least in some cases, led to a return to lower input diverse-shaded coffee production (e.g. Westphal 2008).

Surveys of farmers conducted at the time reveal details of some of these changes. In Guatemala during 2001 and 2002, coffee prices reached lows of between $0.4 and $0.6 USD per pound. In response, farmers reported planting more trees in their coffee fields of all kinds – fruit trees (15–39% of farmers), timber trees (23–43%), and bananas (35% of farmers) – to diversify income. They also planted more shade trees (17–38% of farmers) to reduce fertilizer needs for the coffee and thus input costs (Guharay et al. 2005). As mentioned previously, both relative and absolute income from associated trees increase when coffee or other crop prices fall as farmers plant more trees that yield marketable products. During times of high coffee prices or in systems with high coffee productivity, income from associated products falls to 1–2% of the total system gross income. This is illustrated from a study in Honduras that compared income across years starting from the low coffee prices during 2001–2003 and continuing to the better prices of 2004–2006 (Haggar 2006; Guharay et al. 2005). It appears that coffee farmers had greater presence and productivity from bananas and plantains during the low coffee price years (Table 4). As coffee prices increased, farms reduced the presence of bananas and plantains, presumably with the intention of concentrating on coffee production. During the years of low coffee prices, products from associated trees represented about 11% of income from coffee agroforestry, which then fell to about 2% when prices improved.

These studies show the value of shaded coffee systems in enabling farmers to adapt the management of their coffee according to market conditions: when prices fall, increased shade cover and fruit tree production reduce the need for fertilizer inputs and provide additional product and thus income sources. When prices improve, investment in coffee can be increased, often at the expense of the tree species. From a development perspective, it may seem that the focus should always be on increasing the productivity of the main income-generating species so that yields are sufficient to survive during times of low commodity prices. However, this eventually requires either increased inputs or expanding the land under cultivation. This erodes the viability of the smallholder model, which has been the case for the production of many staple foods and other crops. From a sustainability perspective, however, the flexible strategy of agroforestry systems supports the smallholder model while maintaining other ecosystem services.

Table 4 Trends in income from coffee and associated trees in Honduras, 2001–2005 (Modified from Haggar (2008), (2006), Guharay et al. (2005))

	2001/2002	2003	2004	2005
Coffee price ($ per pound)	0.40	0.38	0.80	0.87
Cost management ($ per ha)		262	305	576
Productivity (no. of 46-kg sacks per ha)	9	10.5	8.7	13.1
Gross income from coffee ($ per ha)	362	402	687	1,142
Income from fruits ($ per ha)	n.d.	1.5	0.3	0.4
Income from fuelwood ($ per ha)	n.d.	4	n.d.	n.d
Income from *Musa* spp. ($ per ha)	46	48	17.5	3.4
Gross income ($ per ha)	408	455	705	1,146
Net income ($ per ha)		194	400	570
% of net income as tree products	11.3	11.7	2.5	0.3

nd not determined

Smallholders generally are reluctant to eliminate trees entirely from their production system, even where there is an increasing emphasis on production of a single cash or commodity crop. In the case study from Guatemala, shifts in tree cover or conversion of coffee to production of tree-based commodities like rubber generally occurred on medium and large farms, between 50 and 500 ha in size. Small cooperative producers did not eliminate any coffee, all of which was shaded. In this case, the smallholders conserved environmental services more effectively than the larger producers (Medina et al. 2005). This aligns with findings from another survey in Nicaragua showing that small-scale farms (less than 50 ha) maintain more trees per hectare, more tree species, and greater above-ground carbon stocks than large-scale farms (Alvarado and Cuadra 2010). The same decline in coffee prices in Nicaragua had a much smaller effect on use of shade or fruit tree production. Only vegetable production was more profitable than coffee, and converting large areas from coffee to vegetable production would not be a viable proposition. Some increases in pasture and food grain production were planned, but this was not at the expense of coffee (Suarez 2002).

5.2   Payment for Ecosystem Services

Implementing conservation measures in production systems to promote both agricultural sustainability and the provision of ecosystem services can have long-term and lasting benefits both on and off the farm. However, smallholders are physically, technically, and financially limited in their capacity to incur the up-front costs and sacrifice short-term productivity to promote these long-term benefits. Traditionally, governments have shared costs, provided low-cost loans, given technical support, and/or subsidized producers for short-term loss of income to promote conservation practices. More recently, payments for ecosystem services (PES) programs have been developed to compensate landowners for the anticipated or accumulated benefits associated with ecosystem services. As with traditional programs, PES can offset the short-term costs and loss of revenue associated with implementing conservation measures (Herrador and Dimas 2000). Such systems are often set up as multi-year contracts (5–20 years) with specific conditions, requirements for compliance monitoring, and the option of extending the contract at the end of the specified period. This provides an ongoing source of revenue tied directly to the level of ecosystem services provided, rather than a single rate or one-time payment based on implementation of specific practices.

Payment for ecosystem service programs exist at multiple government levels, from municipalities to international, United Nations-sponsored programs. In Central and South America, there are many municipal-level watershed conservation and C sequestration PES programs. Wunder and Albán (2008) reported on both types of municipal programs in Ecuador. The programs had a focus on the target ecosystem service, set strong conditions for participation, and selected participants based on their ability to comply with the conditions and the importance of their land for either conserving ecosystem services or the potential to improve them. This resulted in high compliance among producers with contractual obligations and demonstrable benefits in the case of C sequestration. Locatelli (2005) studied the disposition of water users in the city of Retalhuleu, Guatemala, to pay for improved water quality and a more consistent water supply through protection of coffee agroforestry systems in the Ocosito River catchment. The study concluded there was potential for payments of $11 USD per hectare per year to landowners who implemented land uses that contributed to water conservation. A review of actual payment for ecosystem services in Nicaragua by local water authorities (Wheelock-Diaz and Barrios-Jackman 2007) found that schemes were limited in extent, usually only compensating 10–20 producers in the area immediately around the water sources. Payments were about $20 USD per hectare for conservation of forest patches or introduction of soil and water conservation measures. In general, producers considered this too low to compensate for the investments required. In Costa Rica, hydrological services are paid for in a voluntary manner by hydropower companies and some water supply companies or users (e.g. Costa Rican Brewery) to the National Forestry Financing Fund (FONAFIFO). Funds may either be used for specific purposes, such as for forest conservation, or put in the general FONAFIFO fund, which includes compensation for reforestation (Rojas and Aylward 2003). Similar municipal watershed PES programs exist in other Central and South American countries and are not limited to compensating producers. In El Salvador, for example, funding of guards at a national park is paid for through a small fee to water users in the municipality (Herrador and Dimas 2000).

Costa Rica is the model for national PES programs. The FONAFIFO fund makes payments for reforestation, forest conservation and forest management of between about $200 and $550 USD per hectare distributed over 5 years (Rojas and Aylward 2003). Despite these generous payments, Sánchez-Asofeifa et al. (2007) did not find evidence of decreased rates of deforestation at a national scale due to the implementation of this program. They concluded instead that earlier conservation efforts and policies had been successful in generally reducing deforestation rates. The program did appear to provide multiple positive socioeconomic and development benefits (Locatelli et al. 2008). These included the potential for long-term economic benefits, creation of employment opportunities, stronger land tenure, better relationships with local non-governmental organizations, better enforcement of forestry regulations, and greater awareness among producers of the ecosystem service values of forests. There also was evidence of improved forest conservation practices among smallholders. The only negatives were short-term economic losses, especially among poorer landowners, and a tendency toward land consolidation. The FONAFIFO fund also subsidizes the planting of individual trees in coffee plantations at the rate of $0.25 USD per tree for 3 years (FONAFIFO 2005). Access to other payments depends on a capacity to develop proposals that demonstrate the provision of environmental services, such as carbon sequestration. A technical proposal has been developed for shaded coffee to provide compensation for ecosystem services similar to that received by owners of secondary forest (Cabrera 2009), but funding for this payment is still being negotiated.

At the international level, PES programs have developed around C sequestration in both agricultural systems and as a part of forest conservation and reforestation. Provisions of the Kyoto Protocol allow industrialized nations to offset their greenhouse gas (GHG) emissions through C offset payments, clean development mechanisms (CDMs), and investing in GHG reduction projects in developing countries (UNFCCC 2010a). Both C offset payments and CDMs have relevance for smallholder forestry and agroforestry producers. These mechanisms generally pay for either net gains in biomass or soil C or avoided losses that would have occurred without the development of specific conservation systems. Because of the national scale of C offset efforts under the Kyoto Protocol, developing approved projects for smallholders is challenging. As of late 2010, there were fewer than 20 CDM-registered smallholder reforestation or afforestation projects and no certified emission reduction credits (UNFCCC 2010b). Nevertheless, the potential for carbon sequestration from smallholder systems has been recognized in the development of CDMs using a simplified process for "Small-scale Agroforestry, Afforestation and Reforestation" (UNFCCC 2009).

The voluntary carbon market, though smaller, has a greater number of forestry projects (Jindal et al. 2008), accounting for 36% of the carbon market in 2006 (Capoor and Ambrosi 2008). One of the most developed is the Scolel Té project in Chiapas, Mexico, supported by Plan Vivo (URL: www.planvivo.org), which sold over $300,000 US in carbon certificates between 1997 and 2007, benefiting 2,400 families (Scolel Té 2008). The scheme recognizes eight different options over an area of 2,700,000 ha, including reforestation, improved fallows, living fences, and shade coffee. Soto-Pinto et al. (2006) estimated that carbon stocks can be increased by between 50 and 78 Mg C ha^{-1} through planting timber trees in coffee with *Inga* shade, at a net cost of $4–7 Mg^{-1}, but assuming no effect of the increased tree presence on the coffee production. Balderas-Torres et al (2009) found that such projects require a substantial initial investment for design of the project (in this case $850,000

USD), although they consider this could be reduced by up to 50% in future schemes. Operating costs consume about 50% of total payments; the rest goes to the farmers. Reforesting fallow lands generates the greatest gross income, but the cost to producers is considerably higher. Lower carbon sequestration options, such as shaded coffee or living fences, have lower costs to farmers, and thus may be more attractive and enable more rapid scaling-up.

One objective of PES systems is to encourage greater participation by producers, especially poor farmers who often engage in unsustainable practices out of desperation. In general, pricing ecosystem services should be based on the opportunity costs to the producer of converting the land to some other productive use (Martínez et al. 2009) rather than on the potential loss of value in the ecosystem service itself or revenue generated from fees paid by beneficiaries of the ecosystem services. Analyses of potential and actual PES programs suggest they can achieve effective participation rates while improving ecosystem service indicators and improving livelihoods for the poor (Antle and Stoorvogel 2008). Several factors are of importance when designing PES programs for smallholders. These include the size of the payment, the contribution of ES payments to producer livelihoods and the enabling conditions for practicing sustainable agriculture or forestry. Perhaps not surprisingly, Graff-Zivin and Lipper (2008) found that, within the context of conservation agriculture, the adoption rate by poor farmers is highly sensitive to the price of C sequestration payments. Alix-Garcia et al. (2008) report that targeting PES payments to poor communities tends also to result in greater ecosystem service benefits. These communities often exist near forest margins or other areas with high conservation value or high potential for ecosystem degradation (e.g. steep slopes). They also tend to engage in forest degradation and deforestation out of desperation or in response to financial or economic challenges. Providing reliable payments in return for conservation can maximize benefits to the poor as well as encourage conservation of critical areas. For production-oriented systems covered by CDMs, Roshetko et al. (2007) argue that projects should strive for the same enabling and supportive conditions that would make them viable and sustainable in the absence of ecosystem service payments. In the context of smallholder tree plantations of Southeast Asia, these include identifying areas of low biomass or existing tree cover, enrolling producers already interested in or experienced with growing trees, having or developing accessible markets for the intended tree products, and transparency and equality among project partners.

One concern is the ability of the poor to participate in PES programs. Analysis of a PES system in Nicaragua associated with silvopastoral systems found that the poor did participate at similar rates as better-off producers (Pagiola et al. 2008). In this case, the main constraint for the poor was the transaction costs associated with enrolling and participating in the program. Cacho (2009), Jindal et al. (2008) and Roshetko et al. (2007) also emphasized the high transaction costs for smallholders associated with C sequestration payment programs. These costs include measuring C stocks, certifying compliance and sequestration gains, and selling C credits.

Besides increasing C prices, efforts should focus on developing community-based projects, simplifying guidelines, and providing support for monitoring and information-sharing. These will help projects achieve a minimum participating area necessary to make the project attractive to investors, distribute transaction costs among a group of participants, increase compliance and enforcement, and foster better relationships with NGOs and governments involved in establishment, monitoring, and distributing payments.

Another concern in poor communities is the problem of weak property rights and the inability to enforce compliance. Theoretical analysis of PES for forest conservation in Indonesia suggests that weak enforcement in poor communities may undermine compliance (Engel and Palmer 2009). Even where PES conditions can be enforced, this may simply increase the bargaining power of communities with logging firms, allowing them to negotiate better logging deals rather than choosing to participate in the PES program. Jindal et al. (2008) point out that insecure and complex land tenure arrangements in Africa can inhibit participation by individual smallholders because they may lose access to land enrolled in C sequestration projects or may not be able to enforce compliance of project conditions. Throughout the world, but especially in developing regions, the vast majority of forest land is owned by the government (FAO 2005), with varying use rights to producers based on government policies, zoning, and regulations. Although conversion of such land to strict private ownership is not the answer to land tenure challenges, strengthening community-based management and recognizing and supporting their role in conservation through PES programs can achieve the twin goals of protection of ecosystem services and increasing the livelihoods of poor and rural communities.

5.3 Sustainability Certification

Over the past two decades, sustainability certifications have been developed to promote the provision of ecosystem services from forestry and agroforestry systems. While they are not direct payment programs like PES, they do offer producers the ability to sell their products in premium or exclusive markets in exchange for utilizing certain practices or maintaining desirable conditions that support ecosystem services. For coffee production, the original focus was on incorporating shade trees to promote biodiversity and other ecosystem services associated with tree presence and cover. However, most schemes include criteria and indicators that go far beyond simple tree presence or cover. Although all major certification schemes (Rainforest Alliance, Utz Kapeh, CafePractices, Fair-Trade, Nespresso AAA, Sustainable Agriculture Network, Smithsonian Bird-Friendly, 4 C Common Code) have criteria that recognize shaded coffee (i.e., the farm gains points for having shade) and in some cases recognize the diversity of shade-tree species in the coffee (SAN 2010), only the Smithsonian Bird-Friendly certification requires a minimum level of tree species diversity (SMBC 2008). Current estimates of the total amount of organic and sustainably certified coffee is approximately 1.1 million metric tons of green beans (SCAA Sustainability Council 2010), although this does not necessarily represent the total amount sold on the market. The total coffee sold by exporting countries is approximately seven million metric tons (ICO 2010; FAO 2003). Almost half of all sustainably certified coffee is represented by the 4 C Common Code (437 million kg). The Smithsonian Bird-Friendly certification, the oldest and most stringent with respect to shade tree cover and diversity, represents less than 1% of this volume (<4 million kg).

Only a few studies have been published evaluating the actual ecosystem service benefits of sustainable coffee certification. A limited study of farms in a small region of Nicaragua found that certified farms (organic or Rainforest Alliance) had higher tree density, larger carbon stocks in the shade trees and lower pesticide use (Alvarado and Cuadra 2010). Smaller farms (less than 20 ha) also had higher tree density, diversity, cover and greater carbon stocks. Unfortunately, anecdotal evidence suggests that, in some cases, certification promotes farm consolidation and shade simplification. Mas and Dietsch (2004) compared remnant forest shade, polyculture planted shade, and monoculture planted shade coffee systems to nearby native forest in terms of butterfly and bird diversity. Only the remnant forest shade system had comparable diversity to the forest. It also was the only system that met all the sustainability criteria of major certification systems. The monoculture coffee system generally failed most of these criteria. The authors concluded that the certification systems are capable of distinguishing the relative provision of ecosystem services from coffee agroecosystems through their existing criteria, but they differ in the thresholds necessary to achieve certification. The Rainforest Alliance provides sustainability certification for a number of tropical food and other plant products based on expected ecosystem service benefits, but these do not necessarily require the inclusion of trees. Studies currently being conducted will evaluate if the expected social and environmental impacts of certification are more widespread (Giovannucci et al. 2008).

There also are multiple sustainable forestry certification systems; the most widely used is from the Forest Stewardship Council (URL: http://www.fsc.org). These certification systems are based on a number of sustainability criteria encompassing environmental, economic, and social factors, with indicators for each criterion based as much as possible on locally appropriate conditions. If a project passes the evaluation, it receives certification for a 5-year period, and any forest products harvested within the terms of the contract can carry the certifying institutions label. The transaction costs for undertaking a sustainability evaluation and for maintaining certification for multiple 5-year contracts are considerable, but for large operations, the market premium for certified products may be sufficient to cover these costs (Nebel et al. 2005). At a national level, these certification programs can bring much-needed specificity to sustainable forestry guidelines and practices in developing countries (Dennis et al. 2008; Melgarejo et al. 2006). The hope is that certification can promote 'good forestry' in general, reducing forest degradation and deforestation and increasing incentives to maintain productive and healthy forests. However, as with sustainable coffee certification, evaluation of forest certification impacts is quite limited. Nebel et al. (2005) reviewed FSC certification in Bolivia, where approximately 14% of the forest area (one million hectares) has been certified. Five companies control the majority of the certified land. The strict forestry regulations imposed at the national level are mostly FSC-compliant; thus, it is unclear if there were additional environmental benefits to certification. Conservation biologists have questioned the biodiversity benefits of certification, arguing that there is an inherent tradeoff between forest harvesting and biodiversity conservation (Bennett et al. 2001), and that investments in certification may be better spent on more traditional conservation initiatives (Gullison 2003). Analyses of certified projects suggest that changes in management are mainly administrative – better documentation, monitoring, and reporting of forest condition – rather than actual changes in practices (Nebel et al. 2005; Hartsfield and Ostermeier 2003).

The major limitation of FSC certification in Bolivia was that smallholder and community-based projects had trouble obtaining certification. This is a general problem for small and resource-poor producers. The FSC and other major certification systems have programs targeting small or family-owned forest enterprises, but enrollment in these programs is much smaller in terms of numbers and area than large projects. If the goal is to maximize the forest area certified, focusing on large forest landowners and project areas is desirable. However, this ignores the fact that major gains in ecosystem services are more likely to come from improvements in smallholder or community-based forest management (CBFM). Smallholders have the ability to engage in sustainable forest management, but as most producers come from an agricultural background, they often lack specific training or capacity to do so. Sustainability certification is not intended to develop these capacities but rather to evaluate their application to specific projects, so smallholders may be deterred simply by their lack of knowledge and experience in how to achieve certification. One example of successful certification at the community level are the *ejido* CBFM enterprises in Mexico. These enterprises are generally well-organized and managed as production forests, and the communities have clear ownership and use rights. As a result, they have been able to achieve FSC certification in some cases (Ward and Bihun 2001). There are a growing number of CBFM programs in Southeast Asia as well, but these projects are quite variable with respect to the level of community participation in decision-making; control over access and use of the forest; internal capacity to carry out effective forest management; and technical, market, and legal support from state or national governments (Muhammed et al. 2008; Balooni and Inoue 2007). These barriers form the basis of most recommendations for improved CBFM with the goal of improving the internal capacity of communities for sustainable forest management.

The ability of smallholders to obtain and maintain sustainability certification and the potential benefits derived from certification are generally greater for coffee and other agricultural products than for forest products. For coffee, most certification schemes apply standards to individual producers, but often certification efforts are coordinated within existing producer cooperatives. Fair Trade, in particular, targets smallholder producers, utilizing cooperatives as a means to support individual producers and achieve broader community benefits, such as improved workers' rights and working conditions, producer investment in community development, and improved relationships between producers and buyers (Giovannucci and Koekoek 2003; Muradian and Pelupessey 2005). This last potential benefit may be of particular importance for smallholder producers who are often disconnected from markets and pricing information and generally vulnerable to exploitation by unscrupulous buyers (Blowfield 2003). Although community-based forest management is common in tropical and sub-tropical countries, forest product cooperatives are rare for smallholder communities, and the existing certification schemes do not focus on achieving broader community benefits. Some of the challenges for smallholders are similar, however. As with sustainable forestry certification, compliance monitoring for sustainable coffee is usually conducted by a licensed third party and paid for by the producer (Muradian and Pelupessey 2005). In addition, the price premiums assumed to exist for sustainably certified coffee (and other products) are not guaranteed for most systems. Increased demand for sustainable coffee worldwide has actually led to increased competition to produce lower-priced certified coffee (Muradian and Pelupessey 2005). Finally, with the exception of Fair Trade, some of the important social concerns shared by smallholder forestry and agricultural producers, including increased land tenure security and fair prices and timely payment from buyers, are not generally addressed by sustainable certification schemes (Blowfield 2003). That said, the existence of producer cooperatives in agriculture generally allows smallholders to more effectively address their own internal concerns while also improving achievement of environmental standards than in forestry (e.g. Danse and Wolters 2003). Not surprisingly then, strong community-based forest management programs generally provide greater benefits to smallholders while also supporting the conservation of forest cover and related ES (Balooni and Inoue 2007; Bray et al. 2003).

5.4   Ecotourism

Tourism is an activity that may provide additional income to smallholder forestry and agroforestry systems in the tropics and could help ensure the sustainability of these systems. According to Kiss (2004) and Doan (2000), community-based ecotourism (CBET) is a popular means of supporting biodiversity conservation, particularly in developing countries. Most international conservation organizations, the US Agency for International Development, and the World Bank support CBET as a sustainable means of linking conservation with local livelihoods, preserving biodiversity while reducing rural poverty (Kiss 2004). However, the debate about the economic and conservation impacts of CBET began just recently, and much more work is needed.

No universal definition of ecotourism exists. The International Ecotourism Society defines ecotourism as "Travel to natural areas that conserves the environment and sustains the well-being of local people" (TIES 2010). It is not synonymous with nature tourism because of the explicit objectives relating to environmental and social sustainability. Ecotourism incorporates environmentally friendly and culturally protective techniques. It generally caters to a niche market of tourists who desire to visit places where ecosystem services such as biodiversity and conservation are promoted. Niche marketing generally is pursued to increase profitability by attracting a select group of customers who will spend more for the product or services offered. For the ecotourism market, clientele have higher incomes than the average person, thus making a niche marketing approach potentially viable. The criticism has been that ecotourism merely caters to rich people trying to see something rare, or is promoted by large companies trying to use a trend to make money. The result may be a focus on creating an experience for the customer rather than on the provision of ecosystem services.

The environmental and social components of ecotourism vary widely. Some ecotourism operators focus on cost-saving, environmentally friendly business practices and contribute to community projects. Others may be more dedicated to actually improving the environment and therefore devote more resources to addressing environmental concerns. Still others may put more resources into social concerns and move past making contributions to engaging in regular consultations with the community to partial or full community ownership of the operation. Kiss (2004) stresses that while various interpretations of the environmental and social objectives of CBET exist, the ultimate vision for any tourism effort relative to smallholder forestry and agroforestry systems is the development of a component that actively contributes to community-based natural resource management.

Agricultural tourism and volunteer tourism are other niche markets that, like ecotourism, are marketed as being more sustainable than mass tourism products and services (Stronza and Gordillo 2008). While these forms of tourism are receiving attention, researchers have more widely documented the benefits of ecotourism specifically. These include both market and non-market returns. While many researchers recognize the potential value associated with community-based tourism, less work has been done to qualify and quantify these various benefits and link them directly with the community's sustainability goals, such as conservation (Stronza and Gordillo 2008; Kiss 2004). The market benefits, often referred to as economic benefits, are generally measured in terms of employment and income (Walpole and Goodwin 2001; Wunder 2000; Campbell 1999; Gosling 1999). Kiss (2004) concludes that most CBET projects produce modest economic benefits, but these often accrue to a relatively small segment of the community. For some communities even a small increase in economic benefits is welcome, although these may not be considered by community leaders as the most important benefits ecotourism provides (Stronza and Gordillo 2008). Some would argue that maximizing economic benefits from ecotourism should not be the goal of CBET projects because more attractive returns would draw residents away from actual production (Epler-Wood 2002) or intensify social conflicts that result from increasing income disparity (Cousins and Kepe 2004; Ogutu 2002). By contrast, modest returns from tourism, especially if distributed broadly, provide a degree of diversification to help manage production and marketing risks, which promotes the sustainability of existing rural livelihoods (Honey 1999).

A wide array of non-market benefits have been identified and discussed. They include renewing cultural pride (Epler-Wood 2002), catalyzing new development (Weaver 1998), empowering local people (Scheyvens 1999), and protecting biodiversity (Christ et al. 2003). Stronza and Gordilla (2008) discuss more specific benefits to residents such as gaining new skills, broadening experiences in managing people and projects, strengthening abilities to negotiate with outsiders, increasing self esteem, and expanding circles of contracts and support for community efforts. These benefits contribute to sustainability by increasing social capital so that local institutions can better engage in resource management (Jones 2005; Pretty and Smith 2003). The non-market costs include time that must now be devoted to CBET, decreased reciprocity and increased social conflict (Stronza and Gordillo 2008). Participatory approaches to CBET that build social capital facilitate community efforts to develop ecotourism that achieves more sustainable development (Stronza and Gordillo 2008; Bray et al. 2005; Jones 2005).

Ecotourism is sometimes promoted over direct PES because an enterprise approach is assumed to promote self-sufficiency rather than dependency on government payments (Ferraro and Kiss 2002). However, private investment in CBET enterprises may be deterred due to the large amount of investment that is required relative to the returns. Kiss (2004) concludes that the public sector will need to support the building of social capital and conservation beyond what tourists are willing to pay to ensure the sustainability of these enterprises. For example, a thorough analysis of socioeconomic goals and a well-developed business plan are needed to create the conditions for success. Often, smallholders lack the merchandising skills needed to make a tourism operation successful (Cox and Fox 2003). At the same time, external factors such as infrastructure development can be obstacles outside the direct control of the community (Cox and Fox 2003). A realistic assessment of feasibility, cost-effectiveness, social impacts, existing community institutions, and other aspects of sustainability is needed before any strategy is selected (Kiss 2004).

Regardless of whether the niche in question is volunteer, agricultural, or ecotourism, the sustainability of the effort is likely the most pressing issue of concern. Since most tourism operations are run by outside companies rather than smallholders and the surrounding communities directly, often the community bears most of the costs but does not end up with most of the benefits (West and Carrier 2004). Therefore, the overall design of any tourism effort that is a component in a community-based natural resource management plan must ensure that the benefits are channeled to local communities. These benefits are expected to become additional incentives for residents to protect the forestry and agroforestry systems that the tourists pay to see (Ross and Wall 1999).

6   Promoting Adoption and Adaptive Management of Sustainable Smallholder Systems

The complexity and diversity of smallholder forestry and agroforestry systems, potentially consisting of almost any combination of crop plants, trees and livestock, provide smallholders with a considerable potential to meet almost any combination of socioeconomic needs and environmental production conditions. At the same time this represents a considerable challenge to develop technical assistance capacities to work with producers in implementing and managing diverse smallholder systems. Furthermore, forestry and agroforestry systems are long-term investments; smallholders' capacity and interest to invest in them may vary according to changing markets and environmental conditions. Haggar et al. (2001) concluded that participatory design and implementation processes are the most appropriate means of working with smallholders on management of agroforestry systems. A participatory approach facilitates analysis and decision-making among the smallholders, developing their capacity to implement an adaptive management strategy for production systems. The basic principles of participatory design and management of agroforestry systems are:

1. Determine the producer objectives in managing forestry or agroforestry systems (commercial production, food security, investment in future production, etc.).
2. Develop an inventory of the trees present for existing systems or desired for new systems.
3. Evaluate selected crop and tree species based on the production objectives of the producer, their adaptability to the site conditions, and interspecific compatibility within particular spatial or temporal arrangements.
4. Establish an ideal density, spacing and distribution of the prioritized species.

Other criteria such as value for wildlife can be included.

This approach was used in the Yucatan Peninsula of Mexico with Mayan producers to design agroforestry systems according to self-identified needs (Haggar et al. 2004). Some producers combined annual crops, perennial crops, fruit trees and timber trees into a complex system that would provide continuous production as the system developed and result in a complex but high investment multi-strata agroforest. Others just planted native fruit and timber trees that could grow with the natural succession as an enriched fallow. A similar process was developed for new and existing coffee agroforestry systems where producers made an inventory of the tree species present and evaluated the benefits of those species for household use, sale of products and as shade for coffee (Haggar et al. 2001). On this basis producers decided whether they needed to make changes in the shade tree composition or what species to plant for new plantations.

In established agroforestry systems the performance of the component species may be evaluated through the use of agroecological diagnostic tools. These have been most developed for coffee and cocoa agroforestry systems (e.g. Haggar and Staver 2001). These tools enable producers to evaluate the levels of production or incidence of pests and diseases in the perennial crops according to the levels of shade and type of shade from the trees. Then producers can make management decisions as to the level and type of shade most appropriate for their production conditions. Hundreds of extension workers in Central America, from dozens of organizations, have trained some 10,000 coffee smallholders in these decision-making tools (Haggar 2008; Guharay et al. 2005).

Pattanayak et al. (2003) in a study of 120 agriculture and forestry practices, including 32 agroforestry systems, identified five categories of factors that explain adoption by producers:

1. Market incentives or demand for the product or other economic savings from adoption of the practice.
2. Biophysical conditions (soil, climate, etc.) that are appropriate for the crops to be produced.
3. Resource endowments in terms of land, labor and capital.
4. Perceived risk of production or market failures; systems with high perceived risk have very low adoption rates.
5. Household preferences related to culture, gender and economic and social needs.

Many agroforestry systems are long-term production systems, so market signals can often be very uncertain and limit adoption. Also, producers with better endowments are more likely to adopt because they can afford to take greater risks and make larger initial investments. Since producers adopt innovations when clear economic incentives are present and associated risks are manageable, understanding current and historic patterns of adoption and economic incentives are required to develop effective technological and institutional interventions (Scherr 1995).

Even with financial incentives, however, reforestation programs directed at small producers in Central America have often produced low participation rates and uncertain long-term results (Thacher et al. 1997). The authors found that small producers in Costa Rica adopt reforestation schemes primarily for the alternative short-term benefits the household derives from participation in the incentive programs rather than long-term economic production. It is also important to recognize that not all smallholders are in a position to manage complex agroforestry systems that require dynamic learning processes associated with adaptive management. Confirming some of these points, a study in the Yucatan Peninsula found that producers most likely to adopt agroforestry practices were those with greater technical experience, higher incomes and a greater proportion of deforested land (Mercer et al. 2005).

7 Summary and Recommendations

Smallholder forestry and agroforestry systems provide for the subsistence and livelihood needs of millions of households throughout the tropics, especially in rural and poor areas. They contribute significantly to the food, fuel, and fiber supplies of local markets; and many of the world's diversified agricultural products, most notably coffee and cacao, are produced mainly in smallholder systems. Crop productivity in smallholder systems may be lower due to a combination of tree shade and lower rates of water and chemical inputs as compared to high-input open-grown systems. Tree productivity, however, can be greater due to the relatively intensive management provided to crops integrated into agroforestry systems. Smallholder tree plantations provide important supplies of wood for local mills in Indonesia and the Philippines. Rising demand for such wood has actually led to reforestation of degraded lands. In Mexico, there are hundreds of communally managed forests dedicated to timber production. Most of these tree plantations also are sources of important non-timber forest products and may be partially or fully intercropped at establishment to provide for food production until the trees form a closed canopy.

For annual cropping systems, sequential agroforestry designs, such as taungya or shamba systems, generally provide for an optimal trade-off of productivity and ecosystem services. Border or line plantings, especially for windbreaks, living fences, or vegetated terraces, can be appropriate simultaneous designs, but these require regular management of tree density and shade level to maintain crop productivity. Use of multipurpose trees, especially N fixers, has generally been recommended in these situations. For perennial understory crops like coffee or cacao, maintaining shade levels less than 50% is recommended to minimize loss of productivity. Where timber trees are being used for shade, this may reduce total stand volume production due to low densities, although individual tree growth will benefit. For tree plantations, fast-growing tree species have been identified and tested in most areas. Development and dissemination of improved planting stock can increase growth rates of these identified species. As well, timber stand improvement practices, such as early pruning and selective thinning, are needed to improve stem form and average tree size without compromising total stand volume.

The net economic benefits of integrating trees and crops in agroforestry designs have been documented for various types of smallholder systems. Timber and fruit/nut trees in shade coffee systems provide multiple products to minimize risk and stabilize and potentially increase income, especially during times of low crop prices. In rural areas, woodlots and on-farm tree-planting have proven highly beneficial for smallholders dependent upon fuelwood. On-farm woodlots or fuelwood trees offset costs of purchased wood, save on time and labor necessary to gather fuelwood from surrounding forests and often provide an excess of wood that can be sold in local markets. Tree plantations have proven profitable for many smallholders, especially where demand from local mills provides a known and reliable market. Timber from communal forest enterprises in Mexico is an important source of income for rural communities and encourages active conservation and management of forest cover. For smallholders in the Brazilian Amazon, occasional timber sales provide needed income during times of financial stress. For most communally managed forests, non-timber forest products are important for household use and market sales, substituting for purchased products and providing additional revenue.

Smallholder systems generally support higher diversity of native plant species than large, technologically intensive plantations, especially when considered across multiple farms at the landscape scale. Diversified smallholder systems can act as important refugia for native fauna in highly deforested areas; they serve as biological corridors that connect forest fragments; and situated at the margins of intact forests, they expand the suitable habitat of forest-dependent species. Improving support for biodiversity is mainly a function of landscape-scale planning. Smallholder systems cannot fully substitute for large, intact forests to support forest-dependent species, but they can be integrated into forested landscapes to maximize connectivity and expand the range of existing forests. There also are opportunities to integrate smallholder systems in reforestation programs, as long as there is flexibility to accommodate tree harvesting, understory planting or manipulation, and the inclusion of non-native species that are of particular value to smallholders.

Trees in general improve soil and water conservation and contribute to the ecosystem services associated with watershed function, such as water quality, stream habitat, and the health of coastal and near-shore marine ecosystems. Smallholder forestry and agroforestry systems may be relegated or preferentially placed in steeper areas, meaning that proper management is important for conserving these ecosystem services. Maintaining continuous soil cover and minimizing soil disturbance are key for soil and water conservation. Inclusion of trees also generally requires modified management practices that result in greater soil cover and reduced disturbance. Use of trees for vegetated terraces can significantly reduce runoff and erosion in hill-slope cropping systems. Integrating water diversion and capture structures within terrace designs often improves their function. Many communally managed forests in agricultural landscapes are situated in the most steeply sloping areas or in riparian zones. These forests are often conserved because the land is less suitable for cropping; there is active use of the forest for grazing, limited timber harvesting, and NTFP gathering; and there is recognition of the value of the forest for watershed function. Even where communities do not technically own the land, community-based forest management should be supported through official policies and cooperative management planning and decision-making. This provides incentives to conserve forest cover while allowing for traditional uses and community policing of restricted activities.

Associated with soil and water conservation, incorporation of trees in agricultural systems generally improves nutrient cycling and inherent soil fertility. Indeed, shifting cultivation systems rely upon these ecosystem services to renew fertility depleted by annual cropping. The use of N-fixing trees is common for this purpose. Trees and crops can and do compete for soil water and nutrients. This is generally a concern in agroforestry systems where trees are planted in rows or along field borders to minimize shading. Tree roots can extend laterally for many meters, impacting nutrient and water availability beyond the zone where shading would be a more important limitation to crop productivity. However, trees generally improve fertilizer use efficiency and retention and recycling of fertilizer nutrients. Deep-rooted trees can 'mine' base cations from lower soil horizons that may be deficient in surface horizons of highly weathered soils. However, tree-crop competition for nutrients will be intense where nutrients are already lacking. The ability of trees to capture applied nutrients leaching below the crop zone is of greater importance for nutrient use efficiency. Although wood itself has low concentrations of most nutrients, frequent harvesting of trees, such as for fuelwood, can reduce site nutrient capital. Recommendations to maximize nutrient retention and recycling depend largely on the type of system. For shifting cultivation, improved fallows with select N-fixing trees have proven useful for restoring site fertility within a shorter time period. For integrated crop-tree systems, selecting tree species with deeper root systems can minimize root zone overlap with crop species. Pruning of N-fixing trees not only manages shade levels but provides a nutrient-rich source of organic matter for maintaining or improving soil fertility. Where soils are particularly nutrient-poor or water is limiting, a sequential agroforestry system is the best choice for minimizing tree-crop competition and retaining or restoring fertility. Where tree harvesting is frequent or pruned branches and leaves are not returned to the site, supplemental fertilization is recommended.

Conversion of intact forests to production systems generally results in loss of biomass C that is not fully compensated for in tree plantations or agroforestry systems. However, compared to open-grown cropping, smallholder forestry and agroforestry systems can sequester significant amounts of C in both plant biomass and soil organic matter. Retaining large trees in agroforestry systems is generally recommended to conserve biomass C stocks within a target shade level. Maintaining a diversified understory in tree plantations also supplements the C sequestered in overstory trees. There has been only limited quantification of belowground biomass (i.e., roots) in smallholder systems, but various models and allometric equations exist that can be applied to allow for estimation of coarse root C. Carbon sequestration in soil organic matter can be significantly improved in smallholder systems, especially those that include N-fixing trees. Agroforestry practices that increase plant inputs, such as mulch additions of tree pruning residues, may increase soil C over just a few years. Soil C sequestration in tree plantations and agroforestry systems appears to require increased stabilization of soil and soil organic matter as well as increased organic matter inputs. The stability and permanence of this soil C has not been well-studied, but it bears investigation due to the changing vegetation cover and management practices in agroforestry systems and the relatively short rotations of smallholder plantations.

Compensation to smallholders for establishing and maintaining tree cover and biodiversity in agroforestry systems and tree plantations occurs through various markets and support programs. Fruit, nut and timber trees provide flexibility in production and income sources to reduce risks and take advantage of emerging opportunities. Studies have shown that in times of low crop prices, tree cover and diversification actually increase, and tree products become a more important source of household income. Reforestation of degraded areas in Southeast Asia was actually driven by development of local mills that increased the demand for wood supplies. In Africa, declining forest cover increased the value of fuelwood, increasing the incentives for tree planting on farms and unused land.

Payments for ecosystem services (PES) have provided new opportunities to compensate smallholders for diversified forestry and agroforestry systems. The major services included in these programs are carbon sequestration and soil and water conservation. There are many successful examples at national, provincial, and even municipal levels of PES programs for these services. Direct payments to individual smallholders for sequestered C or avoided greenhouse gas (GHG) emissions are not feasible because of their small size relative to the quantity of C or GHG generally packaged and traded on international markets. However, the Clean Development Mechanism (CDM) program through the United Nations has supported the development of community-based projects. The high transaction and monitoring costs of these projects can be a financial disincentive, so the development and support of community-level capacity to plan, implement, and manage these projects is critical to their success.

Sustainability certification systems for forestry, coffee, and other shade-grown agricultural products have proliferated in the last 15 years. These programs provide access to niche markets with the goal of achieving price premiums for certified products. For forestry, certification is generally oriented toward commercial forest operations and large government-owned forest land due to the costs associated with developing a written management plan, monitoring and documentation, and the evaluation process itself. 'Family forest' programs exist for most major forest certification programs, but these are still not feasible for poor smallholders in developing countries. There are multiple sustainability certification systems for coffee production. As with forestry certification, the requirements for a written management plan and monitoring and documentation are impediments for smallholder participation. However, smallholder certification of coffee and other agricultural products has benefited from the existence and promotion of producer cooperatives. The producer cooperative model to support certification of smallholder groups should also be applicable to smallholder tree plantations as well. Providing support to communities and cooperatives for developing written management plans and instituting appropriate monitoring protocols would help existing sustainable forestry certification programs to adapt their standards so the products and services from these systems could be recognized and marketed officially.

The environmental benefits of sustainability certification vary. For forestry, price premiums for sustainably certified wood can cover the costs of certification for large landowners, but for others the benefits are mainly improvements in planning and administration. For coffee and other agricultural products, the price premium for certified products is better developed and can result in net benefits to smallholders. The fair trade system provides a price guarantee and then recovers these costs through increased wholesale or retail prices. Other systems provide reliable access to premium markets or to select retailers and alternative trading organizations. For many smallholders, more sustainable growing practices can reduce management costs and improve the sustainability of production in the long run. Producer cooperatives can also increase bargaining power in the marketplace, support education and training of individual producers, and increase investments in community development. The impact of these certification systems upon actual provisioning of ecosystem services has not been well-studied. For forestry, it appears that many certified operations are already utilizing best management practices. Including more smallholder production systems may actually yield greater improvements in ecosystem services. For coffee certification, production systems that meet the most stringent standards appear to also have the highest provision of ecosystem services, such as wildlife biodiversity.

Ecotourism is an enterprise-based approach to promote ecosystem services. Niche marketing is used to attract tourists interested in experiencing the benefits of nature conservation. Most enterprises are run by outside companies, and most activities occur at the community level. Ensuring equitable distribution of costs and benefits between companies and the community and among smallholders within the community are inherent challenges with this enterprise model. Attracting private investment in community-based ecotourism also is a challenge due to the high costs of initial investment relative to rates of return. As with PES programs, the sustainability of ecotourism depends upon promoting community capacity to define socioeconomic goals, analyze resources and opportunities, develop plans and manage activities as part of a community-based enterprise.

Promoting adoption of smallholder forestry and agroforestry requires building the capacity to adaptively manage these complex and dynamic systems. Participatory approaches can help smallholders identify needs and objectives of various production systems, prioritize compatible mixtures of crops and trees, and design systems that optimize tradeoffs between tree and crop productivity and maximize overall ecosystem services. Agroecological diagnostic tools have been developed that enable producers to evaluate agroforestry system performance based on the type and level of shade. Although market incentives are important for adoption of forestry or agroforestry systems, producers also need sufficient land, labor, and capital to make the required initial investments and take on the associated risk. Not surprisingly then, producers with greater education, income, and resources tend to be more likely to adopt these systems. Thus, while promoted for their long-term benefits for both ecosystem services and socioeconomic development, the pre-existing capacity of smallholders and communities to engage in new and transformative production systems and management practices is key to adoption and success. Promotion, therefore, should be integrated into overall sustainable development programs rather than as a separate initiative. Support is most needed during the planning and adoption stages to boost capacity and provide a supportive environment until the social and economic benefits allow for sustained management by the community or individual producers.

The many successful examples illustrated in this review demonstrate that smallholder forestry and agroforestry can be sustainable and have great potential for expansion throughout the tropics. This is an alternative to further consolidation of agricultural and forest land-holdings and operations or degradation of existing forests and natural areas as a desperate response to poverty, isolation, or lack of domestic or foreign aid. The inherent diversity of existing and potential systems provides for flexibility and adaptability for individual producers, allowing them to be competitive with large, input-intensive monocultural systems. Compensation for the provision of ecosystem services through PES programs can provide important supplementary income to support conservation practices or the development of systems that support long-term sustainability within a smallholder model. Sustainability certification allows for niche marketing of products while providing recommendations and support for conservation practices. Innovative enterprises like ecotourism allow smallholders to tap into niche markets for services, as well as products. Although individual smallholders may be constrained by size and capacity to take advantage of these opportunities, well-organized community-based projects and enterprises can overcome these limitations while taking advantage of the diversity and flexibility of smallholder systems. Supporting community-based enterprises in this way provides community as well as individual household benefits, further strengthening the sustainability of smallholder production and environmental conservation.

References

Akinnifesi F, Kang B, Ladipo D (1998) Structural root form and fine root distribution of some woody species evaluated for agroforestry systems. Agroforest Syst 42:121–138

Akinnifesi FK, Sileshi G, Ajayi OC, Chirwa PW, Kwesiga FR, Harawa R (2008) Contributions of agroforestry research and development to livelihood of smallholder farmers in Southern Africa: 2. Fruit, medicinal, fuelwood and fodder tree systems. Agr J 3:76–88

Alfaia SS, Ribeiro GA, Nobre AD, Luizão RC, Luizão FJ (2004) Evaluation of soil fertility in smallholder agroforestry systems and pastures in western Amazonia. Agr Ecosyst Environ 102:409–414

Alix-Garcia J, Janvry A, Sadoulet E (2008) The role of deforestation risk and calibrated compensation in designing payments for environmental services. Environ Dev Econ 13:375–394

Allen SC, Jose S, Nair PKR, Brecke BJ, Nkedi-Kizza P, Ramsey CL (2004) Safety-net role of tree roots: evidence from a pecan (Carya illinoensis K. Koch)-cotton (Gossypium hirsutum L.) alley cropping system in the southern United States. Forest Ecol Manag 192:395–407

Altieri M (1999) The ecological role of biodiversity in agroecosystems. Agr Ecosyst Environ 74:19–31

Alvarado NUP, Cuadra, MLC (2010) Evaluación de Servicios Ambientales de Café Agroforestal en Fincas en el Macizo de Peñas Blancas, Matagalpa. Undergraduate thesis, Universidad Nacional Agraria, Managua, 50 pp

Amacher GS, Merry FD, Bowman MS (2009) Smallholder timber sale decisions on the Amazon frontier. Ecol Econ 68:1787–1796

Anane SY, Twumasi-Ankrah R (1998) Woodlots: a profitable venture for Ghana's smallholder farmers. Agroforest Today 10:19–21

Antle JM, Stoorvogel JJ (2008) Agricultural carbon sequestration, poverty, and sustainability. Environ Dev Econ 13:327–352

Anyonge CH, Roshetko JM (2003) Farm-level timber production: orienting farmers towards the market. Unasylva 54:48–56 (English ed.)

Arnold JEM, Contreras AH (1979) Economic analysis of forestry projects: case studies. FAO Forestry paper 17, Suppl 1

Babbar L, Zak D (1994) Nitrogen cycling in coffee agroecosystems: net N mineralization and nitrification in the presence and absence of shade trees. Agr Ecosyst Environ 48:107–113

Balderas-Torres A, Marchant R, Lovett J, Smart J, Tipper R (2009) Analysis of the carbon sequestration costs of afforestation and reforestation agroforestry practices and the use of cost curves to evaluate their potential for implementation of climate change mitigation. Ecol Econ 69:469–477

Balooni K, Inoue M (2007) Decentralized forest management in south and southeast Asia. J Forest 105:414–420

Barros E, Neves A, Blanchart E, Fernandes E, Wandelli E, Lavelle P (2003) Development of the soil macrofauna community under silvopastoral and agrosilvicultural systems in Amazonia. Pedobiologia 47:273–280

Beer J (1998) Litter production and nutrient cycling in coffee (Coffea arabica) or cacao (Theobroma cacao) plantations with shade trees. Agroforest Syst 7:103–114

Bennett JA (2000) Pacific forest: a history of resource control and contest in Solomon Island, c. 1800–1997. The White Horse Press, Cambridge

Bennett EL, Cauley HA, Peters CM, Donovan RZ, O'Connor JM, Putz FE, Romero C, Ghazoul J (2001) Timber certification: where is the voice of the biologist? Conserv Biol 15:308–319

Bensel TG (1995) Rural woodfuel production for urban markets: problems and opportunities in the Cebu province, Philippines. Pac Asian J Energy 5:9–28

Bensel T (2008) Fuelwood, deforestation, and land degradation: 10 years of evidence from Cebu province, the Philippines. Land Degrad Dev 19:587–605

Bertomeu MG (2004) Smallholder timber production on sloping lands in the Philippines: a systems approach. World Agroforestry Centre, Lagnua

Bertomeu M (2006) Financial evaluation of smallholder timber-based agroforestry systems in Claveria, Northern Mindanao, the Philippines. Small Scale Forest 5:57–81

Beukema H, Danielsen F, Vincent G, Hardiwinoto S, Andel J (2007) Plant and bird diversity in rubber agroforests in the lowlands of Sumatra, Indonesia. Agroforest Syst 70:217–242

Bhagwat SA, Rutte C (2006) Sacred groves: potential for biodiversity management. Front Ecol Environ 4:519–524

Bhagwat S, Willis K, Birks H, Whittaker R (2008) Agroforestry: a refuge for tropical biodiversity? Trends Ecol Evol 23:261–267

Bigelow SW, Ewel JJ, Haggar JP (2004) Enhancing nutrient retention in tropical tree plantations: no short cuts. Ecol Appl 14:28–46

Binkley D (2005) How nitrogen-fixing trees change soil carbon. In: Binkley D, Manyailo O (eds.) Tree species effects on soils: implications for global change. Springer, Dordrecht, pp 155–164

Binkley D, Ryan MG (1998) Net primary production and nutrient cycling in replicated stands of *Eucalyptus saligna* and *Albizia falcataria*. Forest Ecol Manag 112:79–85

Binkley D, Dunkin KA, DeBell D, Ryan MG (1992) Production and nutrient cycling in mixed plantations of *Eucalyptus* and *Albizia* in Hawaii. Forest Sci 38:393–408

Bisseleua DHB, Missoup AD, Vidal S (2009) Biodiversity conservation, ecosystem functioning, and economic incentives under cocoa agroforestry intensification. Conserv Biol 23:1176–1184

Blowfield (2003) Ethical supply chains in the cocoa, coffee and tea industries. Greener Manag Inter 15–24

Boffa JM, Jourget JG, Turyomurugyendo L, Kindt R, Katumba B (2008) Management of tree diversity in agricultural landscapes around Mabira Forest Reserve, Uganda. Afr J Ecol 46:24–32

Brandt J (1988) The transformation of rainfall energy by a tropical rainforest canopy in relation to soil erosion. J Biogeogr 15:41–48

Bray DB, Merino-Pérez L, Negreros-Castillo P, Segura-Warnholtz G, Torres-Rojo JM, Vester HFM (2003) Mexico's community-managed forests as a global model for sustainable landscapes. Conserv Biol 17:672–677

Bray D, Cornejo M, Cohen S, Beitl C (2005) Community-based ecotourism – a conservation and development strategy in search of an analytical framework. Unpublished paper. Annual meeting of the Society for Applied Anthropology, Sante Fe

Brockington D (2002) Fortress conservation: the preservation of the Mkomazi Game Reserve. James Currey Publishers, Tanzania

Bruijnzeel LA (2004) Hydrological functions of tropical forests: not seeing the soil for the trees? Agr Ecosyst Environ 104:185–228

Burns R, Honkala B (1990) Silvics of North America, vol 2, Hardwoods. U.S. Department of Agriculture/U.S. Forest Service, Washington, DC

Butterfield RP (1995) Promoting biodiversity: advances in evaluating native species for reforestation. Forest Ecol Manag 75:111–121

Cabrera MJ (2009) Propuesta de borrador de decreto para la creación de una modalidad de pago por servicios ambientales a los sistemas agroforestales con café. Fundación Café Forestal, Costa Rica, 57 pp

Cacho O (2009) Economics of carbon sequestration projects involving smallholders. In: Lipper L, Sakuyama T, Stringer R, Zilberman D (eds.) Payment for environmental services in agricultural landscapes: economic policies and poverty reduction in developing countries. Food and Agriculture Organization of the United Nations (FAO), Rome, pp 77–102

Cadisch G, Rowe E, Van Noordwijk M (1997) Nutrient harvesting-the tree-root safety net. Agroforest Forum 8:31–33

Cairns MA, Brown S, Helmer EH, Baumgardner GA (1997) Root biomass allocation in the world's upland forests. Oecologia 111:1–11

Calvo L (2004) Importancia de los cafetales bajo sombra en la conservación de biodiversidad. Centro para la Conservación de Biodiversidad de Guatemala, Guatemala City, 14 pp

Campbell L (1999) Ecotourism in rural developing communities. Ann Tourism Res 26:534–553

Capoor K, Ambrosi P (2008) State and trends of the carbon market 2007. World Bank Institute. http://wbcarbonfinance.org/docs/State_Trends_FINAL.pdf

Cassano CR, Delabie JHC, Bede L, Schroth G, Faria D (2009) Landscape and farm scale management to enhance biodiversity conservation in the cocoa producing region of southern Bahia, Brazil. Biodivers Conserv 18:577–603

CATIE (2002) Sistema geografica sobre area cafetalera de Costa Rica. Informe final al Instituto de Café de Costa Rica. Turrialba, 55 pp

Ceccon E (2005) Eucalyptus agroforestry system for small farms: a 2-year experiment with rice and beans in Minas Gerais, Brazil. New Forest 29:261–272

Cedamon EO, Emtage NF, Suh J, Herbohn JL, Harrison SR, Mangaoang EO (2005) Present tree planting and management activities in four rural communities in Leyte Province, the Philippines. Ann Trop Res 27:19–34

CEPAL (2002) Centroamérica: El impacto de la caída de los precios de café en 2001. Comisión Económica para América Latina y el Caribe, 61 pp

Chacon Leon M, Harvey CA (2006) Live fences and landscape connectivity in a neotropical agricultural landscape. Agroforest Syst 68:15–26

Christ C, Hillel O, Matus S, Sweeting J (2003) Tourism and biodiversity: mapping tourism's global footprint. United Nations Environment Program and Conservation International, Washington, DC

Clough Y, Faust H, Tscharntke T (2009a) Cacao boom and bust: sustainability of agroforests and opportunities for biodiversity conservation. Conserv Lett 2:197–205

Clough Y, Tscharntke T, Pitopang R, Dwi Putra D (2009b) Local and landscape factors determine functional bird diversity in Indonesian cacao agroforestry. Biol Conserv 142:1032–1041

Conant R, Paustian K, Elliott E (2001) Grassland management and conversion into grassland: effects on soil carbon. Ecol Appl 11:343–355

Cousins B, Kepe T (2004) Decentralization when land and resource rights are deeply contested: a case study of the Mkambati eco-tourism project on the Wild Coast of South Africa. Eur J Dev Res 16:41–54

Cox LJ, Fox M (2003) Agriculturally based leisure attractions. J Tourism Stud 14(1):49–58

Danse M, Wolters T (2003) Sustainable coffee in the mainstream: the case of the SUSCOF Consortium in Costa Rica. Greener Manag Inter 43:37–51

Davidson S (2005) Shade coffee agro-ecosystems in Mexico: a synopsis of the environmental services and socio-economic considerations. J Sustain Forest 21:81–95

de Foresta H, Michon G (1996) The agroforest alternative to Imperata grasslands: when smallholder agriculture and forestry reach sustainability. Agroforest Syst 36:105–120

de Miguel Magaña S, Harmand JM, Hergoualc's K (2004) Cuantificación del carbono almacenado en la biomasa aérea y el mantillo en sistemas agroforestales de café en el suroeste de Costa Rica. Quantification of carbon stored in aboveground biomass and the litter layer in agroforestry systems with coffee in the South-west of Costa Rica. Agroforestería Am CATIE 41–42:98–104

de Virginio Filho EM (2005) Evaluacion de los sistemas agroforestales com café em fincas vinculadas al COOCAFE. CATIE, Turrialba, 63 pp

DeClerk F, Escalante M, Philpott S, Sinclair F, Soto-Pinto L, Vaast P (2007) Synthesis of coffee and tree cover for the CORRIDOR Project. CATIE, Costa Rica, 12 pp

DeGryze S, Six J, Paustian K, Morris S, Paul E, Merckx R (2004) Soil organic carbon pool changes following land-use conversions. Glob Change Biol 10:1120–1132

Dennis RA, Meijaard E, Nasi R, Gustafsson L (2008) Biodiversity conservation in Southeast Asian timber concessions: a critical evaluation of policy mechanisms and guidelines. Ecol Soc 13(1):25

Detwiler R, Hall C (1988) Tropical forests and the global carbon cycle. Science 239:42–47

Doan T (2000) The effects of ecotourism in developing nations: an analysis of case studies. J Sustain Tourism 8:288–304

Dommergues YR (1987) The biological role of nitrogen fixation in agroforestry. In: Steppler HA, Nair PK (eds.) Agroforestry-a decade of development. ICRAF, Nairobi, pp 245–271

Eggleton P, Bignell D, Hauser S, Dibog L, Norgrove L, Madong B (2002) Termite diversity across an anthropogenic disturbance gradient in the humid forest zone of West Africa. Agr Ecosyst Environ 90:189–202

Elevitch CR, Idol T, Friday JB, Lepczyk C, Easton Smith V, Nelson SC (2009) Shade-grown coffee for Hawaii. Permanent Agriculture Resources, Holualoa

Elmqvist B, Olsson L, Elamin EM, Warren A (2005) A traditional agroforestry system under threat: an analysis of the gum arabic market and cultivation in the Sudan. Agroforest Syst 64:211–218

Engel S, Palmer C (2009) Designing payments for environmental services with weak property rights and external interests. In: Lipper L, Sakuyama T, Stringer R, Zilberman D (eds.) Payment for environmental services in agricultural landscapes: economic policies and poverty reduction in developing countries. Food and Agriculture Organization of the United Nations (FAO), Rome, pp 35–57

Epler-Wood M (2002) Ecotourism: principles, practices and policies for sustainability. United Nations Publication, New York

Ewel JJ, Hiremath AJ (1998) Nutrient use efficiency and the management of degraded lands. In: Gopal B, Pathak PS, Saxena KG (eds.) Ecology today: an anthology of contemporary ecological research. International Scientific Publications, New Delhi, pp 199–215

FAO (1986) Communal forestry strategies. In: Tree growing by rural people. Forestry paper 64, United Nations, Food and Agriculture Organization, Rome, pp 51–66

FAO (2001) Mean annual volume increment of selected industrial forest plantation species. United Nations Food and Agriculture Organization, Rome

FAO (2003) Medium-term prospects for agricultural commodities. United Nations Food and Agriculture Organization, Rome

FAO (2005) Global forest resources assessment. Food and Agriculture Organization of the United Nations, Rome

Fares A, El-Kadi AI (2008) Coastal watershed management. WIT Press, Southampton

Ferraro P, Kiss A (2002) Direct payments to conserve biodiversity. Science 298:1718–1719

FONAFIFO (2005) Montos pagados en el PSA (Pago por servicios ambientales) en Costa Rica. www.fonafifo.com

Galloway G, Beer J (1997) Oportunidades para fomentar la silvicultura en cafetales en America Central. Serie Tecnica, Informe Tecnico 285, CATIE, Costa Rica, pp 103–126

Garcia-Montiel DC, Binkley D (1998) Effect of *Eucalyptus saligna* and *Albizia falcataria* on soil processes and nitrogen supply in Hawaii. Oecologia 113:547–556

Garrity DP (2004) Agroforestry and the achievement of the Millennium Development Goals. Agroforest Syst 61(62):5–17

Garrity DP, Mercado AR (1993) Reforestation through agroforestry: market driven small-holder timber production on the frontier. International Centre for Research in Agroforestry, SEA Regional Research Programme, Bogor

George A (2006) Estudio comparativo de indicadores de calidad de suelo en fincas de café orgánico y convencional en Turrialba, Costa Rica. M.S. Thesis, CATIE, Turrialba, 118 pp

Giardina C, Binkley D, Ryan M, Fownes J, Senock R (2004) Belowground carbon cycling in a humid tropical forest decreases with fertilization. Oecologia 139:545–550

Gillespie A (1989) Modelling nutrient flux and interspecies root competition in agroforestry interplantings. Agroforest Syst 8:257–265

Gillison A, Jones D, Susilo F, Bignell D (2003) Vegetation indicates diversity of soil macroinvertebrates: a case study with termites along a land-use intensification gradient in lowland Sumatra. Organism Divers Evol 3:111–126

Giovannucci D, Koekoek FJ (2003) The state of sustainable coffee: a study of twelve major markets. The International Coffee Organization, Cali, Colombia

Giovannucci D, Potts J, Killian B, Wunderlich C, Soto G, Schuller S, Pinard F, Schroeder K, Vagneron I (2008) Seeking sustainability: COSA preliminary analysis of sustainability initiatives in the coffee sector. Committee on Sustainability Assessment, Winnipeg

Gosling S (1999) Ecotourism: a means to safeguard biodiversity and ecosystem functions? Ecol Econ 29:303–320

Government of India (2005) Joining the dots: the report of the Tiger Task Force. Project Tiger, Union Ministry of Environment and Forests, New Delhi

Graff-Zivin J, Lipper L (2008) Poverty, risk, and the supply of soil carbon sequestration. Environ Dev Econ 13:353–373

Greenberg R, Bichier R, Sterling J (1997) Bird populations in rustic and planted shade coffee plantations of eastern Chiapas, México. Biotropica 29:501–514

Guharay F, Haggar J, Staver C (2005) Final report on results and impacts: regional program on ecologically based participatory implementation of Integrated Pest Management and Coffee agroforestry in Nicaragua and Central America. Final report to NORAD, CATIE, Managua, 130 pp

Gullison RE (2003) Does forest certification conserve biodiversity? Oryx 37:153–165

Guo LB, Gifford RM (2002) Soil carbon stocks and land use change: a meta analysis. Glob Change Biol 8:345–360

Haggar JP (2006) Report on inception phase: innovation of value chains for quality production and marketing in coffee growing highlands of Nicaragua and Honduras. Report to the Norwegian Embassy by CATIE, Managua, 24 pp

Haggar JP (2008) Informe final de proyecto vinculando pequeños productores Centroamericanos a mercados de cafés especiales. Final report to World Bank, CATIE, Managua, 25 pp

Haggar J, Staver C (2001) Como determinar la cantidad de sombra que disminuya los problemas fitosanitarios de café? Agroforestería Las Am 8(29):42–45

Haggar J, Tanner E, Beer J, Kass D (1993) Nitrogen dynamics of tropical agroforestry and annual cropping systems. Soil Biol Biochem 25:1363–1378

Haggar JP, Wightman KE, Fisher R (1997) The potential of plantations to foster woody regeneration within a deforested landscape in lowland Costa Rica. Forest Ecol Manage 99:55–64

Haggar JP, Briscoe CB, Butterfield RP (1998) Native species: a resource for the diversification of forestry production in the lowland humid tropics. Forest Ecol Manage 106:195–203

Haggar JP, Ayala A, Diaz B, Uc C (2001) Participatory design of agroforestry systems: developing farmer participatory research methods in Mexico. Dev Pract 11:417–424

Haggar J, Rheingans R, Arroyo P, Alvarado B (2003) Benefits and costs of intercropping reforestation in the Atlantic lowlands of Costa Rica. New Forest 25:41–48

Haggar J, Sosa M, Diaz B, Hernandez G, Contreras J, Uc C (2004) Adaptation of agroforestry systems in SE Mexico through the integration of farmer and bioeconomic evaluations. Int J Agric Sustain 2:154–166

Hamilton L, King P (1983) Tropical forested watersheds: hydrologic and soils response to major uses or conversions. Westview Press, Boulder

Hartsfield A, Ostermeier D (2003) Certification: the view from FSC-certified land managers. J Forest 101:32–36

Harvey CA, González Villalobos JA (2007) Agroforestry systems conserve species-rich but modified assemblages of tropical birds and bats. Biodivers Conserv 16:2257–2292

Hassan R, Scholes R, Ash N (2005) Ecosystems and human well-being: current state and trends. Island Press, Washington, DC

Herrador D, Dimas L (2000) Payment for environmental services in El Salvador. Mt Res Dev 20:306–309

Hobbs P (2007) Conservation agriculture: what is it and why is it important for future sustainable food production? J Agric Sci 145:127

Hoch L, Pokorny B, Jong W (2009) How successful is tree growing for smallholders in the Amazon? Int Forest Rev 11:299–310

Hoehn P, Tscharntke T, Steffan-Dewenter I (2010) Relative contribution of agroforestry, rainforest and openland to local and regional bee diversity. Biodivers Conserv 19:2189–2200

Holding C, Carsan S, Njuguna P (2006) Smallholder timber and firewood marketing in the coffee and cotton/tobacco zones of eastern Mount Kenya. In: Wall S (ed.) Small-scale forestry and rural development: the intersection of ecosystems, economics and society. Proceedings of the IUFRO 3.08 conference, hosted by the Galway-Mayo Institute of Technology, National Council for Forest Research and Development (COFORD), Galway, 18–23 June 2006, pp 178–190

Honey M (1999) Ecotourism and sustainable development: who owns paradise? Island Press, Washington DC

Hyman EL (1983) Pulpwood tree farming in the Philippines from the viewpoint of the smallholder: an ex post evaluation of the PICOP project. Agric Adm 14:23–49

ICO (2010) Total production of exporting countries: crop years 2004/05 to 2009/10 International Coffee Organization. http://www.ico.org/prices/po.htm. Last accessed 05 Nov 2010

Imbach A, Fassbender H, Borel R, Beer J, Bonneman A (1989) Modeling agroforestry systems of cacao (*Theobroma cacao*) with laurel (*Cordia alliodora*) and poro (*Erythrina poeppigiana*) in Costa Rica. IV. Water balances, nutrient inputs and leaching. Agroforest Syst 8:267–287

Jagger P, Pender J (2003) The role of trees for sustainable management of less-favored lands: the case of *Eucalyptus* in Ethiopia. For Policy Econ 5:83–95

Jindal R, Swallow B, Kerr J (2008) Forestry-based carbon sequestration projects in Africa: potential benefits and challenges. Nat Resour Forum 32:116–130

Jobbágy EG, Jackson RB (2004) The uplift of soil nutrients by plants: biogeochemical consequences across scales. Ecology 85:2380–2389

Johnson K, Olson E, Manandhar S (1982) Environmental knowledge and response to natural hazards in mountainous Nepal. Mt Res Dev 2:175–188

Jones S (2005) Community-based ecotourism: the significance of social capital. Ann Tourism Res 32:303–324

Jones D, Susilo F, Bignell D, Hardiwinoto S, Gillison A, Eggleton P (2003) Termite assemblage collapse along a land-use intensification gradient in lowland central Sumatra, Indonesia. J Appl Ecol 40:380–391

Kabir ME, Webb EL (2008) Can homegardens conserve biodiversity in Bangladesh? Biotropica 40:95–103

Kabutaulaka TT (2005) Rumble in the jungle: land, culture and (un)sustainable logging in Solomon Islands. In: Hooper A (ed) Culture and sustainable development in the Pacific. Australian National University E Press and Asia Pacific Press, Adelaide, pp 88–97

Kaiser B, Roumasset J (2002) Valuing indirect ecosystem services: the case of tropical watersheds. Environ Dev Econ 7:701–714

Kalinganire A (1996) Performance of *Grevillea robusta* in plantations and on farms under varying environmental conditions in Rwanda. Forest Ecol Manag 80:279–285

Kaonga M, Bayliss-Smith T (2009) Carbon pools in tree biomass and the soil in improved fallows in eastern Zambia. Agroforest Syst 76:37–51

Kessler JJ, Breman H (1991) The potential of agroforestry to increase primary production in the Sahelian and Sudanian zones of West Africa. Agroforest Syst 13:41–62

Kiernan M (2000) The forest *ejidos* of Quintana Roo, Mexico: a case study for shifting the power: decentralization and biodiversity conservation. U.S. Agency for International Development, Biodiversity Support Program, Washington, DC

Kiss A (2004) Is community-based ecotourism a good use of biodiversity conservation funds? Trends Ecol Evol 19:232–237

Klock J (1995) Indigenous woodlot management and ethnobotany in Ifugao, Philippines. Int Tree Crops J 8:95–106

Kotto-Same J, Woomer PL, Appolinaire M, Louis Z (1997) Carbon dynamics in slash-and-burn agriculture and land use alternatives of the humid forest zone in Cameroon. Agric Ecosyst Environ 65:245–256

Lal R (2005) Soil carbon sequestration in natural and managed tropical forest ecosystems. In: Montagnini F (ed.) Environmental services of agroforestry systems. Haworth Press, Binghamton

Lasco R (2002) Forest carbon budgets in Southeast Asia following harvesting and land cover change. Sci China C Life Sci 45:55–64

Lawrence DC (1996) Trade-offs between rubber production and maintenance of diversity: the structure of rubber gardens in West Kalimantan, Indonesia. Agroforest Syst 34:83–100

Lehmann J, Weigl D, Droppelmann K, Huwe B, Zech W (1998) Nutrient cycling in an agroforestry system with runoff irrigation in Northern Kenya. Agroforest Syst 43:49–70

Locatelli B (2005) Factibilidad de un sistema de pago por los servicios ambientales del agroecosistema café en la Cuenca de Ocosito en Guatemala. Report of Sistemas Agroforestales con Café Project, ANACAFE/CIRAD/CATIE, Guatemala, 62 pp

Locatelli B, Salinas Z, Rojas V (2008) Impacts of payments for environmental services on local development in northern Costa Rica: a fuzzy multi-criteria analysis. Forest Policy Econ 10:275–285

Maas B, Tscharntke T, Schulze CH, Clough Y, Putra DD, Waltert M (2009) Six years of habitat modification in a tropical rainforest margin of Indonesia do not affect bird diversity but endemic forest species. Biol Conserv 142:2665–2671

Macdicken KG, Hairiah K, Otsamo A, Duguma B, Majid NM (1996) Shade-based control of *Imperata cylindrica*: tree fallows and cover crops. Agroforest Syst 36:131–149

Magcale-Macandog DB, Visco RG, Delgado MEM (2006) Agroforestry adoption, innovations and smallholder farmers' motivations in tropical uplands of southern Philippines. J Sustain Agr 28:131–143

MAGFOR (2002) Elaboración del mapa de cultivo de café en Nicaragua. Ministerio de Agricultura, Ganadería y Forestal, Managua, 15 pp

Martinez M (2005) Contribución económica del componente forestal en diferentes tipos d fincas cafetaleros en la bocacosta pacifica de Guatemala. M.S. thesis, CATIE, Turrialba, 148 pp

Martínez ML, Pérez-Maqueo O, Vázquez G, Castillo-Campos G, García-Franco J, Mehltreter K, Equihua M, Landgrave R (2009) Effects of land use change on biodiversity and ecosystem services in tropical montane cloud forests of Mexico. Forest Ecol Manag 258:1856–1863

Mas AH, Dietsch TV (2004) Linking shade coffee certification to biodiversity conservation: butterflies and birds in Chiapas, Mexico. Ecol Appl 14:642–654

McGrath DA, Comerford NB, Duryea ML (2000) Litter dynamics and monthly fluctuations in soil phosphorus availability in an Amazonian agroforest. Forest Ecol Manag 131:167–181

Medina B, Aguilar R, Anzuelo F, Haggar J, Locatelli B (2005) Evaluación de Servicios Ambientales en el Agroecosistema Café en Guatemala. ANACAFE-CATIE, Guatemala, 68 pp

Mehta NG, Leuschner WA (1997) Financial and economic analyses of agroforestry systems and a commercial timber plantation in the La Amistad Biosphere Reserve, Costa Rica. Agroforest Syst 37:175–185

Melgarejo O, Ríos F, Colán V, Sabogal C (2006) Status of sustainable forest management in the Peruvian Amazon. Situación del manejo forestal sostenible en la Amazonia peruana. In: Kaimowitz D (ed) Recursos naturales y ambiente. Centro Agronómico Tropical de Investigación y Enseñanza, Turrialba, pp 31–37

Mena-Mosquera VE (2008) Relación entre el carbono almacenado en la biomasa total y la composición fisionómica de la vegetación en los sistemas agroforestales con café y en bosques secundarios del Corredor Biológico Volcánica Central-Talamanca, Costa Rica. M.S. thesis, CATIE, Turrialba, 90 pp

Méndez VE, Gliessman SR, Gilbert GS (2007) Tree biodiversity in farmer cooperatives of a shade coffee landscape of western El Salvador. Agr Ecosyst Environ 119:145–159

Menz K, Grist P (1996) Economic opportunities for smallholders to combine pulpwood trees and food crops. Agroforest Syst 36:221–232

Mercer DE, Haggar JP, Snook A (2005) Agroforestry adoption in Calakmul biosphere reserve, Campeche, Mexico. Small Scale Forest Econ Manag Policy 4:163–184

Miyasaka SC, Habte M (2001) Plant mechanisms and mycorrhizal symbioses to increase phosphorus uptake efficiency. Commun Soil Sci Plant Anal 32:1101–1147

Moguel R, Toledo VM (1999) Biodiversity conservation in traditional coffee systems of México. Conserv Biol 13:11–21

Montagnini F, Cusack D, Petit B, Kanninen M (2005) Environmental services of native tree plantations and agroforestry systems in Central America. J Sustain Forest 21:51–67

Montgomery P (1995) Forestry in the Solomon Islands. Pac Econ Bull 10:74–76

Mosley MP (1982) The effect of a New Zealand beech forest canopy on the kinetic energy of water
drops and on surface erosion. Earth Surf Process Land 7:103–107

Muhammed N, Koike M, Haque F (2008) Forest policy and sustainable forest management in
Bangladesh: an analysis from national and international perspectives. New Forest 36:201–216

Munguia R, Ponce A, Haggar J, Staver C (2007) Biomass residue production and nitrogen cycling
and balance under coffee agroforestry systems in Nicaragua. In: Second international symposium on multistrata agroforestry systems, CATIE, Turrialba, 5 pp

Muñiz-Miret N, Vamos R, Hiraoka M, Montagnini F, Mendelsohn RO (1996) The economic value
of managing the acaí palm (*Euterpe oleracea* Mart.) in the floodplains of the Amazon estuary,
Pará, Brazil. Forest Ecol Manag 87:163–173

Muradian R, Pelupessy W (2005) Governing the coffee chain: The role of voluntary regulatory
Systems. World Dev 33:2029–2044

Murty D, Kirschbaum MUF, McMurtrie RE, McGilvray H (2002) Does conversion of forest to
agricultural land change soil carbon and nitrogen? a review of the literature. Glob Change Biol
8:105–123

Nair VD, Graetz DA (2004) Agroforestry as an approach to minimizing nutrient loss from heavily
fertilized soils: the Florida experience. Agroforest Syst 61–62:269–279

Nair P, Kang B, Kass D (1995) Nutrient cycling and soil-erosion control in agroforestry systems. In:
Juo ASR, Freed RD (eds.) Agriculture and environment: bridging food production and environmental protection in developing countries. Proceedings of an international symposium spon-
sored by Division A-6 of the American Society of Agronomy, Cincinnati, Ohio. ASA Special
Publication No. 60, American Society of Agronomy, Madison, 7–12 Nov 1993, pp 117–138

Nair PKR, Mohan Kumar B, Nair VD (2009) Agroforestry as a strategy for carbon sequestration.
J Plant Nutr Soil Sci 172:10–23

Narain P, Singh RK, Sindhwal NS, Joshie P (1998) Water balance and water use efficiency of
different land uses in a western Himalayan valley region. Agr Water Manag 37:225–240

Nebel G, Quevedo L, Jacobsen JB, Helles F (2005) Development and economic significance of
forest certification: the case of FSC in Bolivia. Forest Policy Econ 7:175–186

Ngetich KA, Birech RJ, Kyalo D, Bett KE, Freyer B (2009) Caught between energy demands and
food needs: dilemmas of smallholder farmers in Njoro, Kenya. J Agr Rural Dev Tropics
Subtropics 110:23–28

Nissen TM, Midmore DJ, Keeler AG (2001) Biophysical and economic tradeoffs of intercropping
timber with food crops in the Philippine uplands. Agr Syst 67:49–69

Norgrove L, Hauser S (2002) Measured growth and tree biomass estimates of *Terminalia ivorensis*
in the 3 years after thinning to different stand densities in an agrisilvicultural system in southern Cameroon. Forest Ecol Manag 166:261–270

Oelbermann M, Voroney RP, Gordon AM (2004) Carbon sequestration in tropical and temperate
agroforestry systems: a review with examples from Costa Rica and southern Canada. Agr
Ecosyst Environ 104:359–377

Oelbermann M, Voroney RP, Kass DCL, Schlönvoigt AM (2006) Soil carbon and nitrogen dynamics using stable isotopes in 19- and 10-year-old tropical agroforestry systems. Geoderma
130:356–367

Ogutu Z (2002) The impact of ecotourism on livelihood and natural resource management in
Eselenkei, Amboseli Ecosystem, Kenya. Land Degrad Dev 13:251–256

Pagiola S, Rios AR, Arcenas A (2008) Can the poor participate in payments for environmental
services? Lessons from the silvopastoral project in Nicaragua. Environ Dev Econ 13:299–325

Palm C (1995) Contribution of agroforestry trees to nutrient requirements of intercropped plants.
Agroforest Syst 30:105–124

Palm C, Woomer P, Alegre J, Arevalo L, Castilla C, Cordeiro D, Feigl B, Haiiah K, Kotto-Same J,
Mendes A (1999) Carbon sequestration and trace gas emissions in slash-and-burn and alterna-

tive land-uses in the humid tropics. ASB Climate Change Working Group Final report, Phase II. ICRAF, Nairobi, 33 pp

Pardini R, Faria D, Accacio G, Laps R, Mariano-Neto E, Paciencia M, Dixo M, Baumgarten J (2009) The challenge of maintaining Atlantic forest biodiversity: a multi-taxa conservation assessment of specialist and generalist species in an agro-forestry mosaic in southern Bahia. Biol Conserv 142:1178–1190

Parrotta JA, Turnbull JW, Jones N (1997) Catalyzing native forest regeneration on degraded tropical lands. Forest Ecol Manag 99:1–8

Pattanayak SK, Mercer DE, Sills EO, Yang J (2003) Taking stock of agroforestry adoption studies. Agroforest Syst 57:173–186

Perfecto I, Rice R, Greenberg R, van der Moort M (1996) Shade coffee: a disappearing refuge for biodiversity. Bioscience 46:598–608

Philpott S, Dietsch T (2003) Coffee and conservation: a global context and the value of farmer involvement. Conserv Biol 17:1844–1846

Post W, Kwon K (2000) Soil carbon sequestration and land-use change: processes and potential. Glob Change Biol 6:317–327

Powers JS, Haggar JP, Fisher R (1997) The effect of overstory composition on understory woody regeneration and species richness in seven year-old plantations in Costa Rica. Forest Ecol Manag 99:43–54

Pretty J, Smith D (2003) Social capital in biodiversity conservation and management. Conserv Biol 18:631–638

Radersma S, Grierson PF (2004) Phosphorus mobilization in agroforestry: organic anions, phosphatase activity and phosphorus fractions in the rhizosphere. Plant Soil 259:209–219

Ramadhani T, Otsyina R, Franzel S (2002) Improving household incomes and reducing deforestation using rotational woodlots in Tabora district, Tanzania. Agr Ecosyst Environ 89:229–239

Resh S, Binkley D, Parrotta J (2002) Greater soil carbon sequestration under nitrogen-fixing trees compared with *Eucalyptus* species. Ecosystems 5:217–231

Rojas M, Aylward B (2003) What are we learning from experiences with markets for environmental services in Costa Rica? A review and critique of the literature. International Institute for Environment and Development, London, 102 pp

Roshetko JM, Delaney M, Hairiah K, Purnomosidhi P (2002) Carbon stocks in Indonesian homegarden systems: can smallholder systems be targeted for increased carbon storage? Am J Altern Agr 17:138–148

Roshetko JM, Lasco RD, Los Angeles MSd (2007) Smallholder agroforestry systems for carbon storage. Mitig Adapt Strateg Glob Change 12:219–242

Roshetko JM, Purnomosidhi M, Purnomosidhi P (2004) *Gmelina arborea*-a viable species for smallholder tree farming in Indonesia. New Forest 28:207–215

Roskoski JP (1982) Nitrogen fixation in a Mexican coffee plantation. Plant Soil 67:283–291

Ross S, Wall G (1999) Ecotourism: toward congruence between theory and practice. Tourism Manag 20:123–132

Samper M (1999) Itinerarios tecnologicos del café en Centroamerica. Ateliers de Caravelle No 13

SAN (2010) Sustainable agriculture standard. Sustainable Agriculture Network, San José, 49 pp

Sanchez P (1995) Science in agroforestry. Agroforest Syst 30:5–55

Sánchez-Azofeifa GA, Pfaff A, Robalino JA, Boomhower JP (2007) Costa Rica's payment for environmental services program: intention, implementation, and impact. Conserv Biol 21:1165–1173

Sandström K (1998) Can forests "provide" water: widespread myth or scientific reality? Ambio 27:132–138

SCAA Sustainability Council (2010) Sustainable coffee certifications: a comparison matrix. Specialty Coffee Association of America, Long Beach

Scherr SJ (1995) Economic factors in farmer adoption of agroforestry: patterns observed in Western Kenya. World Dev 23:787–804

Scheyvens R (1999) Ecotourism and the empowerment of local communities. Tourism Manag 20:245–249

Schibli C (2001) Percepciones de familias sobre el uso y manejo de sistemas agroforestales con café en el norte de Nicaragua. Agroforestaría Am 8:8–14

Schroth G (1995) Tree root characteristics as criteria for species selection and systems design in agroforestry. Agroforest Syst 30:125–143

Schroth G, Lehmann J, Rodrigues MRL, Barros E, Macêdo JLV (2001) Plant-soil interactions in multistrata agroforestry in the humid tropics. Agroforest Syst 53:85–102

Sears RR, Padoch C, Pinedo-Vasquez M (2007) Amazon forestry transformed: integrating knowledge for smallholder timber management in Eastern Brazil. Hum Ecol 35:697–707

Sheng T (1989) Soil conservation for small farmers in the humid tropics: FAO Soils Bulletin 60. United Nations Food & Agriculture Organization, Rome

Shepherd K, Ohlsson E, Okalebo J, Ndufa J (1995) Potential impact of agroforestry on soil nutrient balances at the farm scale in the East African Highlands. Nutr Cycl Agroecosyst 44:87–99

Sherperd G (1992) Managing Africa's dry tropical forests: a review of indigenous methods. Overseas Development Institute, London, 19 pp

Six J, Conant RT, Paul EA, Paustian K (2002) Stabilization mechanisms of soil organic matter: implications for C-saturation of soils. Plant Soil 241:155–176

SMBC (2008) Shade management criteria for "Bird Friendly" coffee. Smithsonian Migratory Bird Center, Washington, DC

Smith NJH, Falesi IC, Alvim PdT, Serrão EAS (1996) Agroforestry trajectories among smallholders in the Brazilian Amazon: innovation and resiliency in pioneer and older settled areas. Ecol Econ 18:15–27

Somarriba EJ, Beer JW (1987) Dimensions, volumes and growth of *Cordia alliodora* in agroforestry systems. Forest Ecol Manag 18:113–126

Somarriba E, Valdivieso R, Vasquez W, Galloway G (2001) Survival, growth, timber productivity and site index of *Cordia alliodora* in forestry and agroforestry systems. Agroforest Syst 51:111–118

Sopandi A, Rule LC (2000) Comparative analysis of four pine plantation management situations in West Java. J Trop Forest Sci 12:656–668

Soto Pinto L, De Jong BHJ, Esquivel Bazán E, Quechulpa S (2006) Potencial ecológico y económico de captura de carbono en cafetales. In: Pohlan J, Soto Pinto L, Barrera J (eds.) El cafetal del futuro: realidades y visiones. Shaker, Aachen, pp 333–360

Spellerberg IF, Sawyer JWD (1996) Standards for biodiversity: a proposal based on biodiversity standards for forest plantations. Biodivers Conserv 5:447–459

Sreedevi TK, Shiferaw B, Wani SP (2006) Adarsha watershed in Kothapally: understanding the drivers of higher impact. SAT eJournal 2:1–19

Steffan-Dewenter I, Kessler M, Barkmann J, Bos M, Buchori D, Erasmi S, Faust H, Gerold G, Glenk K, Gradstein S (2007) Tradeoffs between income, biodiversity, and ecosystem functioning during tropical rainforest conversion and agroforestry intensification. Proc Natl Acad Sci 104:4973–4978

Stronza A, Gordillo J (2008) Community views of ecotourism. Ann Tourism Res 35(2):448–468

Suarez D (2002) Cuantificación y valoración económica del servicio ambiental almacenamiento de carbono en sistemas agroforestales de café en la comarca Yassica Sur, Matagalpa, Nicaragua. M.S. thesis, CATIE, Turrialba, 116 pp

Swamy SL, Puri S (2005) Biomass production and C sequestration of *Gmelina arborea* in plantation and agroforestry system in India. Agroforest Syst 64:181–195

Swanson FJ, Dyrness CT (1975) Impact of clear-cutting and road construction on soil erosion by landslides in the western Cascade Range, Oregon. Geology 3:393–396

Szott L, Melendez G (2001) Phosphorus availability under annual cropping, alley cropping, and multistrata agroforestry systems. Agroforest Syst 53:125–132

Té S (2008) Annual report 2007. San Cristóbal de las Casas, Chiapas, 11 pp

Tefera AT, Rao MR, Mathuva MN, Atta-Krah K (2001) Farmer-participatory evaluation of *Grevillea robusta* in boundary plantings in semi-arid Kenya. Forest Trees Livelihood 11:13–27

Thacher T, Lee DR, Schelhas JW (1997) Farmer participation in reforestation incentive programs in Costa Rica. Agroforest Syst 35:269–289

Thulasidas PK, Bhat KM (2009) Log characteristics and sawn timber recovery of home-garden teak from wet and dry localities of Kerala, India. Small Scale Forest 8:15–24

TIES (2010) What is ecotourism? In: The International Ecotourism Society. http://www.ecotourism. org/site/c.orLQKXPCLmF/b.4835303/k.BEB9/What_is_Ecotourism__The_International_ Ecotourism_Society.htm. Accessed Sept 2010

Townsend AR, Vitousek PM, Trumbore SE (1995) Soil organic matter dynamics along gradients in temperature and land use on the island of Hawaii. Ecology 76:721–733

Trauernicht C, Ticktin T (2005) The effects of non-timber forest product cultivation on the plant community structure and composition of a humid tropical forest in southern Mexico. Forest Ecol Manag 219:269–278

Treacy J, Denevan W (1994) The creation of cultivable land through terracing. In: Miller NF, Gleason KL (eds.) The archaeology of garden and field. University of Pennsylvania Press, Philadelphia, pp 91–110

Tscharntke T, Hoehn P, Tylianakis JM, Sodhi NS, Sekercioglu CH, Dietsch TV (2008) Landscape constraints on functional diversity of birds and insects in tropical agroecosystems (Erratum: June 2008, 89(6): 1775). Ecology 89:944–951

Uezu A, Metzger JP, Beyer DD (2008) Can agroforest woodlots work as stepping stones for birds in the Atlantic forest region? Biodivers Conserv 17:1907–1922

UNFCCC (2009) Approved simplified baseline and monitoring methodology for small-scale agroforestry: afforestation and reforestation project activities under the clean development mechanism. United Nations Framework Convention on Climate Change. Clean Development Mechanism Executive Board AR-AMS0004 / Version 02

UNFCCC (2010a) Mechanisms: Kyoto protocol. United Nations Framework Convention on Climate Change. http://unfccc.int/kyoto_protocol/mechanisms/items/1673.php. Accessed Sept 2010

UNFCCC (2010b) Clean Development Mechanism. Registered Projects Directory. United Nations Framework Convention on Climate Change. http://cdm.unfccc.int/Projects/registered.html. Accessed Sept 2010

Van Noordwijk M, Lusiana B (1998) WaNuLCAS, a model of water, nutrient and light capture in agroforestry systems. Agroforest Syst 43:217–242

Van Noordwijk M, Purnomosidhi P (1995) Root architecture in relation to tree-soil-crop interactions and shoot pruning in agroforestry. Agroforest Syst 30:161–173

Van Noordwijk M, Lawson G, Soumaré A, Groot JJR, Hairiah K (1996) Root distribution of trees and crops: competition and/or complementarity. In: Ong CK, Huxley P (eds) Tree-crop interactions – a physiological approach. CAB International, Wallingford, pp 319–364

Verbist B, Putra AED, Budidarsono S (2005) Factors driving land use change: effects on watershed functions in a coffee agroforestry system in Lampung, Sumatra. Agric Syst 85:254–270

Vitousek PM, Sanford JRL (1986) Nutrient cycling in moist tropical forest. Annu Rev Ecol Syst 17:137–167

Walpole M, Goodwin H (2001) Local attitudes towards conservation and tourism around Komodo National Park, Indonesia. Environ Conserv 28:160–166

Ward JR, Bihun Y (2001) Stewardship of Mexico's community forests: expanding market and policy opportunities for conservation and rural development. In: Bowles IA, Prickett GT (eds.) Footprints in the jungle: natural resource industries, infrastructure, and biodiversity conservation. Oxford University Press, New York, pp 145–167

Weaver D (1998) Ecotourism in the less developed world. CABI, New York

Weerd M, Snelder DJ (2008) Human-altered tree-based habitats and their value in conserving bird and bat diversity in northeast Luzon, The Philippines. In: Snelder DJ, Lasco RD (eds.) Smallholder tree growing for rural development and environmental services – lessons from Asia. Springer Science, Dordrecht, pp 347–377

West P, Carrier J (2004) Ecotourism and authenticity: getting away from it all? Curr Anthropol 45:483–491

Westphal SM (2000) Estrategias agroforestales y socioeconomicas de pequeños productores de café con sombra en el Pacifico Sur de Nicaragua. Report to Project MIP-AF, CATIE, Managua, 18 pp

Westphal SM (2008) Coffee agroforestry in the aftermath of modernization: diversified production and livelihoods in post-reform Nicaragua. In: Bacon CM, Mendez VE, Gliessman SR, Fox JA (eds.) Confronting the coffee crisis: fair trade, sustainable livelihoods and ecosystems in Mexico and Central America. MIT, Cambridge, pp 177–205

Wheelock-Diaz SI, Barrios-Jackman M (2007) Análisis comparativa de experiencias de pago por servicios ambientales en Nicaragua. Cuaderno de Investigación No. 28, Nitlapan, Managua, 54 pp

Wischmeier WH, Smith DD (1965) Predicting rainfall-erosion losses from cropland east of the Rocky Mountains. Agriculture Handbook 282, US Department of Agriculture, Agricultural Research Service, Washington, DC

Witcomb M, Dorward P (2009) An assessment of the benefits and limitations of the shamba agroforestry system in Kenya and of management and policy requirements for its successful and sustainable reintroduction. Agroforest Syst 75:261–274

Wolters V (2001) Biodiversity of soil animals and its function. Eur J Soil Biol 37:221–227

World Commission on Environment and Development (1987) Our common future. Oxford University Press, Oxford

Wunder S (2000) Ecotourism and economic incentives: an empirical approach. Ecol Econ 32:465–479

Wunder S, Albán M (2008) Decentralized payments for environmental services: the cases of Pimampiro and PROFAFOR in Ecuador. Ecol Econ 65:685–698

Youkhana A, Idol T (2009) Tree pruning mulch increases soil C and N in a shaded coffee agroecosystem in Hawaii. Soil Biol Biochem 41:2527–2534

Young A (1997) Agroforestry for soil management. CAB International, Wallingford

Youkhana A, Idol T (2011a) Allometric models for predicting above- and belowground biomass of *Leucaena*-KX2 in a shaded coffee agroecosystem in Hawaii. Agroforest Syst (in press)

Youkhana A, Idol T (2011b) Addition of Leucaena-KX2 mulch in a shaded coffee agroforestry system increases both stable and labile soil C fractions. Soil Biol Biochem 43:961–966

Ziemer RR (1981) Roots and the stability of forested slopes. In: Davies TRH, Pearce AJ (eds.) Proceedings of the international symposium on erosion and sediment transport in Pacific Rim steeplands, Christchurch, New Zealand. International Association for the Hydrological Sciences Publication No. 132, pp 343–361

Zuluaga J (2004) Dinámica de la material orgánica del suelo en sistemas agroforestales con *Erythrina poeppigiana* en Costa Rica. M.S. thesis, CATIE, Turrialba, 97 pp

Sustainable Development of an Agricultural Region – The Case of the Allgäu, Southern Germany

Sabine Weizenegger and Alexander Wezel

Abstract Since the United Nations Conference on Environment and Development in 1992 in Rio de Janeiro, Brazil, much has been written about sustainable development. Nevertheless, information about sustainable development linked explicitly to a particular region is still relatively rare. In this review, we analyse and evaluate the sustainable development of the Allgäu, a rural agricultural region in southern Germany which has high touristic potential. The Allgäu is a typical rural region which had to experience many challenges, and undergo many changes and adaptations during its history, even when living conditions were not easy until the midst of the twentieth century.

From our evaluation we conclude that there is a relatively positive economic, social and ecological development towards sustainability. There exists a good economic and income situation for most people, good ecological conditions with relatively rich biodiversity, a relatively well-established social structure, as well as a certain identity with the region and relatively low social discrepancy. Nevertheless, different actual and future threats exist such as potential negative impacts related to increased development of the tourism sector, intensification or abandonment of agriculture in certain areas, or loss of traditions and customs.

S. Weizenegger
Regionalentwicklung Oberallgäu, Rathausplatz 1, 87452 Altusried, Germany
e-mail: weizenegger@regionalentwicklung-oberallgaeu.de

A. Wezel (✉)
Department of Agroecosystems, Environment and Production, ISARA Lyon,
23, rue Jean Baldassini, Lyon cedex 07 69364, France
e-mail: wezel@isara.fr

W.B. Campbell and S. López Ortíz (eds.), *Integrating Agriculture, Conservation and Ecotourism: Examples from the Field*, Issues in Agroecology – Present Status and Future Prospectus 1, DOI 10.1007/978-94-007-1309-3_6,
© Springer Science+Business Media B.V. 2011

1 Introduction

Quite a lot has been written about sustainable development over the last two decades (e.g. Blewitt 2008; Rogers et al. 2008). The range of topics includes different global aspects of sustainable development, general theoretical considerations (Grober 2010), development of indicators (Job 1996), social dimensions and policies, economics of sustainability or more particular aspects of sustainable development such as that for rural areas (Friedel and Spindler 2009), how income of small-scale farmers could be improved or diversified or how tourism could be organised in a sustainable manner (Weizenegger 2003; Becker et al. 1996; Vorlaufer 1996). Examples of sustainable development linked explicitly to a particular region are still relatively rare (e.g. Gong and Lin 2000; Eglington et al. 1998).

In this review, we provide a comprehensive picture of the development of a region, and whether this development can be evaluated as socially, economically and ecologically sustainable. We carry out a combined qualitative-quantitative assessment where we use quantitative data and indicators when available, and when not available we use literature sources or expert knowledge from the region for a qualitative assessment. Thus, we will be relatively descriptive in certain parts. In this contribution we will describe how the Allgäu region – consisting of complex structures and landscapes, different economic activities and actors – evolved and developed. We will look at how the region used and uses its endogenous natural and cultural assets and its existing structures in order to react to changing conditions and external influences. We will outline the natural and historical contexts of the region which led to certain types of (economic) activities and sometimes to particular adaptations of these activities. For the recent past, we will present some projects, programs and instruments which have been established to foster and direct positive regional development and show that local and regional stakeholders play an important role in those processes. We will assess to what extent those activities and processes can contribute to sustainable development and to what degree we can generally speak of sustainable economic, social and environmental development in the Allgäu region. Finally, we will provide perspectives for potential changes in the future.

1.1 The Case of the Allgäu in the Context of Agroecology

Presently, three main interpretations of agroecology exist world-wide. Agroecology is either seen more as a practice, a movement or a scientific discipline (Wezel and Jauneau 2011; Wezel et al. 2009). Within the latter two, different types of meanings can be distinguished. If we take the case of the Allgäu, this concerns mainly the agroecosystem approach and the food system approach within agroecology as a scientific discipline (more information about the different approaches can be found in Wezel and Jauneau 2011). Regarding the agroecosystem approach, different types of agricultural production, their evolution and their constraints will be presented and evaluated for the Allgäu. In relation to the larger food system approach, interactions and exchanges which take place in an agroecosystem or a region among different stakeholders, in networks, or with society in general, but also influences, impacts and opportunities from policies or the economy are considered. This will be particularly related to regional development of the Allgäu and different projects promoting local food or forestry products. This review of sustainable development in the Allgäu also permits an illustration in more detail about the greater theoretical link between agroecology and rural or territorial development than that presented in Wezel and Jauneau (2011). Although rural development also can be a sub-type within agroecology as a movement, this does not apply to the Allgäu, as no common, goal-oriented movement so far exists.

2 The Study Area

2.1 Delimitation of the Allgäu

The Allgäu is located in the very south of Germany, sharing borders with neighbouring Austria (Fig. 1). Various delimitations of the region exist because different approaches are used such as political or administrative, landscape, or one that involves the identity of the people who live within its boundaries (Fig. 2).

Fig. 1 Location of the Allgäu region in southern Germany

Region of the Allgäu

Fig. 2 Location of the Allgäu in southern Germany with its administrative (*above*) and landscape (*below*) delimitations (Source: Institut für Länderkunde Leipzig 1997)

The administrative delimitation is made with three districts in the Federal State of Bavaria which contain "-allgäu" in their name (Unterallgäu, Oberallgäu, Ostallgäu, standing for Lower, Upper, Eastern Allgäu). A fourth one, formerly Westallgäu, has since 1972, been called Landkreis Lindau (district Lindau). Those four districts and three urban municipalities Kempten, Kaufbeuren and Memmingen comprise an area of approximately 4,650 km². A slightly different grouping for the Allgäu is made with Bavaria's spatial planning regions. Here, the Allgäu consists only of the three districts Lindau, Oberallgäu and Ostallgäu, with the Unterallgäu belonging to a different planning region.

In Germany's National Atlas (Institut für Länderkunde 1997), the Allgäu is displayed as a landscape unit, including parts of Bavaria and a small part of Baden-Württemberg (the other Federal State in southern Germany). Sometimes, even the Austrian exclaves Jungholz or Kleinwalsertal are considered as parts of the Allgäu. In having no fixed borders, the Allgäu belongs to landscapes having

Fig. 3 The prealpine landscape of the Allgäu (*in the foreground*) and the Alps (*in the background*)

flexible delimitations (temporally or topically) with regard to tourism, landscape planning, or administration (Liedtke 1997; Jahn 1989; Klima 1989).

In this review, we follow the landscape delimitation from the National Atlas (Institut für Länderkunde Leipzig 1997) for the Allgäu, but we refer to the administrative units when it comes to statistical data as they are only available for these units. Unless indicated, statistical data refer to the four districts and the three urban municipalities from the Bavarian portion of the Allgäu.

2.2 Geography, Landscapes and Special Features of the Allgäu

Three main types of landscapes can be distinguished for the Allgäu. The most southern part belongs to the Alps, a large mountain range which stretches from southeastern France, over Switzerland and northern Italy, to Austria and Slovenia. The landscape of the prealpine lands and the foothills are located more in the central part (Fig. 3). The most northern part is mainly characterised by plains and smaller areas with gentle undulating lower hills (Fig. 4).

Scholz (1995) describes the geological and geomorphologic processes that led to today's landscapes of the Allgäu. The geological process was the alpine orogenesis (the primary geological mechanism, upheaval, by which mountains are formed on continents) that resulted in peaks, which in the very south of the Allgäu reach

Fig. 4 Landscape of the northern part of the Allgäu (*in the foreground*) and the snow-covered Alps (*in the background*)

altitudes up to 2,649 m above sea level (Hochfrottspitze) with a typical alpine character. The geomorphologic process involved several glacial periods during the Quaternary which shaped the Alps and the glacial landforms in the prealpine landscape which contain moraines, specific valley forms, gentle hills, rivers, and lakes. The glacial periods also were responsible for the creation of gravel plains intermixed in a slightly undulating landscape in the northern part of the Allgäu. This natural landscape has been transformed by man, leading to the cultural landscape of the Allgäu as described later.

The Allgäu is part of the temperate climate zone which is characterized by higher precipitation with around 1,600 mm/year in lower altitudes in the south and west (Oberstdorf, Isny) (Deutscher Wetterdienst 2010; Mühr 2010) and up to 2,500 mm/year in the higher mountains of the south. The northern and eastern parts receive less precipitation; between 900 and 1,200 mm/year (Mindelheim, Kaufbeuren) (Deutscher Wetterdienst 2010; Mühr 2010). Average annual temperatures range from 5.5°C to 7.5°C for the lower altitudes with lower temperatures in the southern portion. The snowy winter period is, in general, relatively long.

Land use in the Allgäu is related to the different landscape types and the climatic situation. The dominant land use in the four Bavarian districts of the Allgäu is agriculture which is practiced over 58% of the area (Bayerisches Staatsministerium für Ernährung, Landwirtschaft und Forsten 2010). Only in the northern portion can a mixture of crops (mainly maize, but also some rape, wheat or potatoes) and permanent grassland be found. The other areas are almost exclusively used as permanent grasslands (pastures, meadows) for livestock production (mainly dairy cows, fewer for meat production, very few goats and sheep for milk and meat). The second major land use is forestry, occupying 29% of the area. The share of forest areas increases to 36% towards the south of the Allgäu. The rest of the area consists of towns, settlements or other urban land and road infrastructure (13%).

The Allgäu is known for its relatively high species and ecosystem diversity because of a highly heterogeneous landscape with different types of forests, bogs, marshes and wetlands, ponds, lakes, grasslands, and alpine ecosystems. About half of the Allgäu is classified by the German Federal Agency for Nature Conservation as an area which should receive certain protection (different protection levels being possible) because it has a high proportion of protected areas as well as endangered species and special biotopes (Bundesamt für Naturschutz 2004). Two out of 42 endemic plant species in Germany are endemic to the Allgäu. Different types of protected areas also are found in the Allgäu, such as nature reserves (8%), landscape protected areas (15%), or NATURA 2000 areas (13%) (LfU 2010). The latter may fall into more than one category, so that at present 24% of the Allgäu is protected under different types of protection status.

In 2008, the transnational Nature Park "Nagelfluhkette" was founded in collaboration with Austria. Parts of the Nature Park are protected areas in the strict sense (as nature reserves or landscape protected areas). The Park as a whole underlies a management plan following the objectives of sustainable development to include protection of nature and landscapes, sustainable agriculture and maintenance of cultural landscapes, sustainable tourism, and sustainable regional development.

2.3   Actual Economic Activity Data (Including Population)

The four districts and three urban municipalities on the Bavarian side of the Allgäu comprise a population of approximately 645,000 people, about one third of them living in one of the urban municipalities. The population density is 139 inhabitants per square kilometre (Destatis 2010).

The labour force is distributed among sectors including agriculture and forestry: 3.7%, industry: 31.6%, and services: 64.6% (Destatis 2010). Compared to other regions, this Allgäu region shows good performance in its economic development. One indicator is the low unemployment rate averaging 4.9% (ranging from 3.3% in Unterallgäu to 7.3% in Kaufbeuren), compared to the rate in Germany (8.2%). This is due to a mixture of agriculture, tourism, crafts and industry. In the southern Allgäu, tourism is dominant, while in the central and northern parts, trade and industry play important roles.

Tourism plays a major role with 2.4 million arrivals and almost 10.5 million overnight stays in 2009 (Allgäu Marketing 2010). More than 2,000 hotels or pensions offer about 70,000 beds. The average length of stay is 4.2 days (compared to 2.9 in Bavaria). While most tourist destinations in Bavaria had a decline in arrivals as well as in overnight stays during the last years, the Allgäu showed a growth of 4.2% for arrivals (the highest rate in Bavaria), followed by the Bavarian capital Munich (3.2%) and a very small decline of 0.2% for overnight stays (Allgäu Marketing 2010).

3   Changes in Land Use Patterns and Structure of Agriculture from the Middle Ages up to the Present

The historical perspective is very important in understanding the appearance of the present cultural landscape of the Allgäu. Different major changes of land use, particularly agricultural land use, occurred in the Allgäu from the Middle Ages up to the present. This often followed structural changes in agriculture such as changes in heritage rules, political systems, and changes in national or global markets.

The first human traces found in the Allgäu date back to 4,000 B.C. (Middle Stone Age). At this time, there were waves of human settlements and the first primitive forms of grain cultivation. Later, forests regained those early settlements. The Late Bronze Age brought men back to the Allgäu, settling first in its northern parts and later proceeding further south. By 1,200 B.C., Illyric farmers (from today's Hungary) founded settlements and brought with them the technique of iron processing. By 15 B.C. the Romans had conquered the region, making it a Roman province. The next 200 years were characterized by Alemann intrusions. During that time a number of settlements were destroyed. Starting from 400 A.D., the Alemanns created many new settlements.

3.1   Middle Age History of Agriculture

When the Franconians arrived in the sixth century A.D., heritage rules were changed. Under the Alemanns, all farmland was inherited as a whole, and this was altered to the land being divided. This meant that a family's land holdings were divided among the beneficiaries in equal terms, resulting in the number of small and smallest plots/fields increasing over generations (van Endert 2000). Fiefdoms also came into existence when the Franconians established royal civil servants (earls) to monitor royal domains and furnished them with lands. In addition, clerical domains developed on lands of dioceses monasteries. Between the ninth and thirteenth centuries, unexploited land belonged to the King, who gave large portions to his clerical and secular sovereigns (earls, dukes, princes). For them, the land was useless as long as it was wilderness. Therefore, settlers were invited, and they were given more rights than usual. At the same time, there were many properties having foreign masters, and towns also gained influence.

This land ownership structure and heritage practice of splitting the land finally led to a degree of fragmentation which did not allow for reasonable cultivation any longer because agricultural land per family became too small from which to make a living, or access to the many very small parcels became difficult or created tensions among the farmers. Nowotny (1984) gives an example of a farmer's 5 ha plot which was divided into 39 different fields in 1550.

The subsequent history of the Allgäu will be presented below in more detail as important changes in land use patterns took place which played important roles in the development of the region up to the present.

3.2   Vereinödung – New Agricultural Structures

Vereinödung stands for changes in settlement patterns, which became necessary due to the existing field structure with many small parcels. Vereinödung means the establishment of dispersed single farmsteads or small hamlets, with the fields (in most cases newly allocated) located next to the farmstead. The Vereinödung has a threefold significance:

– geographical: a place outside the village
– juridical: possession without mutual burdens
– economical: reshaping with or without extensions.

Most known forms of Vereinödung transformation are found in the Allgäu and other parts of southern Germany, in England, and in Scandinavia. As described above, during the Medieval Ages there were generally a lot of rules on how to work fields as well as financial obligations towards the feudal lord. But those rules were not applied to common lands that had been created through forest clearances. Thus, these areas were open to the establishment of new farmsteads.

It is not possible to determine exactly when the process of Vereinödung started in the Allgäu. Most authors refer to a Vereinödung in the village of Sulzberg in the year 1550 (Nowotny 1984; Lochbrunner 1976). In the beginning the process was slow; up to the start of the Thirty Years War in 1618, only 21 cases were reported, all of them near the town of Kempten. But the idea spread, so that up to the end of the nineteenth century more than 1,200 shifts had taken place, most of them between the 1770s and 1830s (Lochbrunner 1976).

Vereinödung did not only mean new allocation of fields, but very often also a displacement of the houses. In some cases it took only one week to break down the house and set it up in another location. Nowotny (1984) claims that the movement of houses had primarily economic reasons – there were no more long distances, a fact that saved time. From the social point of view, the movement brought a disintegration of the families that now had to live outside their social environment.

According to Nowotny (1984), Vereinödung in the Allgäu was a process initiated by the farmers themselves, and not a constraint imposed by the authorities. It was only in 1791 that the authorities began regulating Vereinödung through a legal ordinance – but by then, about two thirds of all Vereinödungen had already been carried out. It was a slow, but well grounded and well reflected movement.

3.3   Transformation from Blue Allgäu to Green Allgäu

The Allgäu had never been well suited for cropping; difficult topography, watery depressions, moory soils, frosts, long winters, and high precipitation were all reasons for not favouring cropping, at least from an economic point of view. Nevertheless, forms of agriculture existed. Alemann settlers had brought the practices into the region but never adapted them to the conditions in their new home, nor did their descendants up to the nineteenth century. They exclusively practiced subsistence farming because of the lack of transport, leading to a highly diversified agriculture (Thierer 1985).

Already in the early Medieval Ages, flax (*Linum usitatissimum* L.) for linen production and hemp (*Cannabis sativa* L.) for yarn production had been cultivated (the blue flower of flax providing the name "Blue Allgäu"), for subsistence and trade. Over time, the region developed a high competence in the textile sector. Beginning in the fifteenth century, however, the climate changed. Average temperatures declined and precipitation increased. Overexploitation of soils had lasted for centuries, and the situation for the farmers became worse, eventually leading to the Peasants' War in 1525. With the colonial era and industrialisation starting in the middle of the nineteenth century, imported industrially processed cotton came into the markets, representing a serious competition for linen, and leading to a decreased income for farmers.

Among other agricultural activities and processing, cheese had always been produced, but on a small scale and mostly for subsistence. When the engagement of farmers and their families in the textile sector started to become increasingly difficult from 1850 onwards, more and more farmers started to produce more conservable types of cheese (cheeses lasting longer periods of time) after the methods of pioneers such as Johann Althaus and Carl Hirnbein, who had learned the techniques in Switzerland and the Netherlands (Krattenmacher 1983). Althaus brought in techniques to make Emmentaler cheese (hard cheese) in 1827, and Hirnbein followed a few years later with Limburger cheese (soft cheese). For the production of hard cheeses like Emmentaler, large amounts of milk were necessary, so that cooperation between the farmers was required, leading to the founding of cooperatives (Thiel 2000; Jahn 1989). The production of soft cheese demanded less milk, and also was easier to produce. The main objective, however, was to produce cheese of better quality, which lasted longer and could therefore be transported.

New means of transport such as the railway allowed for exportation of these products. In his travel guide dating from the middle of the nineteenth century, Buck (1856) mentioned that the production of Backstein cheese had increased recently and was being exported to northern Germany, Holland, and even America. This was the beginning of dairy farming on a larger scale (the colour of the pastures leading to the "Green Allgäu"). The transition from the blue to the green Allgäu lasted until the beginning of the twentieth century when the last flax fields disappeared (Güthler 2006). According to Jahn (1978), the previously mentioned process of Vereinödung was an indispensable precondition for the change from arable cropping to the pasture system, because only the location of the farms amidst larger block parcels could allow for an improved form of use due to the short distances to the pastures.

The changes which took place with flax production and processing, and the beginning of industrialisation in the nineteenth century can be well illustrated with the small town of Immenstadt in the southern Allgäu. Vogel (2000) estimated that half of the population was occupied with weaving and spinning in Immenstadt around 1660. Nevertheless, industrialisation did not take place before 1855, when the first spinning mills were founded (Kamp 2000). In 1867, the town had a population of 2000, with 10% of the population working in the spinning mills. Train connections and the spinning mills transformed the town to an urban industrial centre within a few decades (today it has around 15,000 inhabitants). Some of the spinning companies closed down towards the end of the nineteenth century, and some lasted longer, with the last one closing in 1973. The knowledge acquired in this field during the long time period – many of the workers came from families that had been employed for generations and were proud to work for a successful traditional company – allowed the textile industry to exist further into the twentieth century. Another reason for the success was the aspect of quality (Kamp 2000; Vogel 2000). During the second half of nineteenth century, the Allgäu also became a global player for hemp processing. In the beginning, hemp came from the region, but large quantities were soon imported from Italy or the Soviet Union, and were transformed to different types of yarn which was then exported around the world, even to Australia (Kamp 2000).

3.4   Recent Changes in Land Use and the Landscape of the Green Allgäu

In the middle of the nineteenth century, agriculture was a mixture of fields and pastures that prevailed in the southern part of the Allgäu (towards the Alps) (Güthler 2006). Potatoes, oats, barley and rye were cultivated for local consumption, whereas flax and livestock were produced to generate income. Schaffer and Zettler (1984) analysed the changes of the cultural landscape coming with changes in agricultural structuration in the Unterallgäu (northern part of the Allgäu). While around 85% of the agricultural land in the valleys and smaller plains was used for cropping (flax, hemp, oats and spelt; Konold 1996) around 1840, it changed to 90% grassland at the beginning of the 1980s. Schaffer and Zettler (1984) distinguish three main phases:

– 1840–1910: crises for the farmers and changes between forest clearance and afforestation
– 1911–1949: modernisation and the beginning of a new economic attitude
– 1950–1984: cultural landscape between use and misuse.

From 1840 to 1910, many economic crises appeared which affected the farmers (low cereal prices, famines, high credit costs). Agricultural work was still handwork with simple tools. There had been smaller changes in the agricultural land structure, but in general a highly diversified landscape prevailed with many small landscape structures and many embedded cleared older forest parcels or newly afforested sites.

The modernisation of agriculture with the introduction of machines, the establishment of agrarian markets and the transition from cropping to livestock breeding, led to an initial clearance of smaller landscape structures such as single trees or hedges and to initially smaller relief work (levelling of some areas) from 1911 to 1949.

The period from 1950 to 1984 is signified by a cultural landscape between use and misuse. The large-scale use of machines on larger fields or meadows/pastures, the clearing of fields of many hedges, trees and creeks, the filling up of smaller depressions, hollows or creeks, and the intensification of use (drainage, fertilisation, purchase of fodder from outside the region or on the world market) led toward a homogenisation of the landscape over the entire Allgäu region (Güthler 2006; Schaffer and Zettler 1984). In addition, in the 1990s and 2000s most settlements or towns expanded and some new roads and ski lifts were constructed, as well as the installation of water basins for snowmaking with snow cannons (Güthler 2006). This process also led to a loss of special or rare biotopes such bogs, fruit tree meadows, fish ponds, dry and humid meadows (Güthler 2006; Konold 1996). In contrast, in the southern Oberallgäu, forested areas increased by 13% over the last two decades (Güthler 2006), and a change back from spruce dominated forests to more natural mixed forests (the dominating forest type in former times) also occurred during this period.

4 Agriculture and Forestry Today – Strengths and Weaknesses

4.1 Forestry

Forests cover 29% of the Allgäu (Bayerisches Staatsministerium für Ernährung, Landwirtschaft und Forsten 2010). In particular, the southern part of the Allgäu is characterized by small scale forest ownership, which makes it difficult to reconcile the ecological function (e.g. protection against avalanches in the mountains) of the forests with a sustainable economic perspective for the owners (AELF 2010). In the districts of Oberallgäu and Lindau, 59% of the forest area is owned privately with small forest parcels, 9% is larger private forest areas, 6% is community forest and 26% belongs to the Federal State. The average size of the private woodland ownership is only 2–3 ha, and many of the owners lack technical equipment and skills. In fact, estimates show that only about 60–70% of the timber which could be exploited is actually being used in these private forests.

4.2 Agriculture

More than half of the study area (58%) is under agricultural use (four Bavarian districts) (Bayerisches Staatsministerium für Ernährung, Landwirtschaft und Forsten 2010). In the most southern districts (Oberallgäu and Ostallgäu) the use is almost exclusively (99.9% and 90%, respectively) as permanent grassland (pastures, meadows) for livestock production. Parts of the grasslands, especially in the mountainous southern parts, cannot be cut with machines due to their steep slopes. In the climatically more favourable northern part of the Allgäu (warmer, less precipitation), there is a mixture of cropping (31%) and permanent grassland (69%).

The average farm size is 55 ha, of which up to 35% of the area is leased land (Bayerisches Staatsministerium für Ernährung, Landwirtschaft und Forsten 2010; Destatis 2010). Since the 1980s, both figures have strongly increased due to a focus towards fewer, but larger farms (Bayerisches Staatsministerium für Ernährung, Landwirtschaft und Forsten 2010; Destatis 2010). The average number of large live-stock units per hectare is 1.4 with a total of 8,173 livestock farms in the four Bavarian districts (Bayerisches Staatsministerium für Ernährung, Landwirtschaft und Forsten 2010). Approximately 69% of the farms are full time agricultural enterprises, com-pared to 46% in the Federal State of Bavaria. The highest proportion of part time farmers is found in the most southern districts of Oberallgäu and Lindau (35%) with an average farm size of 18 ha. In these districts and the other southern parts of the Allgäu, the so-called "Alpwirtschaft" (mountain pasture operations) play an impor-tant role as it is practiced on more than half of the agricultural land. This area under-lies often difficult production conditions because of the steep slopes of many pastures, and in certain cases because of the remoteness (low accessibility) of the pastures. Nearly 635 alps (a production unit with several mountain pastures) were registered in the Allgäu in 1972 (Jahn 1978). During the 1930s, butter and cheese were still completely produced directly on the alps, but later production moved to the valleys and even more north to be closer to towns with better traffic infrastruc-ture. Today, in most of the alps, only young cows that have not yet been used for reproduction and therefore do not give milk, are found. During the summer of 2010, there were 689 alps with more than 28,000 young cattle, almost 3,000 milk cows, and a few horses, sheep, goats and pigs, with the numbers being stable for the last several years (Honisch, personal communication, press article in Allgäuer Zeitung 2010). Due to governmental aid programmes, the alps are an agricultural unit in the Allgäu which remained more or less untouched during the generally strong struc-tural changes in agriculture during the last decades, which is reflected by their stable numbers (Güthler, 2010, personal communication).

The majority of the farms in the entire Allgäu produce milk from cows. In the four Bavarian districts of the Allgäu milk production is a bit less than 1.2 million tons per year (Bayerisches Staatsministerium für Ernährung, Landwirtschaft und Forsten 2010). This milk is processed in famous dairy firms (e.g. Champignon, Allgäuland, Edelweiß) and transformed into well-known brand-name dairy prod-ucts (cheese, milk, butter, yoghurt, cream). According to the EU-database DOOR (EU 2010), within the Allgäu two cheese products carry the EU-label Protected Designation of Origin (PDO): "Allgäuer Bergkäse" and "Allgäuer Emmentaler". This label covers agricultural products and foodstuffs which are produced, pro-cessed and prepared in a given geographical area using recognised methods. In Germany, only four types of cheese carry the label. For a comparison in the same cheese category, France has 45 listings and Italy 38.

Organic agriculture is relatively well developed in the Allgäu. The share of organic agriculture enterprises over all agricultural enterprises is 4.7% (5.1% for agricultural land) for the district Unterallgäu, 8.1% (7.9%) for Ostallgäu, and 11.3%

(9.9%) for Oberallgäu, the latter two being clearly higher than the average share of 5.1% in the southwestern part of the Federal State of Bavaria (10 districts) (Bayerisches Staatsministerium für Ernährung, Landwirtschaft und Forsten 2010; Bio-Ring Allgäu 2010).

The importance of agriculture in the Oberallgäu, besides the production of milk, meat and cattle breeding, is increasingly founded on the maintenance and conservation of the cultural landscape. Without extensive agricultural use, the economically important tourism sector in the southern Allgäu would be unthinkable (see also Güthler 2006).

More and more, agricultural enterprises are trying to increase their income by renting guest rooms or holiday apartments. In particular, in the touristically more attractive southern Allgäu, many agricultural enterprises gain indispensable additional income. For example, 23% of all agricultural enterprises in the district Oberallgäu offer "Holidays on a farm". They provide almost 5,500 beds and have nearly 500,000 overnight stays per year.

5 Tourism

Tourism has a long tradition in the Allgäu. As early as 1856 a handbook for travellers was published (Buck 1856). The author not only recommended never undertaking a hike into the mountains without a reliable guide and some wine or rum or similar beverage, but also mentioned that all over, even in the most remote villages, cheap and good accommodations could be found. There were already a wide range of activities available. For example, in a mountain lake just underneath the peak Großer Daumen at an altitude of about 2,000 m above sea level, boat tours were offered. Also, a visit to a 'Viehscheid' (the cattle returning to the valleys after a summer on the alpine pastures) was proposed to the visitor. Even today, the Viehscheid is one of the most appealing events for visitors.

Tourism continues to play an important role for the Allgäu. Its location close to the Alps and its landscape diversity attract overnight tourists as well as excursionists. There are potentials for many activities during all seasons, the more popular ones are skiing and hiking, followed by mountain biking, swimming in lakes, or visits to traditional festivities. Centers for tourism are located in the southern parts of the Allgäu, within well known places like Schwangau, where the famous Neuschwanstein castle is located, or Oberstdorf, where international competitions in ski-jumping or cross-country skiing take place. The district Oberallgäu strongly profits from tourism, but also is relatively independent of it with a share of 14% of the total economic income generated in the district. In single communities in the southern part of the Allgäu, the share of total income generated by tourism is much higher; the Bavarian average is 3.4%. Tourism intensity (nights per 1,000 inhabitants) in the Allgäu is 12,000 with a range from 1,400 in the town of Kaufbeuren to 35,000 in the district of Oberallgäu. Private hosts play an important role in host structure. For the Oberallgäu, approximately one third of all overnights take place in private houses. Farm holidays are particularly important for farmers in providing additional income by renting rooms or apartments to tourists.

In 1985, the concept of the community of Bad Hindelang for the integration of agriculture and tourism, later known as "Eco-Model Hindelang", was launched. Its objective was to maintain or renew the diverse cultural landscape and at the same time produce income for the mostly small-structured farms in order to keep them in existence. The idea is simple: farmers receive economic compensation for maintaining the cultural landscape, which is of great importance for tourism. The model received an award as a "recognised international world-wide project" at EXPO 2000 in Hannover, Germany, and was acknowledged by the scientific community (Hemmer 1997; Maier 1996), and even by researchers from Japan (Kureha 2002). Yet, realisation of the eco-model proved to be difficult. Today, Bad Hindelang follows a concept with softer criteria. For example, a historic car race that had been stopped for several years because it would not fit into an ecological concept (Maier 1996) is once again in operation. The present concept is based on ski tourism and health, as well as landscape tourism experiences.

6　Current Regional and Rural Development Programs and Project Examples in the Allgäu

Rural areas all over Europe are facing similar problems. The absence of job options and perspectives, especially for young people, leads to a steady out-migration or, in the worst case, to rural exodus. Declining numbers of inhabitants leads to a reduction of infrastructures, which in turn results in a loss of quality of life. Structures in agriculture also often change. Rural areas that want to react to those processes need a development strategy suitable for their specific situation and problems. Here we present a number of projects that have been started in the Allgäu in order to promote the products of the region. All of them have received funding for an initial phase or for the first years of the projects (the latter three through the LEADER-Programme of the European Union) and were then continued completely on a private basis.

6.1　Von Hier – Products (From Here)

An interesting marketing strategy in the Allgäu was developed with the "Von Hier" products (From Here) (Fig. 5). The label was developed by a regional German supermarket chain (Feneberg) for promoting and selling food products, which are both from the region and organic. The supermarket chain was founded in 1947 and has about 3,800 employees. It is run by the family Feneberg (2010b). In 2009, the chain consisted of 73 supermarkets and 9 larger department stores (Siegel 2010).

The Von Hier project started in 1997, with co-funding from the Federal State of Bavaria and the EU for the years 1999–2001 (E. Wirthensohn, 2010, personal

Fig. 5 The Von Hier (From Here) label for the regional supermarket chain in the Allgäu

communication) and the label was established in 1998 (Kippes 2008). Today about 300 different Von Hier products exist including cheese, dairy products, eggs, vegetables, fruits, meat, sausage, flour, bread, fruit juice and beer (Feneberg 2010a). Most products purchased are dairy (55%), followed by meat and sausage (18%), and fruits and vegetables (9%) (Siegel 2010). Most customers of the supermarket (84%) buy Von Hier products, in particular people older than 30 years of age and who have been living for several years in the Allgäu.

The regional aspect of Von Hier products is very important. Only products which are produced within a radius of 100 km around the central Feneberg site in the town of Kempten are considered under the label (Feneberg 2010a). The area within the 100 km radius completely includes the Allgäu, but also covers about the same area beyond it. At present, more than 600 organic farmers and 23 processing enterprises deliver their products to Feneberg (Siegel 2010). All organic farms are certified under the European Union Bio-Certification 2092/91.

Under the Von Hier strategy, an added regional value has been established in linking regional farmers and processing enterprises via the supermarket to the consumers. To strengthen this regional link, information about the different organic farmers and their addresses are provided on the internet or in brochures in the supermarkets with the objective of making the organic producers visible to the consumers (Feneberg 2010c; Kippes 2008). A yearly turn-over of 16 million Euro for Von Hier products was calculated for 2009 providing an added value of 3.22 million Euro for the contracted organic farmers (on average, 5,000 Euro per year and enterprise; added value = price difference paid to the organic farmers compared to prices of conventional products) (Feneberg 2010c; E. Wirthensohn, 2010, personal communication). Approximately 500 additional jobs have been created by the Von Hier program.

Over the last years, the Von Hier program has won different awards (Kippes 2008). Among them is second place in a competition during 2007, financed by the German Ministry of Nutrition, Agriculture and Consumer Protection, for regional initiatives promoting regional economic structures in rural areas (DVL and BMELV 2007). The success of the Von Hier concept also is reflected by the fact that it has been copied to other regions in north-eastern Germany (Kippes 2008), reflecting the many positive economic, social and ecological objectives that have been fulfilled by the program for the farmers, consumers and the supermarket chain.

Another example of a supermarket chain which promotes a special label (Rhöngut – regional and organic meat products) is from a mountainous area in central Germany (Roep and Wiskerke 2006). This production has made a positive contribution on the revitalisation of the region and rural development by providing a positive economic, social and environmental profile.

6.2   The LEADER Program of the European Union

LEADER stands for Liaisons Entre les Actions de Développement de l'Economie Rurale (Links between rural economy and development actions) and is a Programme of the European Union. The basic ideas of LEADER are:

- Territorial approach: specific features of the respective regions have a high value – regions are to discover and develop their individual profiles.
- Bottom-up approach: local actors are to participate in the development of a strategy, in the realisation of projects and in decisions on how the EU-money is spent.
- Regional development strategy: different actors of a region develop a strategy for a positive development of the region.
- Integrated approach: actors from different sectors cooperate in planning and realising a project.
- Networks of actors: actors shall learn from each other, exchange information, integrate applied knowledge and cooperate.
- LEADER Action Groups (LAGs): LAGs are regional federations of public and private people and institutions that coordinate and foster the development process, bring the different actors together, and are the contact point for project ideas and proposals.

In each of the four districts on the Bavarian side of the Allgäu, a LEADER Action Group exists. Within the LAG Regionalentwicklung Oberallgäu, 34 projects have been realised that brought nearly 2.4 million Euro of aid money into the region between 2003 and 2006. Compared to LAGs in other regions, the Allgäu has a very high quota of cooperative projects (projects with two or more LAGs involved). This is because the identification of people with their region made it quite easy for the four LAGs to cooperate. In addition, there are other structures responsible for the Allgäu as a whole, such as the Allgäu GmbH (former Allgäu Initiative and Allgäu Marketing; responsible for economic and tourism promotion).

6.2.1   Project: LandZunge – The Taste of the Countryside

LandZunge (LandTongue) is one example of a project starting within the LEADER program (Fig. 6). LandZunge is a network of producers and restaurant chefs promoting and using regional products in restaurants according to their own criteria (Landzunge 2010). LandZunge was initiated by people who care about the region where they grew up and currently live, among them a journalist and a restaurant owner.

Fig. 6 LandZunge label

The initiative started in 2001 in the district of Ravensburg on the Baden-Württemberg side of the Allgäu, was extended to the Bavarian side about 2 years later, and has continued to develop ever since. Each restaurant in the network is obliged to offer at least three meals with beef or cheese from certified producers of the region. Beer and certain soft drinks must be produced within the region as well. One of the problems and a reason why such projects often do not work is logistics; a restaurant chef cannot shop in all places where producers sell their quality products. Hence, from the beginning of the LandZunge project, strong partners were sought. Feneberg, the food supermarket chain described above, is one of them. Given the presence of a journalist in the group, the initiators also were aware that good publicity is a precondition for the success of such a project. Therefore, they founded their own magazine, in which the participating restaurants and partners are presented. Topics comprise not only the ecologic and economic aspects of such a project, but also the social ones; appreciation for clean structures in the villages, a restaurant as a meeting point for local people, or quality of life. Much effort was put forward to convince the local and regional politicians of the idea, and to provide support for the project.

In the beginning of 2010, almost 80 restaurants participated in the network, and many of these purchased all their beef and cheese from regional producers, so that a superior category, LandZunge Plus, was developed for them. The project unifies the different local breweries, which normally are in competition with each other. Quality control has been handed over to an external institute, and a LandZunge Foundation was founded. Restaurant owners report that clients are ready to pay a bit more if they know about their contribution to the development of the region.

6.2.2   Project: Allgäuer Alpgenuss

"Alpen" (singular: Alp) is not only the name for the mountain chain, but also for the mountain huts in which the herdsmen stay during summer months when the cattle graze in the alpine pastures in the mountains. It is quite common that small scale gastronomy is attached to those Alpen, where visitors can get dairy products and basic meals.

When it turned out that more regional products such as bread, meat, or even dairy products were purchased at discount markets, the project "Allgäuer Alpgenuss" was

Fig. 7 Allgäuer Alpgenuss label

Fig. 8 Label for the forestry project in.Silva

initiated (Fig. 7). Its objective was to bring more regional products to the menus in order to stimulate added value for the region – very similar to the LandZunge approach.

Currently, a pool of more than 40 Alpen have formed an association, in which producers and other partners also are represented. They have defined a set of criteria for participation in the network, and the association initiated a cooperative agreement with the regional newspaper to provide public attention for their activities.

6.2.3   Project: in.Silva – Clustering of Wood Supplies from Small Forest Owners

As mentioned above, agriculture and forestry are important for the Allgäu, in particular for maintaining the landscape, which is vital for the regional tourist industry. The region is characterised by small scale forest ownership, which makes it difficult to reconcile the ecological function of the forests with a sustainable economic perspective for the owners.

For this reason the project in.Silva was started in 2005 with support funding from the LEADER programme (Fig. 8). The project brings together regional wood producers into a cooperative that can meet the demands of large scale wood processing industries. The established organisational structure acts on behalf of numerous micro-producers, is innovative and adds significant value to the production cycle. It shows how a major deficiency in terms of competitiveness in a particular sector can be overcome through the development of a new regional governance framework. Regional sawmills are still supplied by members of the cooperative individually, but given the increase in quality and quantity of wood products provided by working together, the cooperative aims to supply major national and international clients.

Most members work according to agreed standards for sustainable forestry practices provided by the PEFC (Programme for Endorsement of Forest Certification Schemes) (H. Wirthensohn, 2010, personal communication).

The project has given the regional wood producers the capacity to act independently and govern themselves after the funding period has finished, and is therefore a good example of how programmes for rural development such as the LEADER-Programme of the European Union can initiate sustainable development. The project aims to create a 'win-win' situation, where major wood processing industries are attracted to the quantity and quality of supply, and at the same time diversifies and stabilises demand to ensure long term revenue for the producers.

6.3   Effects and Success Factors

A success factor common to all these projects is that an individual or group of people see the project as 'their' own. The necessity of networking also has proved to be an important factor for the success of such projects.

A comparison of different development activities in the Allgäu and in Vorarlberg (a neighbouring region in Austria) showed that regional development is a slow process in which results are not necessarily seen quickly. Therefore, one of the success factors is that the long duration of regional development processes leads to a fruitful exchange of knowledge, ideas and experiences (Mayer et al. 2008). Actors learn that they have the ability to find innovative solutions, and networking and cooperation improve through the projects – although the ability to cooperate may be needed even more in the future.

The general problem with most subventions from the European Community (direct payments from the Common Agricultural Policy) is that, in most cases, it slows down or inhibits innovation in agriculture. There are several interesting innovative projects or initiatives in the Allgäu that are linked with agriculture or forestry (From Here or other local and regional products, In.Silva, ecotourism, farm holidays, projects from other sectors with indirect links to agriculture), but many farmers still depend enormously on subventions and the market prices for milk and cereals.

Even the LEADER-Programme, which seeks to foster innovation, motivate actors and create networks, has its limits, due to its focus on specific or pilot projects. Yet, even this can serve as a best practice, although not often solving problems of a whole sector such as agriculture. An Allgäu label is being created, but even in this project the situation for the agriculture sector is difficult. In contrast to tourism or the general marketing of the Allgäu region, no well-established structures or appropriate authorities exist in the region to establish the above mentioned programs on a larger scale. The problem is a lack of common strategy. As mentioned by Lightfoot et al. (2001), there must be a high importance for collective visioning to sustainably develop an agroecosystem or a region.

7 Allgäu – A Development Towards Sustainability?

In general, it is very difficult (or even impossible) to completely answer the question of whether a certain region is developing towards sustainability (i) due to lack of sufficient reliable data to evaluate most of today's potential indicators for the three pillars of sustainability (economic, social, ecological), and (ii) an evaluation simply reflects the present situation which can change rather quickly within a few years. Nevertheless, in the case of the Allgäu, we provide a comparatively good picture of the present situation in conducting a mixed qualitative-quantitative evaluation based on selected indicators. For the evaluation we also consider the knowledge of regional or local experts (see acknowledgments) on certain issues. We also list the positive and negative points for the economic, social and ecological dimensions with the overall evaluation presented in the conclusion.

7.1 Sustainable Economic Development

In general, many positive points can be mentioned which show that the economic situation for many people has improved over the last decades (and centuries if we look at regional history) as most people have sufficient income to make their livelihoods in different activity sectors:

+ Existence of a broad diversity of employment in the primary, secondary and tertiary sectors.
+ Relatively low unemployment rate in the districts of the Allgäu.
+ Fewer farmers abandoning agriculture compared to other regions in Germany because of good possibilities for receiving a diversified income from agriculture (from conventional and, in particular, organic agriculture), from direct or regional sales of their products, from tourism (accommodations or part time engagement in the tourist sector), from full-time or part-time engagement of family members in local industries or tertiary sectors, and from financial support payments by the Federal State or the EU.
+ A stable or even slightly increasing number of farmers producing milk and livestock on alps, the alpine summer pastures. This type of production is strongly supported by different financial aid programs.
+ An increasing tourism sector, although initiatives and concepts for eco- or sustainable tourism still remain quite limited. A lot of the money remains in the region because of the many family enterprises engaged in tourism and a highly diversified small-structured tourism market.
+ Positive income evolution for organic farmers through higher prices for their products and a relatively stable market (example of the added value for Von Hier farmers).
+ A rural region with an increasing population in contrast to most situations elsewhere in Germany and Europe. This does not *per se* mean an improvement of the

economy of a region, but reflects the attractiveness of the Allgäu to stay or move there, which is not the case for other rural areas in Germany or Europe.

+ High potential for local or regional food products (dairy and cheese products in general, From Here products in particular). For example, quite a lot of local breweries have existed for decades, and still find their consumers.
+ Existence of local dairies of which most work profitably and independently of national and global milk prices.

However, these positive points also are accompanied by some negative points:

− Some farmers, in particular smallholders, have large problems in generating sufficient income, and have partly or completely abandoned their work over the last years.
− Low milk prices over the last years, and probably also over the coming years, will prevail among the farmers who focus on milk production, and this is the highest proportion of all farmers in the Allgäu. They will be pushed to complete abandonment, to look for alternative sources of income, or to intensify or rationalise production (more land, more fertilisation, extra fodder, more cows).
− The last remnants of the textile industry changed to new products, but have difficulties in continuing except for very specialised production (e.g. traditional costumes for locals and tourists).
− Many of the smaller tourist accommodation facilities such as pensions (bed and breakfast) need modernisation to meet today's tourist demands for quality.

7.2 Sustainable Social Development

Similar to economic development, some positive points can be mentioned for social development:

+ Less social discrepancy and fewer social problems because of the relatively low unemployment rate.
+ Many family farms remain with the possibility to generate income from other types of work which allows for the maintenance of traditional family structure.
+ The population of the Allgäu has a generally positive image and identity, and is proud of its region (Siegel 2010; Sieth 2006).
+ The continuing existence of local traditions (e.g. traditional festivities and clothes; local or regional customs).
+ The LEADER projects and other local programs have improved the network of stakeholders as well as provided a more common vision on different topics. These social networks have good potential for sustainability.
+ Healthy social structures and involvement lead to a high sense of community and voluntary commitment.

As well, there are negative points to be considered:

− A trend of younger people is to leave the smaller villages to live and work in the towns. Although this trend exists, it is much less pronounced in the Allgäu compared to many rural areas elsewhere. In contrast, in many larger villages there is a positive trend of many young families staying in the village, or even new installations of young families from outside the Allgäu. In many communities in the southern Allgäu, an increasing number of retired people have settled in the villages because of the attractiveness of the landscape, leading to an aging of the village population.

- Ground-breaking projects like Eco-Model Hindelang do not lead automatically to successful implementation because of diverging interests of stakeholders and lack of a defined common vision.

7.3 Sustainable Ecological Development

Although some negative points exist, the Allgäu region still has high ecological value and will likely maintain or even improve it in the future.

+ Relatively high species and ecosystem diversity exist because of a highly heterogeneous landscape with different types of forests, bogs, marshes, ponds, lakes, grasslands, alpine ecosystems and occurrence of endemic species.
+ High share of protected areas in the region.
+ Low intensive livestock breeding created and still preserves many alpine and lowland pastures and meadows with a diverse fauna and flora. Some of them might be abandoned in the future because of insufficient profitability for farmers if they are not remunerated for this activity either via agri-environmental scheme payments or direct payments from the communities to keep the cultural landscape open for tourism (attractiveness of the landscape).
+ No nitrate or pesticide pollution of the ground water (and drinking water) because low-input agriculture dominates, and lakes have good water quality.
+ A high share of organic agriculture, which has a lower impact on the environment and is seen as favouring higher biodiversity compared to conventional agriculture.
+ Increasing forested areas with a more natural composition of tree species (mixture of broadleaf and needle trees) and higher non-woody species diversity.
+ Some negative trends such as expansion of ski lifts and skiing areas, and damage to natural vegetation by hikers has been reversed over the last two decades in certain areas, but only in smaller spatial areas. Also, some rare species such as the beaver and Golden eagle have increased in numbers over the last years.
+ Existence of initiatives/projects for sustainable development such as EZA (Energy and Environment Centre Allgäu) or Allgäuer Moorallianz (an initiative for the conservation of bogs that was among five winners in a national contest).

As well, the negative points to be considered are:

- A loss of valuable ecological areas has occurred over the last two decades, brought about by intensification of agriculture, construction of new roads, ski lifts and water basins for snowmaking and expansion of settlements. Although this concerns larger spatial areas in only a few cases, this loss can be found in many areas of the Allgäu.

– The intensification of land use in valleys or lowland areas as a consequence of low milk prices often leads to less diverse vegetation. The abandonment of the use of some summer alpine pastures can lead to bush and forest encroachment (Güthler 2006). In addition, these vegetation changes decrease the attractiveness of the landscape as flower-rich open areas for tourists because they are replaced by forests and bushland, or by less diverse grasslands.
– Some rare species such as grouse (a group of birds) and the Moorland Clouded Yellow (a butterfly) decreased in numbers over the last years because of increasing landscape homogeneity (less landscape structure, spruce cultivated forests) in certain areas.
– In the past, larger flooding events occurred in some valleys which caused much damage to many areas including human settlements which are in some areas too close to major rivers and lack sufficient natural water retention areas. After the last flooding, such natural water retention areas have been re-established to allow "natural flooding" during peak flood events.

8   Conclusions and Perspectives for the Future

Given that 100% sustainability (meaning that all indicators or parameters show positive trends) is for us illusionary, do we find sufficient evidence for sustainable development in the Allgäu? We would say yes, because many points can be listed for positive economic, social and ecological development in the region. We are aware that many negative points exist as well and some conditions should be improved; thus, some people may not agree with our judgment. Barring catastrophe over the coming years, the overall good economic and income situation for most people, the good ecological conditions and relatively rich biodiversity, the relatively well-established social structure, as well as the identity of the people with the region and relatively low social discrepancy indicate that sustainability for the region can be maintained and improved. If we consider ecological sustainability, for example, it is obvious that some crucial points still exist, but it appears that the Allgäu still has high ecological value and probably can maintain or even improve it in the future. The ecological situation was probably better some decades ago with more pastures or meadows having high species diversity and less use of manure. Yet, the negative trends have been much reduced or even nonexistent in certain areas of the Allgäu compared to many other rural areas in Germany or Europe.

Although we find relatively positive development in the Allgäu, different actual and future threats exist. These are (i) potential negative impacts related to the tourism sector such as construction of new infrastructure in natural areas, increasing traffic and mass tourism in certain areas that degrade habitats and reduce species richness, (ii) intensification of agriculture in certain areas, but also abandonment of agriculture in other areas, (iii) loss of traditions and customs, and (iv) declining numbers of smallholders, in particular those with alpine pastures. The objective is to minimise these negative impacts to assure the sustainable development of the Allgäu in the future. If, for example, the negative impact from tourism (construction of infrastructure, traffic, etc.) continues with the same dynamic as over the last 10 years, future development will probably not be sustainable anymore. Clearly then, there are limits or thresholds for such improvements, after which the effects begin to work against the original intentions.

The Allgäu is a typical rural region which had to undergo many changes and adaptations in its history, and where living conditions were not easy until the middle of the twentieth century. Yet, in contrast to other regions in Europe, several factors have helped such that today there are relatively favourable conditions for sustainable development. One factor was that transition from the Blue to the Green Allgäu was not abrupt, but gradual. The textile industry did not collapse suddenly (within few years), so farmers more easily shifted their practices from cropping to livestock maintenance. This transition was enormously supported and fostered by pioneers such as Carl Hinbein and Johann Althaus who established innovative cheese making techniques in the Allgäu.

Another factor is that since the nineteenth century a certain amount of industry has always existed, so that at least some part of the population did not completely rely on agricultural production. This industry seems to have been able to adapt to new economic realities in changing from textiles to a more diverse and modern set of industries (e.g. high-tech firms, dairy farms, packaging industries, textile industries, machine construction). In addition, tourism became more and more important to the region, which generated a great deal of work in this sector either as full-time jobs, or in combination with agricultural activities. This combination of agriculture (in creating and maintaining the largest part of the present landscape) and tourism, in profiting from the high tourism potential of the region during summer and winter, seems for us to be one of the most important factors in determining that the Allgäu is better-off than other rural landscapes.

Another important factor is that with the change from the Blue to the Green Allgäu, the new agricultural products (in particular milk) were processed in the region and created a dairy 'industry' which is well-established today. These quality cheese and dairy products from the Allgäu have found a good market niche even beyond the region, but the producers are presently struggling with low milk prices. To help compensate, some of the dairy and cheese products produced and processed on the farms or in local dairies can be sold directly to tourists or the tourism sector (e.g. restaurants, festivities). Historically, and in most cases, the different alternatives that have fortunately been available for adaptation in the Allgäu have been indispensable in its continued drive for improved sustainable development.

Acknowledgements We thank very much Andreas Güthler, Michael Honisch, Susanne Kappes, Christine Räder, Johanna Siegel, Julia Wehnert, Henning Werth, Ernst Wirthensohn, Hugo Wirthensohn and Simone Zehnpfennig for their information, expert knowledge and critical discussion of this review. Also highly acknowledged are the corrections and comments on the review by W. Bruce Campbell.

References

AELF (Amt für Ernährung, Landwirtschaft und Forsten) (2010) Forest statistics Allgäu. Kempten

Allgäu Marketing (2010) Statistical data for tourism in Bavaria and in the Allgäu. Allgäu Marketing, Kempten

Bayerisches Staatsministerium für Ernährung Landwirtschaft und Forsten (2010) Agrarpolitik – Daten und Fakten. http://www.stmelf.bayern.de/agrarpolitik/daten_fakten/18646/karten_2008. pdf. Accessed June 2010

Becker C, Job H, Witzel A (1996) Tourismus und nachhaltige Entwicklung. Wissenschaftliche Buchgesellschaft, Darmstadt, 185 pp

Bio-Ring Allgäu (2010) Statistical data for organic agriculture in the Allgäu. Kempten

Blewitt J (2008) Understanding sustainable development. Earthscan, London/Sterling, 288 pp

Buck J (1856) Handbuch für Reisende im Algäu. Lechthal und Bregenzerwald. Verlag Tobias Dannheimer, Kempten (3rd edition 2008)

Bundesamt für Naturschutz (2004) Daten zur Natur. Bundesamt für Naturschutz, Bonn, 474 pp

Destatis (2010) Statistisches Bundesamt Deutschland. ims.destatis.de. https://www.regionalstatistik. de/genesis/online/logon. Accessed June 2010

Deutscher Wetterdienst (2010) Klimadaten. http://www.dwd.de/klimadaten. Accessed May 2010

DVL (Deutscher Verband für Landespflege), (BMELV, Bundesministerium für Ernährung, Landwirtschaft und Verbraucherschutz) (2007) Natürlich Regional. 56 starke Initiativen für die Zukunft des Ländlichen Raumes. Deutscher Verband für Landespflege, Ansbach, 14 pp

Eglington A, Israel R, Vartanov R (1998) Towards sustainable development for the Murmansk region. Ocean Coast Manage 41(2–3):257–271

Endert D van (2000) Die Besiedlung des Allgäus in vor- und frühgeschichtlicher Zeit. In: Kettemann, Otto (ed) "Droben im Allgäu, wo das Brot ein End' hat". Zur Kulturgeschichte einer Region. Druckerzeugnisse des Schwäbischen Bauernhofmuseums Illerbeuren 14. Kronburg-Illerbeuren, pp 43–46

EU (2010) EU agricultural product quality policy. Geographical indications and traditional specialities. European Commission, Brussels. http://ec.europa.eu/agriculture/quality/schemes/ index_en.htm. Accessed May 2010

Feneberg (2010a) VonHier – Informationen für den landwirtschaftlich Interessierten. http://www. feneberg.de/index.php?id=131. Accessed May 2010

Feneberg (2010b) Unternehmensprofil. http://www.feneberg.de/index.php?id=191. Accessed May 2010

Feneberg (2010c) Bauernliste. www.feneberg.de/fileadmin/feneberg_upload/VonHier/Bauernliste_09. pdf. Accessed May 2010

Friedel R, Spindler EA (eds.) (2009) Nachhaltige Entwicklung ländlicher Räume. VS Verlag, Wiesbaden, 501 pp

Gong J, Lin H (2000) Sustainable development for agricultural regions in China: case studies. For Ecol Manage 128(1–2):27–38

Grober U (2010) Die Entdeckung der Nachhaltigkeit. Verlag Antje Kunstmann, München, 299 pp

Güthler A (2006) Allgäu im Wandel. Verlag J, Eberl KG, Immenstadt, 96 pp

Hemmer I (1997) Das Ökomodell Hindelang: eine Erfolgsstory. Geographie heute 151:12–15

Institut für Länderkunde Leipzig (ed.) (1997) Atlas Bundesrepublik Deutschland. Pilotband. Institut für Länderkunde Leipzig, Leipzig

Jahn W (1978) Strukturanalyse des Allgäus. Das Allgäu als Raumeinheit: Geofaktoren – Entwicklungsprozesse – Grenzen. In: Ortsausschuss des 16. Deutschen Schulgeographentages Augsburg (ed) Exkursionen in Schwaben und angrenzenden Gebieten. Kiel, pp 187–197

Jahn W (1989) Der Landschaftsname Allgäu im Wandel der Zeit: Reichweiten-Ambivalenz als Folge sich ändernden Sinngehaltes. Mitteilungen der Geographischen Gesellschaft in München 74:108–162

Job H (1996) Modell zur Evaluation der Nachhaltigkeit im Tourismus. Erdkunde 50(2):112–132

Kamp M (2000) Wir müssen auf gute Waare sehen, wenn wir gutes Garn erzeugen wollen. Zur Kulturgeschichte der Hanfverarbeitung in Immenstadt. Immenstadt

Kippes S (2008) Regionalität als Vermarktungskonzept im Lebensmittelsektor. Eine empirische Untersuchung des "VonHier"-Projekts des Unternehmens FENEBERG und seiner vertraglich gebundenen Produzenten. Masters thesis, University of Erlangen-Nürnberg, Germany, 169 pp

Klima A (1989) Das Abbild der Raumvorstellung "Allgäu" als Facette des Regionalbewusstseins einer heimattragenden Elite. Berichte zur deutschen Landeskunde 63(1):49–78

Konold W (1996) Von der Dynamik einer Kulturlandschaft. Das Allgäu als Beispiel. In: Konold W (ed.) Naturlandschaft – Kulturlandschaft. Die Veränderung der Landschaften nach der Nutzbarmachung durch den Menschen. Landsberg, pp 121–136

Krattenmacher M (1983) Prädikat Allgäufrisch – über Milch und was man alles aus ihr machen kann. Merian 36(5):142–143

Kureha M (2002) System for the relationship of symbiosis between tourism and agriculture in Hindelang, the German Alps. Science reports, vol 23, Institute of Geoscience, University of Tsukuba, Japan, pp 75–90

Landzunge (2010) Das Projekt LandZunge. www.landzunge.info/. Accessed June 2010

LfU – Bayerisches Landesamt für Umwelt (2010) Fachinformationssystem Naturschutz. http://www.lfu.bayern.de/natur/daten/fis_natur/. Accessed June 2010

Liedtke H (1997) Landschaften in Deutschland und ihre Namen. In: Institut für Länderkunde Leipzig (ed) Atlas Bundesrepublik Deutschland. Institut für Länderkunde Leipzig , Pilotband, pp 34–35

Lightfoot C, Fernandez M, Noble R, Ramírez R, Groot A, Fernandez-Baca, Shao F, Muro G, Okelabo S, Mugenyi A, Bekalo I, Rianga A, Obare L (2001) A learning approach to community agroecosystem management. In: Flora CB (ed) Interactions between agroecosystems and rural communities. CRC Press, Boca Raton, pp 131–155

Lochbrunner W (1976) Kemptener Vereinödungen. In: Bayerisches Staatsministerium für Ernährung, Landwirtschaft und Forsten, Berichte aus der Flurbereinigung 24(1976): 20–26

Maier J (1996) Hindelang, ein Modell ökologischer Dorfentwicklung – Zwischenbilanz einer Evaluation auch im Sinne nachhaltiger Entwicklung? In: Arbeitskreis Freizeit und Fremdenverkehrsgeographie. Institut für Tourismus der Freien Universität Berlin, Berichte und Materialien Nr. 14. Berlin, pp 67–78

Mayer M, Metzler K, Job H (2008) Das LEADER+ Programm als Motor einer endogenen Regionalentwicklung – Oberallgäu und Vorarlberg im Vergleich. In: Maier Jörg (ed.) Entwicklungsperspektiven ländlicher Räume Teil 1: Ziele und Strategien einer aktuellen Politik für periphere ländliche Räume in Bayern, pp 142–163

Mühr B (2010) Klimadiagramme weltweit. www.klimadiagramme.de/Deutschland/deutschland2. html. Accessed May 2010

Nowotny P (1984) Vereinödung im Allgäu und in den angrenzenden Gebieten. Verlag für Heimatpflege Kempten im Heimatbund Allgäu e.V

Roep D, Wiskerke H (eds.) (2006) Nourishing networks. Fourteen lessons about creating sustainable food supply chains. Reed Business Information, Wageningen University, Wageningen, 176 pp

Rogers PP, Jalal KF, Boyd JA (2008) An introduction to sustainable development. Earthscan, London, 416 pp

Schaffer F, Zettler L (1984) Kulturlandschaft im Wandel der Landwirtschaft. Fallstudie Unterallgäu. In: Der ländliche Raum in Bayern. Fallstudien zur Entwicklung unter veränderten Rahmenbedingungen. (=Veröffentlichungen der Akademie für Raumforschung und Landesplanung: Forschungs- und Sitzungsberichte, vol 156). Hannover, pp 275–289

Scholz H (1995) Bau und Werden der Allgäuer Landschaft. Schweizerbart'sche Buchgesellschaft, Stuttgart, 305 pp

Siegel S (2010) Eine Untersuchung zu den Motiven, regionale Produkte zu kaufen, am Beispiel der Marke "VON HIER". Masters thesis, Ludwig Maximilians University München, München, 129 pp

Sieth B (2006) Darstellung der Markenpolitik von Regionen und Entwicklung eines Konzeptes der Dachmarke Allgäu. Masters thesis, University of Applied Science Kempten, Kempten, 224 pp

Thiel G (2000) "Weißes Gold". Aspekte zur Entwicklung der Allgäuer Milchwirtschaft. In: Kettemann, Otto (ed) "Droben im Allgäu, wo das Brot ein End' hat". Zur Kulturgeschichte einer Region. Druckerzeugnisse des Schwäbischen Bauernhofmuseums Illerbeuren 14, Kronburg-Illerbeuren, Germany, pp 209–218

Thierer M (1985) Zur Milchwirtschaft im Württembergischen Allgäu. In: Jentsch C, Kulinat K, Moll P (eds.) Beiträge zur Angewandten Geographie an Beispielen aus dem südwestdeutschen Raum. Mannheim, pp 309–327

Vogel R (2000) Von der Leinenweberei zur Textilindustrie. In: Kettemann O (ed.) "Droben im Allgäu, wo das Brot ein End' hat". Zur Kulturgeschichte einer Region. Druckerzeugnisse des Schwäbischen Bauernhofmuseums Illerbeuren 14, Kronburg-Illerbeuren, Germany

Vorlaufer K (1996) Tourismus in Entwicklungsländern. Möglichkeiten und Grenzen einer nachhaltigen Entwicklung durch Fremdenverkehr. Wissenschaftliche Buchgesellschaft, Darmstadt, 257 pp

Weizenegger S (2003) Akteursorientiertes Großschutzgebietsmanagement. Münchner Studien zur Sozial- und Wirtschaftsgeographie, Band 44, Kallmünz/Regensburg

Wezel A, Jauneau JC (2011) Agroecology – interpretations, approaches and their links to nature conservation, rural development and ecotourism. In: Campbell WB, López Ortíz S (eds.) Integrating agriculture, conservation and ecotourism: examples from the field. Issues in agroecology – present status and future prospectus, vol 1, Springer, Dordrecht, pp 1–29

Wezel A, Bellon S, Doré T, Francis C, Vallod D, David C (2009) Agroecology as a science, a movement or a practice. A review. Agron Sust Dev 29:503–515

Index

W.B. Campbell and S. López Ortíz (eds.), *Integrating Agriculture, Conservation and Ecotourism: Examples from the Field*, Issues in Agroecology – Present Status and Future Prospectus 1, DOI 10.1007/978-94-007-1309-3, © Springer Science+Business Media B.V. 2011